DEVELOPMENTS IN QUATERNARY SCIENCES 13
SERIES EDITOR: JAAP J.M. VAN DER MEER

THE MÝRDALSJÖKULL ICE CAP, ICELAND. GLACIAL PROCESSES, SEDIMENTS AND LANDFORMS ON AN ACTIVE VOLCANO

Developments in Quaternary Sciences
Series editor: Jaap J.M. van der Meer

Volumes in this series

1. The Quaternary Period in the United States
   Edited by A.R Gillespie, S.C. Porter, B.F. Atwater
   0-444-51470-8 (hardbound); 0-444-51471-6 (paperback) – 2004

2. Quaternary Glaciations – Extent and Chronology
   Edited by J. Ehlers, P.L. Gibbard
   Part I: Europe ISBN 0-444-51462-7 (hardbound) – 2004
   Part II: North America ISBN 0-444-51592-5 (hardbound) – 2004
   Part III: South America, Asia, Australasia, Antarctica ISBN 0-444-51593-3 (hardbound) – 2004

3. Ice Age Southern Andes – A Chronicle of Paleoecological Events
   By C.J. Heusser
   0-444-51478-3 (hardbound) – 2003

4. Spitsbergen Push Moraines – Including a translation of K. Gripp: Glaciologische und geologische Ergebnisse der Hamburgischen Spitzbergen-Expedition 1927
   Edited by J.J.M. van der Meer
   0-444-51544-5 (hardbound) – 2004

5. Iceland – Modern Processes and Past Environments
   Edited by C. Caseldine, A. Russell, J. Hardardóttir, Ó. Knudsen
   0-444-50652-7 (hardbound) – 2005

6. Glaciotectonism
   By J.S. Aber, A. Ber
   978-0-444-52943-5 (hardbound) – 2007

7. The Climate of Past Interglacials
   Edited by F. Sirocko, M. Claussen, M.F. Sánchez Goñi, T. Litt
   978-0-444-52955-8 (hardbound) – 2007

8. Juneau Icefield Research Project (1949–1958) – A Retrospective
   By C.J. Heusser †
   978-0-444-52951-0 (hardbound) – 2007

9. Late Quaternary Climate Change and Human Adaptation in Arid China
   Edited by David B. Madsen, Chen Fa-Hu, Gao Xing
   978-0-444-52962-6 (hardbound)-2007

10. Tropical and Sub-Tropical West Africa – Marine and Continental Changes During the Late Quaternary
    By P. Giresse
    978-0-444-52984-8 – 2008

11. The Late Cenozoic of Patagonia and Tierra del Fuego
    Edited by J. Rabassa
    978-0-444-52954-1 – 2008

12. Advances in Quaternary Entomology
    By S.A. Elias
    978-0-444-53424-8 – 2010

13. The Mýrdalsjökull Ice Cap, Iceland. Glacial Processes, Sediments and Landforms on an Active Volcano
    Edited by A. Schomacker, J. Krüger, K.H. Kjær
    978-0-444-53045-5 – 2010

For further information as well as other related products, please visit the Elsevier homepage (http://www.elsevier.com)

Developments in Quaternary Sciences, 13
Series Editor: Jaap J.M. van der Meer

THE MÝRDALSJÖKULL ICE CAP, ICELAND. GLACIAL PROCESSES, SEDIMENTS AND LANDFORMS ON AN ACTIVE VOLCANO

Edited by

Anders Schomacker

Institute of Earth Sciences, University of Iceland, Iceland

Johannes Krüger

Department of Geography and Geology,
University of Copenhagen, Denmark

Kurt H. Kjær

Natural History Museum of Denmark,
University of Copenhagen, Denmark

ELSEVIER

Amsterdam • Boston • Heidelberg • London • New York • Oxford
Paris • San Diego • San Francisco • Singapore • Sydney • Tokyo

Elsevier
Radarweg 29, PO Box 211, 1000 AE Amsterdam, The Netherlands
Linacre House, Jordan Hill, Oxford OX2 8DP, UK

First edition 2010

British Library Cataloguing in Publication Data
A catalogue record for this book is available from the British Library

Library of Congress Cataloging-in-Publication Data
A catalog record for this book is available from the Library of Congress

ISBN: 978-0-444-53045-5
ISSN: 1571-0866

Printed and bound by CPI Group (UK) Ltd, Croydon, CR0 4YY

Transferred to Digital Print 2012

Contents

List of Contributors

Ívar Örn Benediktsson, Institute of Earth Sciences, University of Iceland, Reykjavík, Iceland

Simon J. Carr, Centre for Micromorphology UoL, Department of Geography, Queen Mary, University of London, London, UK

Robert Duller, Department of Earth Sciences and Engineering, Imperial College, Royal School of Mines Building, London, UK

Páll Einarsson, Institute of Earth Sciences, University of Iceland, Reykjavík, Iceland

Halldór Geirsson, Physics Department, Icelandic Meteorological Office, Reykjavík, Iceland

Andy Hooper, Department of Earth Observation and Space Systems, Delft University of Technology, Delft, The Netherlands

Ólafur Ingólfsson, University of Iceland, Institute of Earth Sciences, Reykjavík, Iceland

Kurt H. Kjær, Natural History Museum of Denmark, University of Copenhagen, Copenhagen K., Denmark

Knud Erik S. Klint, The Geological Survey of Denmark and Greenland (GEUS), Copenhagen K., Denmark

Johannes Krüger, Department of Geography and Geology, University of Copenhagen, Geocenter Copenhagen, Copenhagen K., Denmark

Guðrún Larsen, Institute of Earth Sciences, University of Iceland, Reykjavík, Iceland

Nigel P. Mountney, Earth Sciences, School of Earth and Environment, University of Leeds, Leeds, UK

Hreggviður Norðdahl, University of Iceland, Institute of Earth Sciences, Reykjavík, Iceland

Benedikt G. Ófeigsson, Institute of Earth Sciences, University of Iceland, Reykjavík, Iceland

Halldór Ólafsson, Nordic Volcanological Centre, University of Iceland, Reykjavík, Iceland

Niels Richardt, Rambøll – Energy & Environment, Virum, Denmark

Andrew J. Russell, School of Geography, Politics & Sociology, Newcastle University, Newcastle Upon Tyne, UK

Anders Schomacker, Institute of Earth Sciences, University of Iceland, Reykjavík, Iceland

Freysteinn Sigmundsson, Nordic Volcanological Centre, University of Iceland, Reykjavík, Iceland

Oddur Sigurðsson, Icelandic Meteorological Office, Reykjavík, Iceland

Erik Sturkell, Department of Earth Sciences, University of Gothenburg, Gothenburg, Sweden and Nordic Volcanological Centre, University of Iceland, Reykjavík, Iceland

Jaap J.M. van der Meer, Centre for Micromorphology UoL, Department of Geography, Queen Mary, University of London, London, UK

Preface

The access to the Mýrdalsjökull region, especially its northern part, has improved up through the last decades. For this and improved technical abilities, research activities have increased both in the marginal areas and on the ice cap itself, generating a huge dataset. Hitherto, the results of these studies have not been compiled or presented in the context of the entire ice cap and its interaction with the Katla Volcanic System. In addition, there have been strong indications that the Katla central volcano is heating up and might be near an eruption. Thus, it is highly attractive to compile data and publications and present the most recent results in a review – research type book. Therefore, the scope of this volume is first to review and update existing knowledge on the glacial and volcanic environments of the Mýrdalsjökull region before the next eruption, and second to produce a standard reference work for studies of Mýrdalsjökull and Katla. Another consequence of the better access to the area is that the interest from the scientific community, i.e. international excursions and field courses with a didactic purpose, has increased immensely in the Mýrdalsjökull area. We therefore hope that this volume will be useful as an advanced guide as well.

The editors of this volume have worked continuously in the Mýrdalsjökull area. *Johannes Krüger* initially took the initiative to carry out research about the glacial geology and geomorphology of Mýrdalsjökull, and he has been PI for numerous research projects since 1977. The large amount of data collected during these projects is reflected in many of the chapters in this volume. The idea for this book also originates from these research projects. *Kurt H. Kjær* has carried out research at Mýrdalsjökull since 1995, and *Anders Schomacker* since 2001. The editors therefore have in-depth experience and knowledge on the state-of-the-art with regards to research in different geosciences. The contributors to this volume are internationally acknowledged scientists and make up a highly qualified group with extensive research experience from the Mýrdalsjökull region and southern Iceland.

After a general introductory chapter (Johannes Krüger, Anders Schomacker and Kurt H. Kjær) follow two chapters about Katla, one chapter dealing with the Katla and Eyjafjallajökull volcanoes (Erik Sturkell, Páll Einarsson, Freysteinn Sigmundsson, Andy Hooper, Benedikt G. Ófeigsson, Halldór Geirsson and Halldór Ólafsson) and another chapter about the eruption history of Katla and the tephrochronology (Guðrún Larsen). The fourth chapter describes the deglaciation and Holocene glacial history of Iceland and the Mýrdalsjökull area (Ólafur Ingólfsson, Hreggviður Norðdahl and Anders Schomacker). Then follows a chapter on the fluctuations of Mýrdalsjökull during postglacial times (Oddur Sigurðsson). The geomorphic and structural genesis of marginal moraines produced along the front of the most prominent outlet glaciers of the Mýrdalsjökull ice cap, such as Kötlujökull, Sléttjökull and Sólheimajökull, are described in Chapter 6 (Johannes Krüger, Anders Schomacker and Ívar Örn Benediktsson). Chapter 7 is devoted to dead-ice environments and a landsystems model for a debris-charged stagnant lowland glacier margin as seen in the terminus region of Kötlujökull (Johannes Krüger, Kurt H. Kjær and Anders Schomacker). Then follows a chapter about sediments and landforms produced in subglacial environments and with examples from Sléttjökull, Kötlujökull and Sólheimajökull (Anders Schomacker, Kurt H. Kjær and Johannes Krüger). Chapter 9 is a case study dealing with evidence for subglacial deformation and deposition during a complete advance-stagnation cycle at Kötlujökull (Knud Erik S. Klint, Niels Richardt and Johannes Krüger). The following chapter treats the micromorphology of glacial sediments (Jaap J.M. van der Meer, Simon Carr and Kurt H. Kjær). The final chapter describes volcanogenic glacier outburst floods (jökulhlaups) from Mýrdalsjökull and their impacts on proglacial environments (Andrew Russell, Robert Duller and Nigel P. Mountney).

We are most grateful to the referees for their constructive suggestions for improvements: Ívar Örn Benediktsson, Svend Funder, Per Möller, Heidi Soosalu and Stefan Wastegård. Furthermore, A.S. acknowledges the financial support from Nordvulk, University of Iceland, Landsvirkjun, and the Carlsberg Foundation for research in Iceland in the recent years.

Anders Schomacker, Johannes Krüger, Kurt H. Kjær,
Reykjavík and Copenhagen

1

The Mýrdalsjökull Ice Cap, Iceland: Glacial Processes, Sediments and Landforms on an Active Volcano: An Introduction

Johannes Krüger[1,*], Anders Schomacker[2] and Kurt H. Kjær[3]

[1]*Department of Geography and Geology, University of Copenhagen, Øster Voldgade 10, DK-1350 Copenhagen K, Denmark*
[2]*Institute of Earth Sciences, University of Iceland, Askja, Sturlugata 7, IS-101 Reykjavík, Iceland*
[3]*Natural History Museum of Denmark, University of Copenhagen, Øster Voldgade 5-7, DK-1350 Copenhagen K, Denmark*
[*]*Correspondence and requests for materials should be addressed to Johannes Krüger (e-mail: jk@geo.ku.dk)*

Glacier–volcano interactions are among the most dynamic landforming processes in Iceland. Volcanic and geothermal activities may trigger dramatic events, such as glacier surges, volcano–glacial jökulhlaups, and tephra deposition on glacier surfaces. Studies of processes related to glacier–volcano interaction and sedimentation in Iceland are highly relevant, because such processes may seriously affect populated areas, the infrastructure, engineering projects and the environment in a wider perspective. Furthermore, investigations of volcanic and glacial sedimentary processes, especially at Icelandic lowland glaciers, have provided depositional models for landscapes in former glaciated areas both in Iceland and elsewhere. Therefore, process–sediment–landform studies at Icelandic glacier margins are highly useful both for reconstructing the impact of volcanic events on the glacial environment and for generating depositional models for glacial sediments. In most reconstructions of glacial and volcanic events in Iceland, establishment of a high-resolution geochronology is realistic, because of tephrochronology, the abundance of organic material for ^{14}C dating and the emerging field of cosmogenic exposure dating. This makes Iceland an attractive 'laboratory' for geoscientists working with the intimate interplay between volcanism and glacier dynamics.

1.1. The Unique Geological Qualities of the Mýrdalsjökull Area

The Mýrdalsjökull ice cap is a geologically highly dynamic area demonstrating the interaction between 'fire, ice and water'. The ice cap is located on the Katla central volcano, and wide areas beneath and around the ice cap have shown intense volcanic activity throughout the Holocene. Epicentres of earthquakes have been located beneath the ice cap, and a large caldera, 100 km^2, situated under the ice cap includes the Katla eruption site, the second most active volcano in Iceland. It has been estimated that more than 30 km^3 of basaltic material has been produced beneath Mýrdalsjökull during the Holocene. Eruptions from the Katla volcanic system occur on average twice a century; between 1580 and 1918 there have, on an average, been 42 years between Katla eruptions, but since 1918 no major eruption has occurred.

Katla eruptions have caused great damage in south Iceland, not only with lava flows and ash falls, but they also triggered the most voluminous and catastrophic glacier bursts known to occur on Earth in our times, flooding large areas of Mýrdalssandur. During these 'Katlahlaups' the peak discharge may exceed some 100,000 m^3/s. This equals the discharge from the river Amazon. After the last Katla eruption in 1918 many blocks of ice, 40–50 m high, were left on the outwash plain west of Hafursey, and masses of glacier ice formed a floating barrier some kilometres offshore showing the effects of the glacier burst accompanying the eruption. As present evidence of glacier bursts, large areas of the upper parts of the Mýrdalssandur are strewn with boulders, 0.5–3 m in size. The heavy glacier bursts were furthermore accompanied by fast shoreline progradation, for example before AD 1300 the table mountain Hjörleifshöfði was a promontory cliff washed by the sea, but now the distance from this mountain to the shore is 3 km.

For many scientists the Mýrdalsjökull ice cap has served as an excellent 'ice-age' laboratory, and an increasing number of publications about glacial processes, sediments and landforms from this region are found in most major peer-reviewed scientific journals. From the ice cap proper, several outlet glaciers, such as Sólheimajökull, Kötlujökull, Sandfellsjökull, Öldufellsjökull, Sléttjökull and Entujökull, terminate in forefields of different nature and demonstrate different conditions of ice activity and glacial environments coupled to volcanic activity. For instance, Sléttjökull provides suitable conditions of studying geomorphological and sedimentological characteristics resulting from subglacial deposition and frontal glacier retreat with a well-defined ice margin and dynamically active ice. In contrast to this, the

DEVELOPMENTS IN QUATERNARY SCIENCES
VOLUME 13 ISSN 1571-0866

terminus region of Kötlujökull consists of stagnant, debris-covered ice providing suitable conditions of studying de-icing processes, supraglacial mass-movement processes and development of dead-ice moraines. In the 1980s, however, the front of Sléttjökull became roughly stationary, while Kötlujökull was rapidly advancing during the same time. These changes in glacier behaviour opened up possibilities of studying currently forming ice-marginal moraines.

1.2. General Geology and Topography

The hardrock geology of the Mýrdalsjökull area can be divided into three main formations: the Pleistocene Grey Basalt Formation, the Pleistocene Hyaloclastite Formation and the Holocene Lava Formation. The Grey Basalts, which were formed during interglacials, or in supraglacial environments, occur only southwest of Mýrdalsjökull and in small isolated areas south and east of the ice cap. The main part of the area surrounding Mýrdalsjökull, however, is covered by the Hyaloclastite Formation, which consists mainly of pillow lavas, breccias and brownish tuffs produced subglacially during the Pleistocene. These rocks are often capped by the Holocene Lava Formation, which formed subaerially when the subglacial volcanoes protruded through the ice cover. The Holocene Lava Formation, which belongs to the Katla volcanic system, is partly covered by the Mýrdalsjökull ice cap. It is exposed especially north and east of the ice cap. In the areas east and northeast of Mýrdalsjökull, several lava flows originating from individual eruptions are interpreted as evidence of major volcanic events, for instance the AD 934 Eldgjá eruption.

The Katla massif capped by Mýrdalsjökull is a southern spur of the central highland of Iceland. To the west the Katla massif joins the ridge-shaped mountainous, volcanic massif capped by Eyjafjallajökull which rises to 1,666 m a.s.l. During the Little Ice Age, Eyjafjalljökull and Mýrdalsjökull were joined, but in the mid-twentieth century they became separated into the two present ice masses due to a general glacier retreat. North of the Eyjafjallajökull massif, the Markarfljót river valley pushes inland an isolated finger of lowland almost to the northwest region of Mýrdalsjökull.

The topography around Mýrdalsjökull can be divided into three morphological regions: the inland plateau, 500–600 m a.s.l., a marginal zone of the plateau rising up to 600–850 m a.s.l. and deeply cut by valleys, and the coastal lowland below the 100–200 m level. The inland plateau north of Mýrdalsjökull appears as an open plain known as Mælifellssandur, which mainly consists of denudational products as well as glacial, glaciofluvial and aeolian deposits. To the north, the plain is bounded by a differentiated mountain area reaching heights of 900–1,250 m a.s.l. and capped by a small plateau glacier, Torfajökull. Mælifellssandur is diversified by ridges running in straight lines, SW–NE, in general alignment with the tectonic direction of this region and steep-sided, isolated table mountains created by the subglacial volcanism during Pleistocene. Among the ridges is the summit line including Mælifell, 791 m a.s.l. The Holocene volcanism has formed similar linear basalt features in the shape of crater rows, for example the eruption fissure with the Eldgjá crater row which can be traced running northeastwards from beneath Mýrdalsjökull.

To the south and southeast, the ice-capped highland is more or less bounded by abrupt slopes, probably to some extent old sea cliffs of the inner coastal plain. All mountain slopes facing the lowland south of Mýrdalsjökull and the Markarfljót river valley to the west are irregular in outline and incised with numerous steep-sided valleys. Many of them are old offshoot valleys formed by distributary glaciers when outlet glaciers from the Mýrdalsjökull ice cap were more numerous and more extended than at present.

Most of the coastal lowland fringing the inland plateau is outwash plains, for example the Skógarsandur, Sólheimasandur, and the extensive Mýrdalssandur. The latter separates the southeast front of Mýrdalsjökull from the coast, some 15–25 km to the south and southeast. Here and there, table mountains belonging to the Pleistocene Hyaloclastite Formation rise isolated or in groups from the surface of Mýrdalssandur and break the monotony of the plain, for example Hafursey reaching 582 m a.s.l.; Hjörleifshöfði, 221 m a.s.l.; and Sandfell, 597 m a.s.l. Around Rjúpnafell, 342 m a.s.l., and southwards to the seashore, solid rock, such as the Álftaver lava, the Hólmsá lava, and the Rjúpnafell lava, appears sporadically as morphological components which break the monotony of the extensive Mýrdalssandur.

Morphologically, the Mýrdalsjökull ice cap covering around 600 km^2 can be divided into two regions: the ice cap proper, where the ice forms an unbroken cover down to the 1,000–1,300 m level, and below that level, the peripheral zone, where the ice cap splits up into separate outlet glaciers. The ice cap proper culminates in two gently rising summits, 1,480 m a.s.l. in Goðabunga and 1,450 m a.s.l. in Háabunga, both of them situated in the southern part of the ice cap. The two cupolas are separated by a wide depression down to the 1,345 m level. In the higher parts of the peripheral zone of the ice cap the ice is penetrated by small nunataks here and there, and in between the ice forms several depressions which are the upper parts of a radial system of outlet glaciers. In the south, valley-type outlet glaciers descend over rock steps, for example Sólheimajökull which flows southwards down to about 100 m a.s.l. Smaller, steeper ice tongues drain towards the west. The outlet glacier Kötlujökull extending towards the southeast expands on the coastal plain into a piedmont shaped glacier and terminates around 220 m a.s.l. Sandfellsjökull and Öldufellsjökull are outlet glaciers descending from the ice cap to the east. Except for the outlet glaciers Merkurjökull and Entujökull extending towards the northwest, the northern margin of the ice cap is lobate and terminates at 550–600 m a.s.l. The ice cap proper merges continuously into this relatively gently sloping and 20 km broad glacier lobe named Sléttjökull, which is bounded by the nunatak Enta in the west and Öldufell in the east.

Meltwater issuing from the western part of the northern terminus region of the ice cap is drained off towards the northwest by Bláfjallakvísl, which flows into the river Markarfljót, while the eastern part of the northern terminus region is drained northeastwards by the river Brennivínskvísl, which carries the meltwater into the river Hólmsá. The drainage from the eastern outlet glaciers is concentrated into four main systems: Jökulkvísl collecting meltwater from Öldufellsjökull, and the braided river Leirá, which drains Sandfellsjökull and the northeastern part of Kötlujökull; both of them join the river Hólmsá. Þverkvísl drains the central part of Kötlujökull and the river Múlakvísl drains the extreme south of the glacier.

1.3. Aim of This Book

The present volume contains chapters covering a wide range of geoscientific topics related to Katla and Mýrdalsjökull. A chapter can be read as an individual paper, whereas the broad overview is given by the complete volume. Each chapter reflects the data, interpretations and conclusions of the individual authors. Thus, there are contrasting viewpoints and conceptions between some chapters – highlighting that there are still many questions to be answered from future research about Katla and Mýrdalsjökull.

The specific aims of study are given in each chapter. Some of the overall aims of this volume are to answer the following research questions: How is the current status of the Katla central volcano and the Mýrdalsjökull ice cap on top of it? What is the eruption history of Katla and the glacial history of Mýrdalsjökull? How does the Katla volcano affect the glacial processes? The last part of the book contains chapters with focus on geomorphology, sedimentology and processes. Their overall aim is to investigate the genesis of sediments and landforms in modern glacial and glacio-volcanic environments in order to provide depositional models for ancient deposits. Many of these chapters build on the conception of Mýrdalsjökull as an 'ice age laboratory', where geological processes can be directly observed.

2

Katla and Eyjafjallajökull Volcanoes

Erik Sturkell[1,2,*], Páll Einarsson[3], Freysteinn Sigmundsson[2], Andy Hooper[4], Benedikt G. Ófeigsson[3], Halldór Geirsson[5] and Halldór Ólafsson[2]

[1]*Department of Earth Sciences, University of Gothenburg, Gothenburg, Sweden*
[2]*Nordic Volcanological Centre, University of Iceland, Askja, Sturlugata 7, IS-101 Reykjavík, Iceland*
[3]*Institute of Earth Sciences, University of Iceland, Askja, Sturlugata 7, IS-101 Reykjavík, Iceland*
[4]*Department of Earth Observation and Space Systems, Delft University of Technology, Delft, Netherlands*
[5]*Physics Department, Icelandic Meteorological Office, Reykjavík, Iceland*
**Correspondence and requests for materials should be addressed to Erik Sturkell (e-mail: erik.sturkell@gvc.gu.se)*

2.1. Introduction

The volcanic system of Katla is one of the most active ones in Iceland with at least twenty eruptions within the central volcano (Larsen, 2000) and one in its fissure swarm during the past 1,100 years. The volcano is covered mostly by the Mýrdalsjökull ice cap (Fig. 2.1); consequently, eruptions within the Katla central volcano are phreato-magmatic and capable of producing glacial bursts, that is, jökulhlaups. One of the most voluminous eruptions ($\sim 18.6\,km^3$, Thordarson and Larsen, 2007) in historical times since AD 874, the Eldgjá eruption AD 934–940, originated from the Katla fissure swarm extending from under the Mýrdalsjökull ice cap towards the north-east. Eruptions within the Katla caldera are much smaller with an upper limit of volume of approximately $2\,km^3$. The neighbouring volcano, 25 km to the west, is Eyjafjallajökull (also referred to as Eyjafjöll). It is currently much less active, with two eruptions during the past 1,100 years, occurring in tandem with Katla eruptions in 1612 and 1821–1823. The erupted volumes from the Eyjafjallajökull volcano have been negligible in historic times and range in the $0.1\,km^3$ scale. However, they can be extremely hazardous, as farms are located at the foot of the volcano, which rises 1.5 km from the agricultural plain.

As most eruptions are confined to the ice-covered Katla caldera area, huge amounts of meltwater are released in the beginning of an eruption. The path of these outburst floods depends critically on the location of the eruption site within the caldera, in relation to the present water divides under the ice within the caldera. In historical time, floods have on most occasions drained to the east and at a few instances also to the south. Jökulhlaups and fallout of tephra are primary hazards during a Katla eruption, and the proximity of the volcano to populated areas and to international flight paths makes it a potent threat.

Different chemistry and mineralogy distinguish the two volcanoes. FeTi basalt dominates volcanism of the Katla system. The Eyjafjallajökull volcanic system, on the contrary, has produced a suite of alkalic rocks ranging from ankaramites to hawaiite and minor silicic rocks (Jakobsson, 1979).

This compiled overview of geology, seismicity and crustal deformation is based principally on two research articles: Sturkell *et al.* (2003, 2008), the time series have been extended to include the collected data up to year 2008.

2.2. Tectonic Setting

The Katla and Eyjafjallajökull volcanoes are located outside the main zones of divergent plate motion. In south Iceland, the plate boundary is in a state of transition (e.g. Einarsson, 1991; Sigmundsson *et al.*, 1995; Sigmundsson, 2006). Plate divergence is presently taken up by two parallel rift zones: the Western Volcanic Zone (WVZ) and the Eastern Volcanic Zone (EVZ). Latitude-dependent variations in spreading rate are observed along both zones, with the EVZ accommodating 40–100% (Fig. 2.2) of the relative motion between the North American and Eurasian plates (LaFemina *et al.*, 2005). The spreading rate decreases southwards and active rifting terminates south of the Torfajökull volcano (Figs. 2.1 and 2.2), where the EVZ rift zone meets the transform boundary of the South Iceland Seismic Zone.

Large volcanoes characterize the segment of EVZ south of its intersection with the transform. Rifting structures on the surface are inconspicuous in this area. From the area between Torfajökull and Mýrdalsjökull to the Vestmannaeyjar islands, the influence of the EVZ is primarily seen in the decreasing alkalinity of rocks with increasing distance from the tip of the rift zone (Óskarsson *et al.*, 1982). The arguments are mainly geochemical and structural. FeTi volcanism, characteristic for propagating rifts, is found within this area, beginning 2–3 Myr ago (Jóhannesson *et al.*, 1990). Voluminous volcanism has created a plateau of FeTi basalt south of the Torfajökull area towards the south coast in Mýrdalur (Fig. 2.1). Therefore, Katla and the neighbouring volcanoes Eyjafjallajökull and Vestmannaeyjar (including

DEVELOPMENTS IN QUATERNARY SCIENCES
VOLUME 13 ISSN 1571-0866

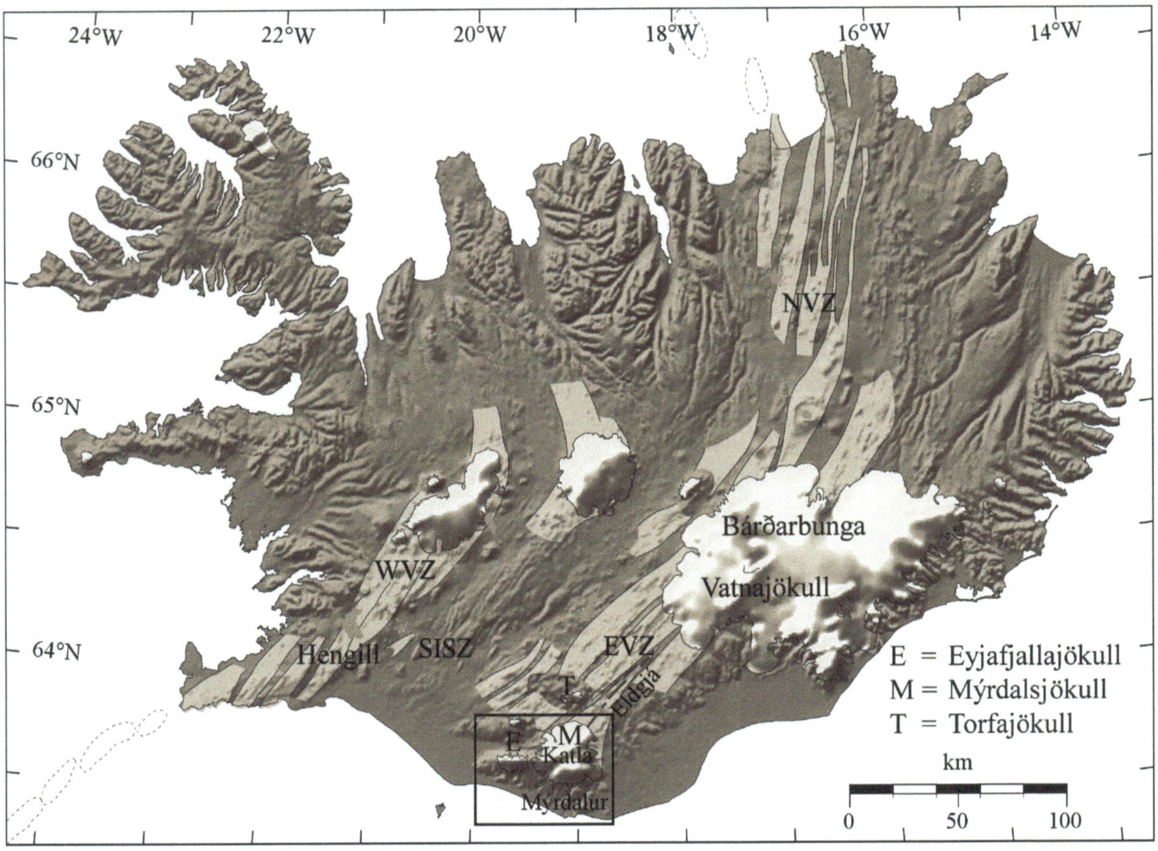

Fig. 2.1. Map of Iceland, showing the neo-volcanic zone consisting of individual volcanic systems coloured in yellow. Dashed lines mark submarine volcanic system and white areas are glaciers. The plate boundary in Iceland is expressed by different segments and is divided into the Northern Volcanic Zone (NVZ), the Western Volcanic Zone (WVZ), and the Eastern Volcanic Zone (EVZ) (Einarsson and Saemundsson, 1987). The EVZ and the WVZ are connected by the South Iceland Seismic Zone (SISZ). The area including Eyjafjallajökull and Katla is outlined with a box. The Eldgjá fissure is marked in red.

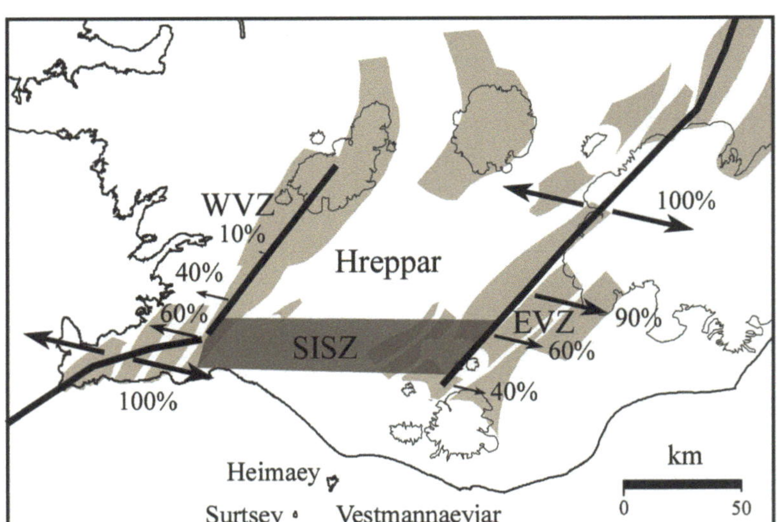

Fig. 2.2. Partition of plate spreading between the Western Volcanic Zone (WVZ) and the Eastern Volcanic Zone (EVZ). The spreading rate increases from north to south along the WVZ progressively from zero to 60% of spreading at the Hengill volcano and the opposite pattern is valid for the parallel EVZ. The surface structures associated with a rift in the EVZ terminates just north of the Mýrdalsjökull ice cap. The Hreppar microplate is located between the parallel spreading zones. Plate spreading data are taken from LaFemina et al. *(2005).*

the Heimaey and Surtsey eruption sites, see Fig. 2.2) have spreading velocities consistent with a location on the stable Eurasian plate (LaFemina *et al.*, 2005; Geirsson *et al.*, 2006). Hence, Katla can be classified as an intraplate volcano, in spite of its occasional connection with rifting in the EVZ, exemplified by the AD 934 Eldgjá eruption (Fig. 2.1).

2.3. Structure and Eruptive Products of the Volcanoes

2.3.1. Katla

The Katla volcano hosts a 600- to 750-m-deep caldera filled with ice (Björnsson *et al.*, 2000). The caldera rim is

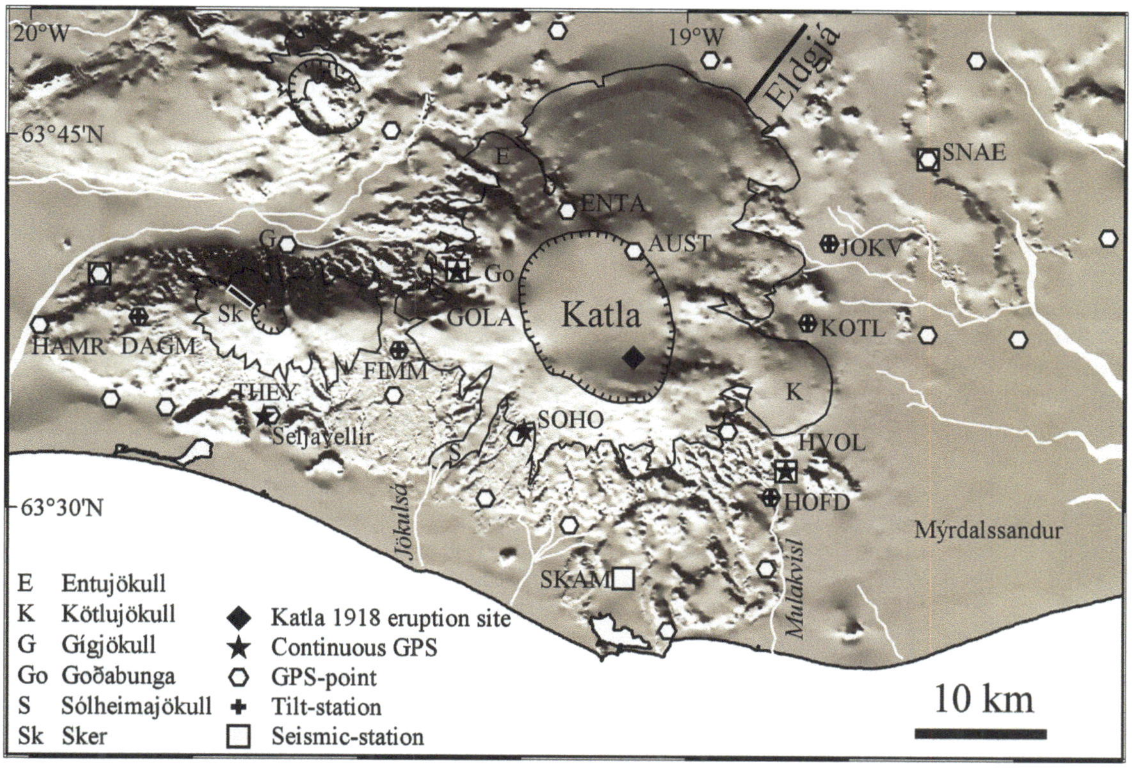

Fig. 2.3. The geodetic and seismic networks around Eyjafjallajökull and Katla (Mýrdalsjökull) showing campaign GPS, continuous GPS, tilt and seismic stations. On Mýrdalsjökull glacier, the hachured line is the Katla caldera, and the hachured line on Eyjafjallajökull marks the summit crater.

breached in three places, to the south-east, north-west and south-west. These gaps in the caldera rim provide outflow paths for ice in the caldera to feed the main outflow glaciers, Kötlujökull, Entujökull and Sólheimajökull (Fig. 2.3). Apart from the large Eldgjá flood lava eruption AD 934–940 (Thordarson *et al.*, 2001), all historical eruptions of the Katla volcanic system have occurred within the caldera (Larsen, 2000).

Volcanic products of Katla are primarily bimodal in composition, comprising alkali basalt and mildly alkalic rhyolites (Lacasse *et al.*, 2007). Intermediate rocks, mostly basalt-rhyolite hybrids and occasional hawaiite, are very subordinate (Lacasse *et al.*, 2007). Volumetrically, FeTi-rich basalt with aphyric appearance dominates. This is ascribed to rapid segregation of material from a large mantle source beneath a propagating rift (Sinton *et al.*, 1983). Silicic volcanism is an important component of Katla activity. At least twelve silicic tephra layers are known from Katla during the Holocene, between 1700 and 6600 BP (Larsen *et al.*, 1999; Larsen, 2000), and almost all known outcrops of Katla at the caldera rim and immediately outside of it are silicic (Jóhannesson *et al.*, 1990; Lacasse *et al.*, 2007). Katla tephras show that phreato-magmatic eruptions have taken place throughout the Holocene (Óladóttir *et al.*, 2005). The relatively homogeneous chemical composition of the products of large basaltic eruptions of Katla during the past 1,000 years seems to indicate that they have not had time to evolve in a shallow magma chamber (Óladóttir *et al.*, 2005, 2008). The wide range of composition for Katla magmas in general (Lacasse *et al.*, 2007), however, points to a complicated magma plumbing system, which may even change significantly on a time scale of thousands of years (Óladóttir *et al.*, 2008).

Seismic undershooting within the Katla caldera has revealed a zone where P-wave velocities are reduced and S-waves are absent; this anomaly is interpreted as evidence of a magma chamber (Guðmundsson *et al.*, 1994). Moreover, results from an aeromagnetic survey indicate the presence of a non-magnetic body within the region of the postulated magma chamber (Jónsson and Kristjánsson, 2000).

2.3.2. *Eyjafjallajökull*

The Eyjafjallajökull volcano is an elongated, flat cone of about 1,600 m height. The Eyjafjallajökull glacier, up to 200 m thick (Guðmundsson and Högnadóttir, 2005), covers the volcano and its elliptical 2.5-km-wide summit crater (Fig. 2.3). The outlet glacier Gígjökull originates from the crater and flows towards the north through an opening in the crater rim. The most recent eruption in 1821–1823 occurred within the crater close to its southern rim and produced intermediate to acid tephra (Thoroddsen, 1925).

The Eyjafjallajökull volcano has an alkaline composition, similar to other off-rift volcanoes in Iceland. This type of volcano generates relatively small amounts of material; typically 0.1 km^3, during each eruption. Most eruptive fissures and crater rows at Eyjafjallajökull are E-W orientated, but occasional radial fissures are observed around the summit of the volcano. The most conspicuous radial eruptive fissure is Sker (Fig. 2.3). In the area SSE of the summit crater and NE of the Seljavellir farm (Fig. 2.3), Jónsson (1998) reports the presence of highly altered rocks that are cut by numerous dikes and veins. This area is interpreted to be the oldest part of the Eyjafjallajökull volcano, with a suggested age of more than 0.78 Myr. If this age estimate is accurate, Eyjafjallajökull is one of the oldest active volcanoes in Iceland. The most pronounced expression of geothermal activity at Eyjafjallajökull is confined to its south flank, in the area around Seljavellir (Fig. 2.3). This area of geothermal activity correlates with the location of recently formed intrusions.

2.4. Recent Unrest in Katla

2.4.1. The 1918 Eruption

The latest large eruption of Katla began on 12 October 1918, lasting for some three weeks (Jóhannsson, 1919; Sveinsson, 1919). It was basaltic in composition, and the eruption site was near the southeast rim of the Katla caldera, beneath about 400 m of ice. A debris-laden jökulhlaup was seen propagating over Mýrdalssandur sand plains (Fig. 2.3) a few hours from the onset of the eruption. The 1918 eruption of Katla was large, but estimates of the total volume of erupted material vary. The amount of tephra fallout is estimated to be 0.7 km^3 (Eggertsson, 1919), and the volume of water-transported material is estimated at between 0.7 and 1.6 km^3 (Larsen, 2000). The dense-rock equivalent may have been as high as 1 km^3.

2.4.2. The 1955 Event

It is possible that a short-lived subglacial eruption took place in 1955 on the eastern rim of the Katla caldera; however, no tephra erupted into the atmosphere. Instead, two shallow ice cauldrons formed on the surface of Mýrdalsjökull and a small jökulhlaup drained from Kötlujökull glacier tongue (Rist, 1967; Thórarinsson, 1975).

2.4.3. The 1999–2005 Episode

In July 1999, an unexpected jökulhlaup drained from Mýrdalsjökull. Its timing is marked in Fig. 2.4 and is followed by elevated earthquake activity. This short-lived jökulhlaup was preceded several hours earlier by earthquakes and pulses of low-frequency tremor that originated from Mýrdalsjökull. Inspection of Mýrdalsjökull revealed a newly formed surface depression (ice cauldron) near to the epicentres of the earthquake swarm that preceded the jökulhlaup. The latest Katla eruption to break the ice surface took place in 1918. During the weeks following the 1999 jökulhlaup, increased geothermal activity was detected along the caldera rim, as manifest by the deepening of pre-existing ice cauldrons (Guðmundsson *et al.*, 2007).

From 2000 until 2004, uplift at the Global Positioning System (GPS) point AUST occurred at a rate of 1.7 cm/yr (Figs. 2.3 and 2.5), but interferometric measurements of deformation from radar satellites (InSAR) indicate no simultaneous uplift of the volcano flanks. Figure 2.6A shows the derived displacement vectors from benchmarks on and around Mýrdalsjökull, relative to stable Eurasia. Vertical uplift and radial displacement from the caldera centre were observed.

Using a point source model, these data place the centre of the magma chamber at 4.9 km depth beneath the northern part of the caldera. However, this depth may be overestimated because of a progressive decrease in the mass of the overlying ice cap, in reality, the depth may be only 2–3 km. About 0.01 km^3 of magma has accumulated between 1999 and 2005. This value is considerably less than the estimated 1 km^3 of material erupted during the latest eruption of Katla in 1918.

2.5. Recent Unrest in Eyjafjallajökull

Since 1994, the ice-capped Eyjafjallajökull volcano, situated 25 km west of Katla, has generated by far the largest amount of crustal deformation (Sturkell *et al.*, 2003). In 1994, and again in 1999, magma intrusion was detected under the southern slopes of Eyjafjallajökull. These intrusions had their centre of uplift approximately 4 km southeast of the summit crater of the volcano (Sturkell *et al.*, 2003; Pedersen and Sigmundsson, 2004, 2006) and were associated with considerable seismic activity (e.g. Dahm and Brandsdóttir, 1997). After the intrusion event in 1999, crustal deformation and earthquake activity at Eyjafjallajökull (Fig. 2.4) have remained low (Sturkell *et al.*, 2006).

This is, however, not the only known case of simultaneous unrest of Eyjafjallajökull and Katla. The two volcanoes apparently erupted in 1612, and the Eyjafjallajökull eruption of 1821–1823 was immediately followed by an eruption of Katla (Thoroddsen, 1925). These are the only known eruptions of Eyjafjallajökull in historic times.

2.6. Earthquakes

2.6.1. Seismic Activity of the Katla and Eyjafjallajökull Region

Katla stands out among Icelandic volcanoes for its high and persistent seismic activity. The only other volcanoes that compare to Katla in this respect are Hengill and Bárðarbunga (Fig. 2.1) (Einarsson, 1991; Jakobsdóttir, 2008), both of which are located on main branches of the plate boundary. As Katla is located at some distance from the main deformation zone of the plate boundary, its high seismicity is somewhat unique.

The seismicity of the Mýrdalsjökull region forms three distinct groups (Figs. 2.4 and 2.7), within the Katla caldera, at Goðabunga on the western flank of Katla and beneath Eyjafjallajökull. Seismic characteristics of

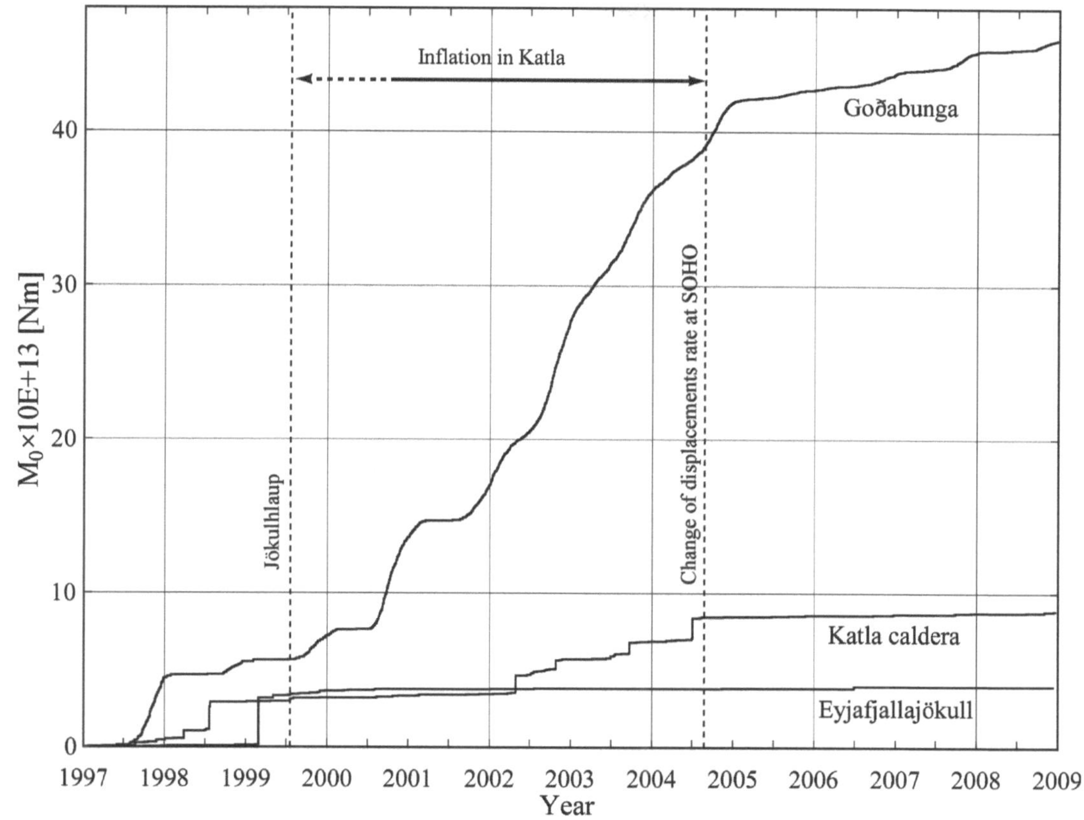

*Fig. 2.4. Cumulative seismic moment released in the Katla caldera, the Goðabunga area and Eyjafjallajökull region, for the time period from 1 January, 1997, to 1 January, 2009. All earthquakes larger than magnitude 0.5*M_{LW} *are used in the moment-magnitude relation* $\log M_0 = 1.5M_{LW}+9.1$, *where* M_0 *is the seismic moment (Nm) and* M_{LW} *the local moment magnitude. Of the three regions, Goðabunga is by far the most active with continuous activity. Between 2002 and the beginning of 2005, earthquakes took place year-round within the Goðabunga region. Note that a new ice cauldron (Sturkell et al., 2008) was formed before the onset of sustained earthquake activity in the Goðabunga region. The character of the seismicity within the Katla caldera and in the Eyjafjallajökull volcano is more swarm-like. The 1999 swarm in Eyjafjallajökull was accompanied by a shallow intrusion. The vertical line in July 1999 represents the jökulhlaup that drained from Mýrdalsjökull (emanating from the Sólhemajökull). The vertical line in late August 2004 represents the change of displacements rate at SOHO continuous GPS-station.*

these groups are quite distinct and different from one another.

Seismicity at Eyjafjallajökull is episodic by nature (Fig. 2.8A). In fact, most of the time, Eyjafjallajökull is almost aseismic. Three episodes of seismicity occurred in the 1990s, in 1994, 1996 and 1999 (Jakobsdóttir, 2008; Dahm and Brandsdóttir, 1997). The 1994 and 1999 episodes were associated with intrusive activity beneath the SE flank of the volcano (Sturkell *et al.* 2003; Pedersen and Sigmundsson, 2004, 2006); the 1996 episode was much smaller and may have been related to the development of a conduit feeding the shallow intrusions (Fig. 2.8A). The beginning of each of the active periods seems to indicate that the semi-vertical conduit is located beneath the northern flank, whereas the magma was subsequently intruded southwards as sills at 4–6 km depth beneath the south-eastern flank. Most of the earthquakes at Eyjafjallajökull are high-frequency events of low magnitude but with sharp P- and S-waves. They have only once reached magnitude 3, preceding the beginning of the 1999 intrusion.

The Katla caldera epicentral cluster is not quite concentric with the caldera: its centre is displaced slightly to the NE with respect to the caldera centre (Fig. 2.7). The largest magnitude earthquakes of the whole area take place within this cluster. The largest event recorded so far was that of June 2, 1977 (m_b 4.9; M_S 5.0, body-wave magnitude and surface-wave magnitude, respectively). There is still some controversy regarding the depth of the caldera earthquakes. Arrival time data for most of the events are consistent with a shallow source, 0–5 km, but deeper sources are not excluded (Vogfjörð and Slunga, 2008). The epicentral cluster is located in the same general area as the low-velocity, high-attenuation body detected by Guðmundsson *et al.* (1994) and interpreted as a magma chamber at a shallow level. It is therefore plausible that the earthquakes are related to stress changes around and above this chamber induced by pressure changes. The appearance of the earthquakes on the seismograms is consistent with this interpretation. They are mostly high-frequency events with some events that would probably classify as hybrid events. The focal mechanism studies that have been conducted on the Katla caldera events so far seem to indicate a large component of reverse faulting (Einarsson, 1987; Sturkell *et al.*, 2008). It is possible that some of the seismic activity is related to the extensive geothermal

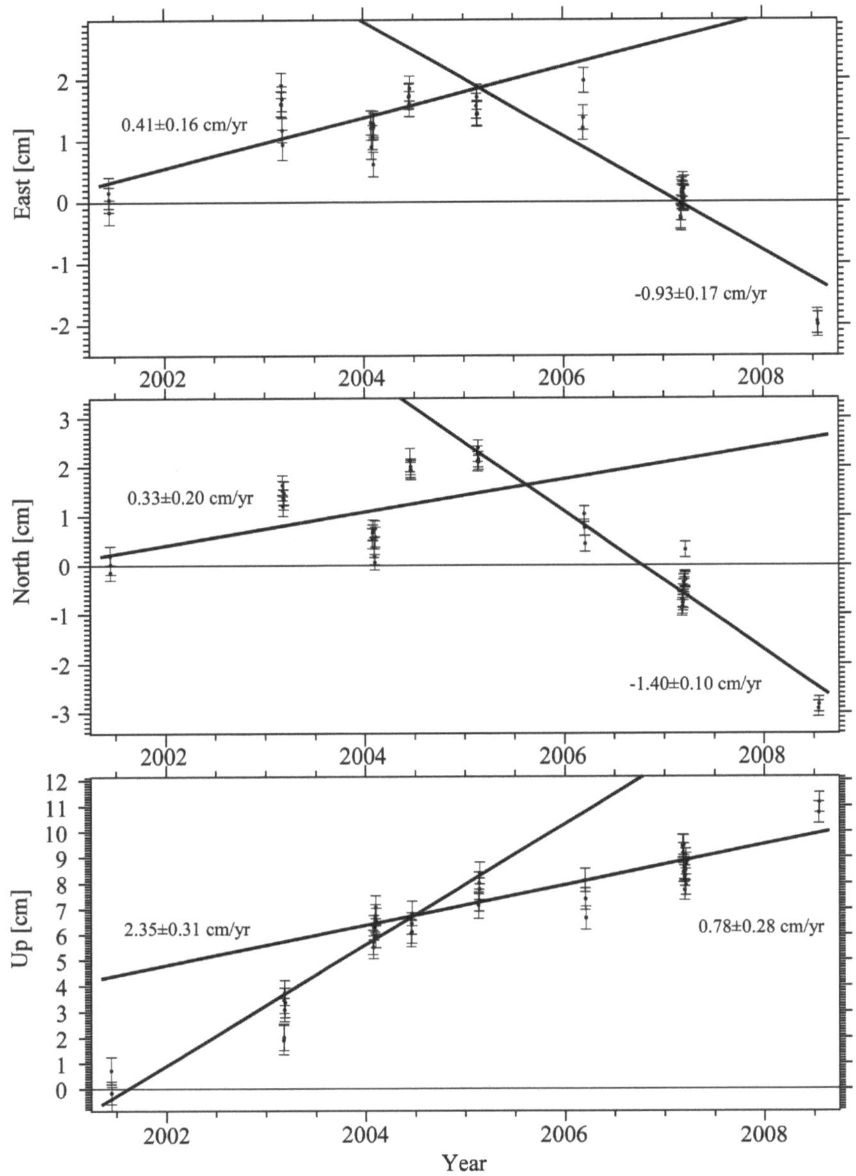

Fig. 2.5. Displacement time series for the campaign GPS site AUST from 2001 to 2008. Horizontal displacements are relative to stable Eurasia but the uplift relative to ITRF 2005 reference frame. The uncertainty is the square of covariance weighted from the least square fit. The sloping lines here, as well as in Fig. 2.11, represent the best line fit before and after August 2004. The displacement rate changed in late August 2004, and this is suggested to represent the termination of the inflation episode.

activity of the Katla caldera region as expressed by the many cauldrons in the ice surface, particularly along the caldera rims (Guðmundsson *et al.*, 2007). The geothermal water systems cool the crust as heat is mined from the magma chamber roof. The stress changes associated with the cooling and contracting crustal rocks may lead to earthquakes.

The Goðabunga earthquake cluster provides some of the most unusual aspects of the seismicity at Mýrdalsjökull. The earthquakes are almost exclusively of the low-frequency type, and this area is one of few areas in Iceland that consistently produce such earthquakes (Soosalu *et al.*, 2006a, 2006b). The earthquakes are often poorly recorded and typically have emergent P-waves, an unclear S-wave and a long low-frequency coda. This makes them difficult to locate. Furthermore, the magnitude scale for these earthquakes is problematic. Because of the long coda, they have a low amplitude/duration ratio, very different from that of 'normal' earthquakes. Magnitudes that are based on maximum amplitude are therefore likely to underestimate the size of the event. Another unusual aspect of the Goðabunga earthquakes is their pronounced seasonal correlation, first described by Tryggvason (1973). The correlation was later confirmed by Einarsson and Brandsdóttir (2000) who point out that both seasonal ice load change and resulting pore pressure change at the base of the ice could induce the modulation of the earthquakes. They argue that the pore pressure effect may be greater than the load effect, which is also consistent with the phase lag of the earthquakes with respect to the time of maximum deloading. The peak in the earthquake activity usually occurs during the fall months, 2–3 months after the maximum in the rate of deloading (Jónsdóttir *et al.*, 2007).

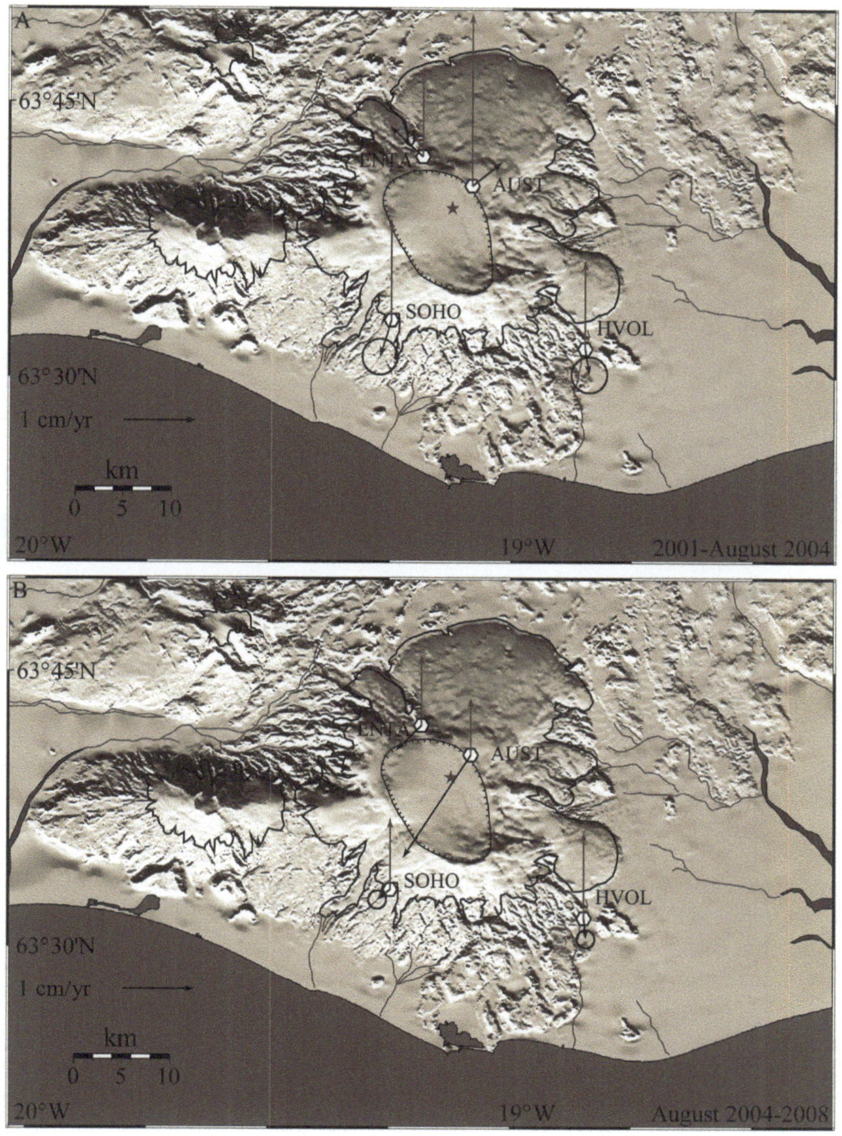

Fig. 2.6. (A) Observed horizontal and vertical displacements at Katla from 2001 to late August 2004, measured by GPS. Black arrows show horizontal displacement, and red arrows show vertical movements. Horizontal displacements are relative to stable Eurasia and vertical displacements are relative to the ITRF2005 reference frame. This uplift and outward displacement from a centre inside the Katla caldera (marked by a red star) suggests magma inflow in a magma chamber, which is possibly located at 3–5 km depth. (B) Observed horizontal and vertical displacements at Katla from the termination of the magmatic uplift in late August 2004 until end of 2008, measured by GPS. Black arrows show horizontal displacement, and red arrows show vertical movements. Horizontal displacements are relative to stable Eurasia and vertical displacements are relative to the ITRF2005 reference frame.

It has been suggested that the Goðabunga cluster is an expression of a rising cryptodome (Einarsson *et al.*, 2005; Soosalu *et al.*, 2006a), that is, a diapir-like volume of viscous magma beneath the surface. Several lines of arguments support this interpretation:

1. The low-frequency characteristics of the earthquakes are typical for dome activity.
2. The deformation field around Goðabunga is very localized. No deformation is detected at nearby tilt and GPS stations. This is one of the characteristics of rising domes (e.g. Poland and Lu, 2008).
3. The earthquake activity is remarkably persistent. There are fluctuations on a months-to-years time scale, but the day-by-day activity is steady.
4. The earthquake cluster is very tight and localized.
5. Geographical position of the cluster with respect to the Katla caldera is the same as for a number of acidic extrusive bodies. They appear to form an aureole around the caldera. Similar relationship is observed at some other central volcanoes in Iceland, such as Krafla (Jónasson, 1994).

Arguments have also been presented for icequakes as the source of the Goðabunga activity (Jónsdóttir *et al.*, 2008). Icequakes are known from several places in the glaciated areas of Iceland, in particular, along the edges of active glaciers and along the largest ice stream of Iceland, the Skeiðarárjökull tongue of Vatnajökull (e.g. Brandsdóttir and Menke, 1990; Roberts *et al.*, 2006).

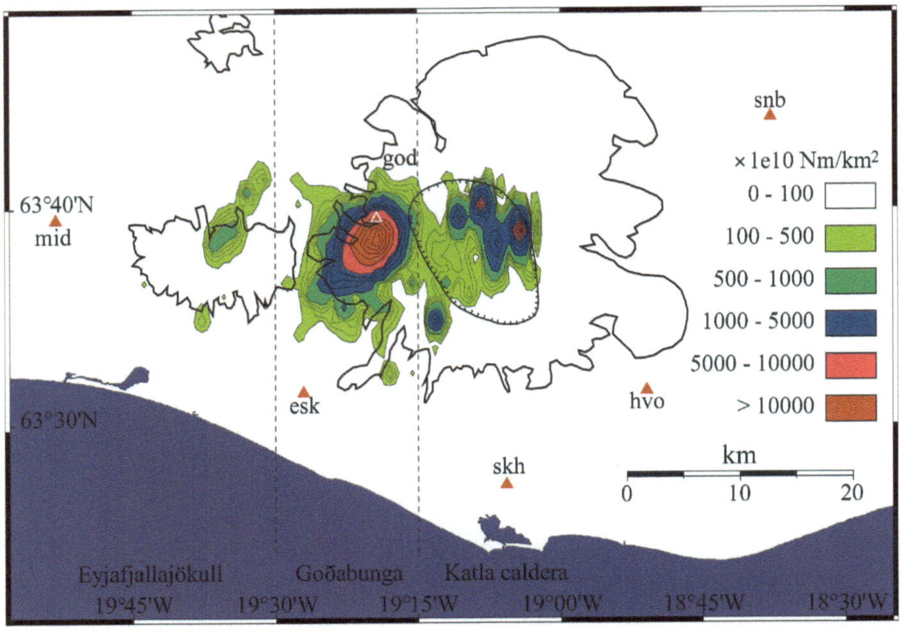

Fig. 2.7. Earthquake activity recorded from 1 January, 1997, to 1 January, 2008, in Eyjafjallajökull and Katla. Contours depict the cumulative seismic moment per square kilometre for the period. Seismic moment was calculated using the moment-magnitude relation $\log M_0 = 1.5M_{LW}+9.1$. *Earthquake activity is focused in three regions, separated by dashed lines: two in the Katla volcano; the area within the Katla caldera (2,176 events) and the Goðabunga region (11,951 events), separated by the dashed line, and in Eyjafjallajökull (396 events). Triangles mark locations of seismic stations of the national network, called with the acronym SIL.*

These events are, however, considerably smaller than the events at Goðabunga. Furthermore, it is difficult to see why icequakes would be concentrated in a tight cluster and be more intensive at Goðabunga than at other glacier edges in Iceland.

Although the seismicity of the Katla area is generally high and persistent, it is by no means constant. Episodes of enhanced activity are separated by longer periods of less intense seismicity. Notable were the two episodes of greatly elevated seismicity in 1967 and 1976–1977 and an episode of unrest that began in 1999 with a flash flood from Sólheimajökull (Fig. 2.3) (Sturkell *et al.*, 2008). Even these episodes are different from each other and probably reflect different processes in the evolution of Katla volcano. The relative importance of the two clusters was comparable during the 1976–1977 episode (Einarsson and Brandsdóttir, 2000), whereas it has changed considerably throughout the 1999–2004 episode (Fig. 2.4). The relative importance of the Goðabunga cluster increased during the episode, and the Katla caldera activity decreased. Also, the seasonal correlation at Goðabunga has changed noticeably. The correlation was strong in 1995–2001, with almost total seismic quiescence during the early part of the year (Fig. 2.4). Then, the activity increased and extended into the quiet periods until the activity became continuous throughout the year. Seasonal correlation was still discernible, but the level of activity was higher than before. Then, in 2004, the activity subsided to the previous level. The maximum magnitudes appear to have diminished, but the quiet periods have not returned (see Fig. 4D of Jakobsdóttir, 2008). At the time of writing (March 2009), the activity level was only moderate, the seasonal correlation was visible, but the activity had continued through the year.

Historical eruptions of Katla have often been preceded by felt and sometimes even damaging earthquakes (e.g. Thoroddsen, 1925; Björnsson and Einarsson, 1981). The precursor time is of the order of half an hour to two hours. Comparing these events to the present activity at Katla, it is quite clear that the historical earthquakes must be either considerably larger than the events that have occurred lately or originate in a different part of the volcano. They demonstrate that a change in the seismicity pattern is to be expected before the outbreak of the next eruption of Katla.

2.7. Crustal Deformation

2.7.1. Tilt

In 1967, three levelling lines, 500–600 m in length (HOFD, KOTL and JOKV in Fig. 2.3), were installed to determine changes in tilt (so-called dry tilt) caused by pressure changes in the Katla volcano. The inflation in 1999–2004 is suggested to be caused by a pressure increase of a source at 4–5 km depth with its apex in the northern region of the caldera (Figs. 2.3 and 2.6A). These stations are located at least 10 km from the Katla caldera rim and even further away from the centre of inflation 1999–2005 and are insensitive to inflation rates of 2 cm/yr during the 1999–2005 pressure build-up in Katla. The uncertainty in tilt obtained from levelling lines is 2–3 μrad. To generate tilt of 2 μrad at the nearest station by pressure increase at 5 km depth, an uplift of about 40 cm is required. The nearest station is KOTL (Fig. 2.3) at 15 km distance from the centre of inflation in 1999–2005. The three tilt stations (KOTL, HOFD and JOKV,

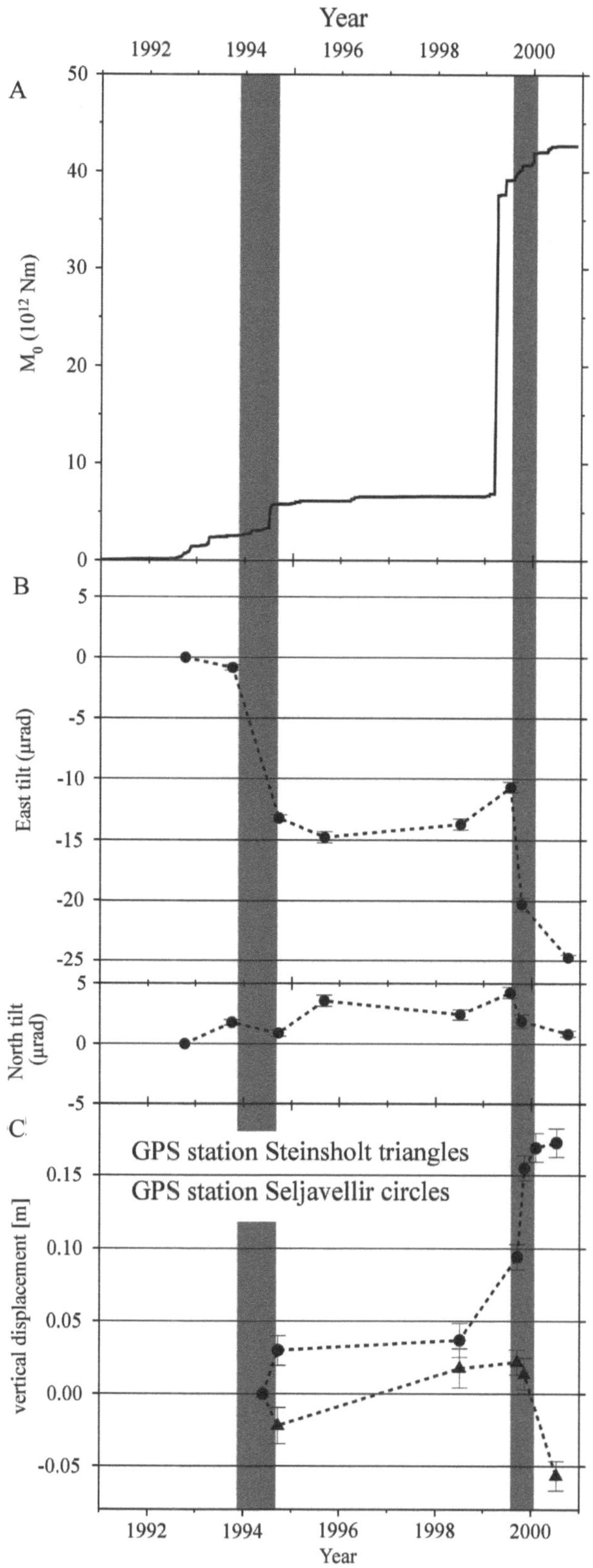

Fig. 2.8. (A) Cumulative seismic moment of earthquakes in the Eyjafjallajökull volcano (from 1 January, 1991, to 1 January, 2002). Seismic activity intensified in the beginning of 1999, and this swarm continued into 2000. Thereafter, the earthquake activity has settled at a higher background level. The 1994 and 1999 periods of crustal deformation (intrusions) are shown as shaded bars (modified after Sturkell et al., 2003). (B) Observed ground tilt (1992–2002) at the optical levelling (dry) tilt station at Fimmvörduháls (installed in 1992), located between Mýrdalsjökull and Eyjafjallajökull (FIMM in Fig. 2.3). The strongest tilt signals are in an eastward direction in 1999. GPS data show that the tilt is a result of uplift under the southern slope of Eyjafjallajökull. (C) Time series of the vertical displacements at SELJ (Seljavellir) and STEI (Steinsholt) GPS stations (see Fig. 2.9) during the 1999–2000 intrusion event, the later subsidence of STEI probably is related to the closeness to the feeder channel. The tilt and GPS-measurements show clearly when the time periods of crustal deformation begin and end. They are shown as shaded bars. The 1999 crustal movement episode is well-constrained by deformation measurements and lags the seismic activity (modified after Sturkell et al., 2003).

Fig. 2.3) are located too far away to be useful in detecting the current pressure changes in the inferred shallow magma camber in Katla.

A completely different story applies for the two tilt stations at each side of Eyjafjallajökull (Fig. 2.3), as they showed two inflation events (Fig. 2.8B). The crustal deformation caused by the two intrusion events (1994 and 1999) in Eyjafjallajökull gave distinct tilt signals at the two stations (DAGM and FIMM) flanking the volcano (Fig. 2.3). The station FIMM gave a clear signal of upward tilt pointing in the direction of the uplift (Figs. 2.8B and 2.9). The large tilt signal observed help to bracket the duration of the intrusions. The DAGM station was installed in 1994 after the first intrusion event. This station gave a clear signal, which was caused by the 1999 intrusion event and narrowed down the area of maximum uplift (Fig. 2.9), which was about half in size compared with the signal at FIMM.

The tilt station FIMM is favourably located for the detection of the deformation bulge on the south flank of Eyjafjallajökull, due to both the small distance and the size of the bulge (35 cm in 1999). A potential shallow deformation source at Goðabunga, on the contrary, does not produce a detectable signal at FIMM in spite of smaller distance. If the source at Goðabunga is located at 1.5 km depth, as the precise depth determinations of earthquakes seem to indicate (Soosalu *et al.* 2006a), the deformation field diminishes rapidly with distance, and the uplift must be of the order of 1 m to give a tilt of 2 μrad at FIMM. This indicates that the FIMM station will not be useful with only minor deformation at Goðabunga. However, the continuous GPS (CGPS) at GOLA (Fig. 2.3) is very sensitive to the localized deformation.

2.7.2. GPS

2.7.2.1. GPS processing

All data since 2000 have been processed with the Bernese software, version 5.0, using double-difference-based analysis with quasi-ionosphere-free (QIF) resolution strategy (Dach *et al.*, 2007). The final network solution is a minimum constraint solution, realized by three no-net-translation conditions imposed on a set of reference

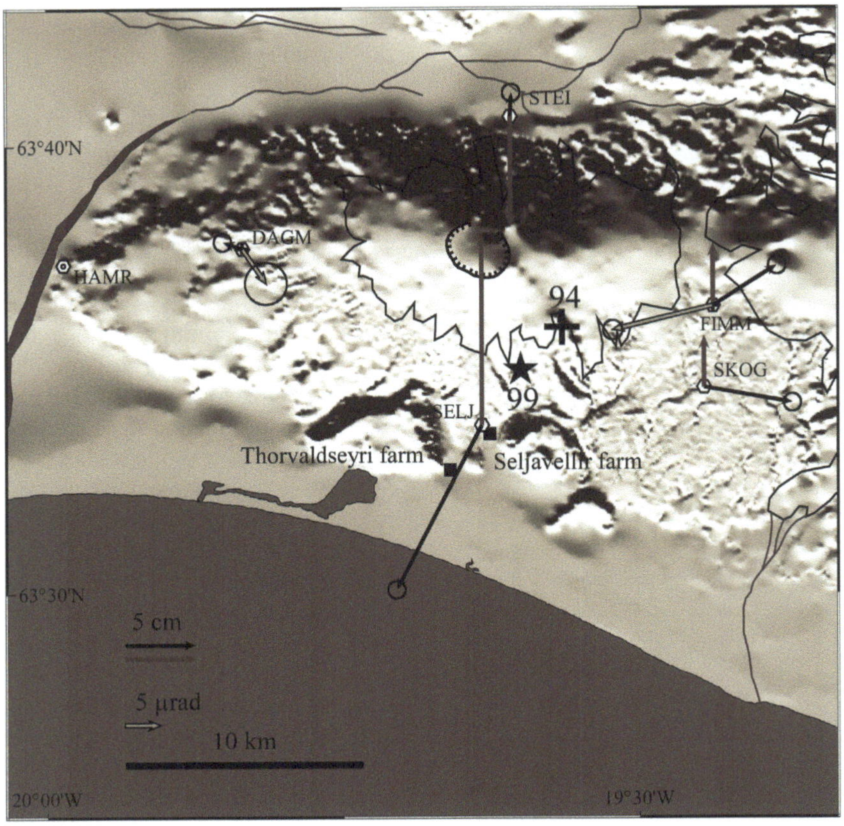

Fig. 2.9. Observed horizontal and vertical displacements, relative to the GPS point HAMR, during the period from July 1998 to July 2000 for the GPS and tilt network around Eyjafjallajökull. Black arrows show horizontal displacement, and red arrows show vertical movements. Shaded arrows from station FIMM and DAGM show the tilt. The black star denotes a best-fit point source location for the 1999 episode, based on horizontal displacements. The suggested centre of uplift of the 1994 event is indicated with a plus sign. Black boxes indicate the location of the Thorvaldseyri and Seljavellir farms (modified after Sturkell et al., 2003).

coordinates of GPS stations derived by the International GNSS Service (IGS). The set of IGS stations used includes stations in North America, Iceland and Scandinavia (REYK, HOFN, ALGO, ALRT, ONSA, TROM, MADR and WES2). The reference coordinates that are used are termed IGS05 and are a realization of the ITRF2005 reference frame (Altamimi *et al.*, 2007). The GPS-derived velocity field has then been plotted relative to stable Eurasia using the ITRF2005 absolute rotation pole for Eurasia (lat, lon, omega) = (56.330 deg, –95.979 deg, 0.261 deg/Myr) (Altamimi *et al.*, 2007). The same procedure is applied for both campaign and CGPS.

2.7.2.2. Campaign GPS

The campaign GPS network, which was directly aimed to follow the crustal deformation of the Katla volcano, was initiated in 1992. The GPS net was expanded in steps and today comprises about 25 stations (Fig. 2.3). Crustal deformation in the Katla region started in the neighbouring volcano Eyjafjallajökull in 1994. This led to the extension of the Katla GPS net to cover that volcano as well. As the GPS network around Eyjafjallajökull was in place, the intrusion stopped. However, the next intrusion in Eyjafjallajökull in 1999 had a good coverage of GPS sites, and the net was extended further.

All various GPS campaigns in the Katla and Eyjafjallajökull area, from 1992 to mid-2000, are listed in Sturkell *et al.* (2003). As the intrusive activity under Eyjafjallajökull ceased in the beginning of year 2000, inflation of the shallow magma chamber in Katla continued until the autumn of 2004. This transferred the focus of the campaigns to Katla. Since the termination of the inflation and reduction in earthquake activity in 2005, the GPS measurements at the Katla volcano have been limited to the sites installed on the two nunataks in Mýrdalsjökull (AUST and ENTA in Figs. 2.3 and 2.5). The complete network was measured in one survey for the first time in July 2000. Also, points in the Icelandic Land Survey network (ISNET) were included to provide ties to this network. The efforts concentrated on semi-annual surveys of the sites on the nunataks, as they were located close to the centre of uplift. The deformation rates turned out to be moderate and the deformation field did not reach far outside the Mýrdalsjökull ice cap. The two CGPS stations (SOHO and HVOL) together with the two campaigns sites boxed the deformation centre in Katla (Fig. 2.6A). With the geometry of the net and closeness to the inflation, the inferred location of the shallow magma chamber could be modelled (Fig. 2.6A).

From 2001 until 2003, constant uplift at AUST occurred at a rate of 1.7 cm/yr (Fig. 2.5), and radial displacement from the caldera centre was observed

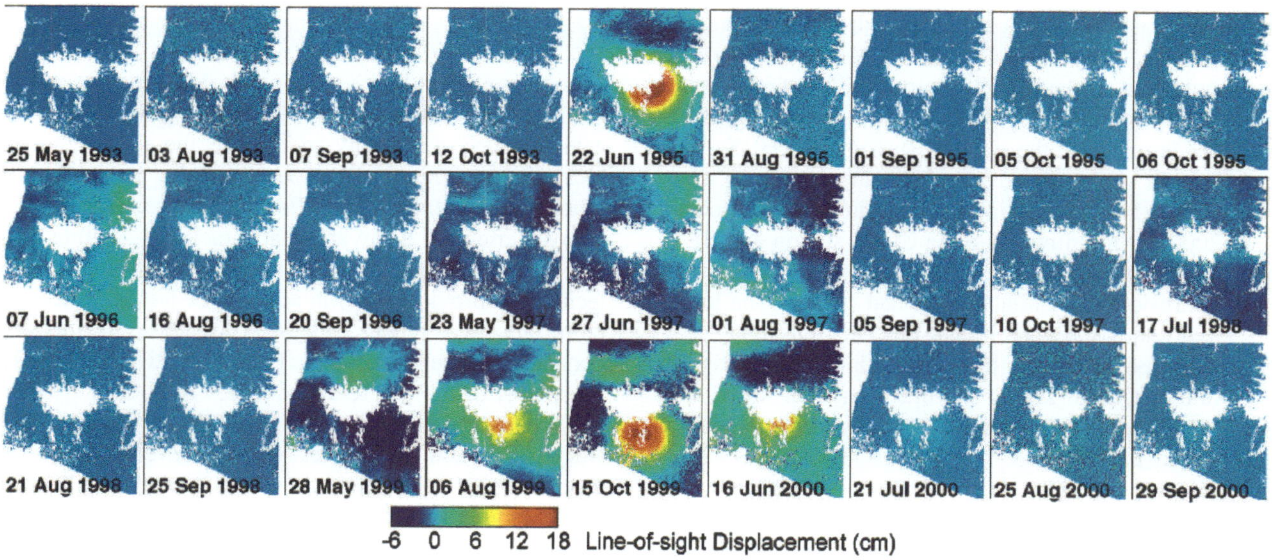

Fig. 2.10. Time series of the deformation field for Eyjafjallajökull volcano 1993–2000 based on InSAR data. The maps show the change in the line-of-sight to the radar satellites. The white area in the centre of each image corresponds to the outline of Eyjafjallajökull ice cap. Each image represents the incremental displacement since the time of the previous image, with reference to pixels in the northwest corner. Spatially correlated nuisance terms have been estimated and removed by spatial and temporal filtering (following Hooper et al., 2007), except for the 1999 images, where the temporal sampling is not high enough with respect to the deformation rate.

(Fig. 2.6A). Figure 2.6A shows the derived displacement vectors from benchmarks on and around Mýrdalsjökull in the Eurasia reference frame. A Mogi-point pressure model was used to calculate the best-fitting location of a point source. In the forward modelling, the horizontal displacements were given the largest weight, as those are much less sensitive to the glacio-isostatic readjustments, compared to the vertical component. GPS points on nunataks near the Katla caldera rim were displaced upwards and horizontally away from the centre at a rate of 1–2 cm/yr. The modelling gave a best-fitting point source at a depth of 4.9 km with an uplift amount of 12 cm at the centre of uplift.

In an attempt to monitor crustal deformation in the zone of maximum earthquake activity in the Goðabunga region (Fig. 2.3), the GPS benchmark (GOLA) was installed in 2004 to the immediate west of the ice cap. Within a six-month interval, 2.0 cm of uplift was detected at GOLA. From the summer 2006, a CGPS station occupies the same benchmark as the campaign measurements used.

Currently, campaign GPS is carried out once or twice per year, in March and if possible in June. The reason for performing the measurements at these times is logistical, as travelling on the glacier is easy during the spring and early summer. With the large annual variation in snow load, which gives a peak-to-peak amplitude of up to 1.2 cm, it is preferable to try to do the measurements during the same time each year to minimize the effect.

The campaign GPS measurements around the intrusion 1999–2000 in Eyjafjallajökull presented here use the Hamragarðar (HAMR, Fig. 2.3) benchmark as fixed in the processing. For a description of the processing procedure, see Sturkell *et al.* (2003). The displacement vectors for an intrusive event point outward from an area on the south slopes of Eyjafjallajökull, indicating inflation (Fig. 2.9). The best-fitting Mogi source gave an estimate for the displacement of the reference station HAMR, resulting in the vector pattern. The 'best-fitting' Mogi model gives a displacement vector of 1.1 cm at 282° for the reference station. A maximum amount of vertical displacement, 35 cm, directly over the point source at a depth at 3.5 km, was inferred by modelling. The star denotes the best fit for the point source location in the 1999 episode by using the GPS and tilt data (Fig. 2.9). The InSAR data gave a much better model of the geometry of the intruding body (Fig. 2.10).

2.7.2.3. Continuous GPS

Crustal deformation is currently recorded at four CGPS sites around Mýrdalsjökull and Eyjafjallajökull (Fig. 2.3), to monitor volcanic deformation, co-seismic displacements and glacio-isostatic deformation (Geirsson *et al.*, 2006). At the end of 2008, over 60 CGPS stations were in operation in Iceland, and most of them are deployed in the plate boundary deformation zone (Arnadottir *et al.*, 2008). The CGPS data are automatically downloaded and processed on a daily basis with the result available on the Icelandic Meteorological Office website for public viewing (http://www.vedur.is). The time series from the CGPS stations are mostly dominated by plate spreading, but deviations are observed at stations close to individual volcanoes (Geirsson *et al.*, 2006). The CGPS stations give excellent time resolution of volcanic activity, but due to high installation and operating costs, their spatial coverage is necessarily much sparser than that provided by campaign GPS.

In response to the unrest in Katla, two CGPS were installed, one at Sólheimaheiði (SOHO) and a second at Láguhvolar (HVOL), south and south-east of Mýrdalsjökull, respectively, in late 1999 (Fig. 2.3). As most deformation in the late 1999 occurred in Eyjafjallajökull, the station Thorvaldseyri (THEY) was set up (Fig. 2.3). The benchmark of the CGPS station at THEY was installed in February 2000 and measured in that campaign, while the continuous measurements started in May 2000. At the time when the CGPS station at THEY became operational in May 2000, crustal deformation due to the intrusion in Eyjafjallajökull had ceased. The long-time trend shows no significant deformation signal that can be related to magma movements in Eyjafjallajökull. The fourth CGPS station in the area was Goðaland (GOLA) installed in 2006 over an existing campaign benchmark (Fig. 2.3).

The time series of displacement at SOHO relative to stable Eurasia (Fig. 2.11) have been corrected for the annual variation caused by the snow load. The station SOHO (and to some extent HVOL) shows southward displacement in excess of the plate movement since 2000, which we attribute to a local source of inflation in the Katla volcano. Applying a point source model for the inflation gives an outward horizontal displacement nearly three times larger than the vertical signal at the SOHO station (see Table 1 in Sturkell *et al.*, 2008). The vertical signal is a combination of glacio-isostatic uplift and the pressure increase in an inferred magma chamber, with the vertical signal caused by the magma chamber only marginal. The same applies for the HVOL station even to a greater extent. The horizontal signal caused by the pressure increase in the modelled shallow magma chamber under the Katla caldera is about a half of what is recorded at SOHO. During the period until early autumn 2004, the north component of SOHO has a rate of 0.5 cm/yr (Fig. 2.11). As the pressure increase apparently ceased in the later part of 2004, the rate of the north-south component slowed down to 0.14 cm/yr. No obvious changes in the displacement rates of the east and vertical components can be observed (Fig. 2.11). The ratio between the vertical and horizontal rates since 2004 is 4.2 (0.97 cm/yr of uplift and 0.23 cm/yr of movement outward from the caldera). Following Pinel *et al.* (2007), this implies that the current deformation signal at SOHO is probably dominated by glacio-isostatic response to the rapid ice-sheet reduction, which takes place at Mýrdalsjökull. A decrease of the seismic activity correlates with the change of deformation rate of the horizontal component at SOHO (Figs. 2.3 and 2.11).

The vertical time series at SOHO shows a seasonal variation, with a peak-to-peak amplitude close to 1.2 cm. In a study by Grapenthin *et al.* (2006), a correlation between the seasonal variation and the annual snow load in Iceland could be established. The model, based on the available CGPS data in Iceland, predicted a peak-to-peak amplitude seasonal displacement of 1.1 cm, at SOHO, which is in fair agreement with our time series, which were calculated in a slightly different manner than in Grapenthin *et al.* (2006). The deformation rate of the outward displacement at SOHO decreased in 2004 at approximately the same time as the seismic intensity decreased (Figs. 2.4 and 2.11). We suggest this marks the termination of the inflation event in Katla volcano, which started in 1999.

2.7.3. InSAR

Spaceborne InSAR is a valuable tool for measuring surface deformation because of the high spatial resolution achieved and the ability to acquire the data remotely. Nevertheless, significant issues arise due to changes in scattering properties of the Earth's surface, variations in atmospheric path delay and inaccuracies in satellite orbit and surface elevation determination. Time series InSAR techniques provide a way to address the issues. Currently, there are two broad categories of these techniques, persistent scatterer methods (e.g. Ferretti *et al.*, 2001; Hooper *et al.*, 2004; Kampes, 2005) and small baseline methods (e.g. Berardino *et al.*, 2002; Schmidt and Bürgmann, 2003). Hooper (2008) developed a new algorithm that combines the two approaches to maximize the spatial coverage of useful signal and allow more reliable estimation of integer phase cycle ambiguities present in the data.

In Hooper *et al.* (2009), the new algorithm was applied to 27 images acquired over Eyjafjallajökull volcano by ERS-1 and ERS-2 satellites in a descending look direction between May 1993 and September 2000 (Fig. 2.10). Two intrusive episodes could be detected, the first of which occurred in 1994 and the second of which began in mid-1999 and ended in early 2000. The deformation caused by the two episodes is clearly centred at different locations, which was also the conclusion of Pedersen and Sigmundsson (2006) from conventional InSAR analysis. However, the overall inferred deformation pattern for each episode is more regular than deduced from conventional InSAR due to the removal of digital elevation model and atmospheric artefacts and more accurate phase unwrapping in the time series analysis. In addition, the displacements during the 1999 episode are decomposed into four sequential steps by the time series processing. The data for both episodes can be explained by an intrusion of a circular sill with a uniform overpressure. Seismicity is interpreted to be mostly associated with a narrow magma feeder channel from depth that does not cause noticeable deformation.

The 1999 increase in seismicity at Eyjafjallajökull was associated with significant inflation of the volcano. The deformation data are modelled using InSAR by Pedersen and Sigmundsson (2006), Hooper *et al.* (2007), Hooper (2008) and by GPS and tilt by Sturkell *et al.* (2003). The inflation was centred to the southern flank of the volcano approximately 4 km south of the summit crater. Maximum uplift of the model is about 0.35 m. The application of Synthetic Aperture Radar (SAR) interferometry greatly supplements other, sparse surface deformation measurements from the Eyjafjallajökull. At Eyjafjallajökull, two deformation events in 1994 and 1999 are consistent with the intrusion of approximately circular, horizontal sills. For the latter event, we are able to resolve the evolution of the deformation in time; it appears the growth history of the sill is not, however, simple. The two intrusion events were associated with elevated earthquake activity. The earthquakes are not spatially constrained to the centre of uplift on the southern slope but are also to a large extent located below the northern slope of Eyjafjallajökull. Almost all earthquakes in 1994 occurred in the north; this

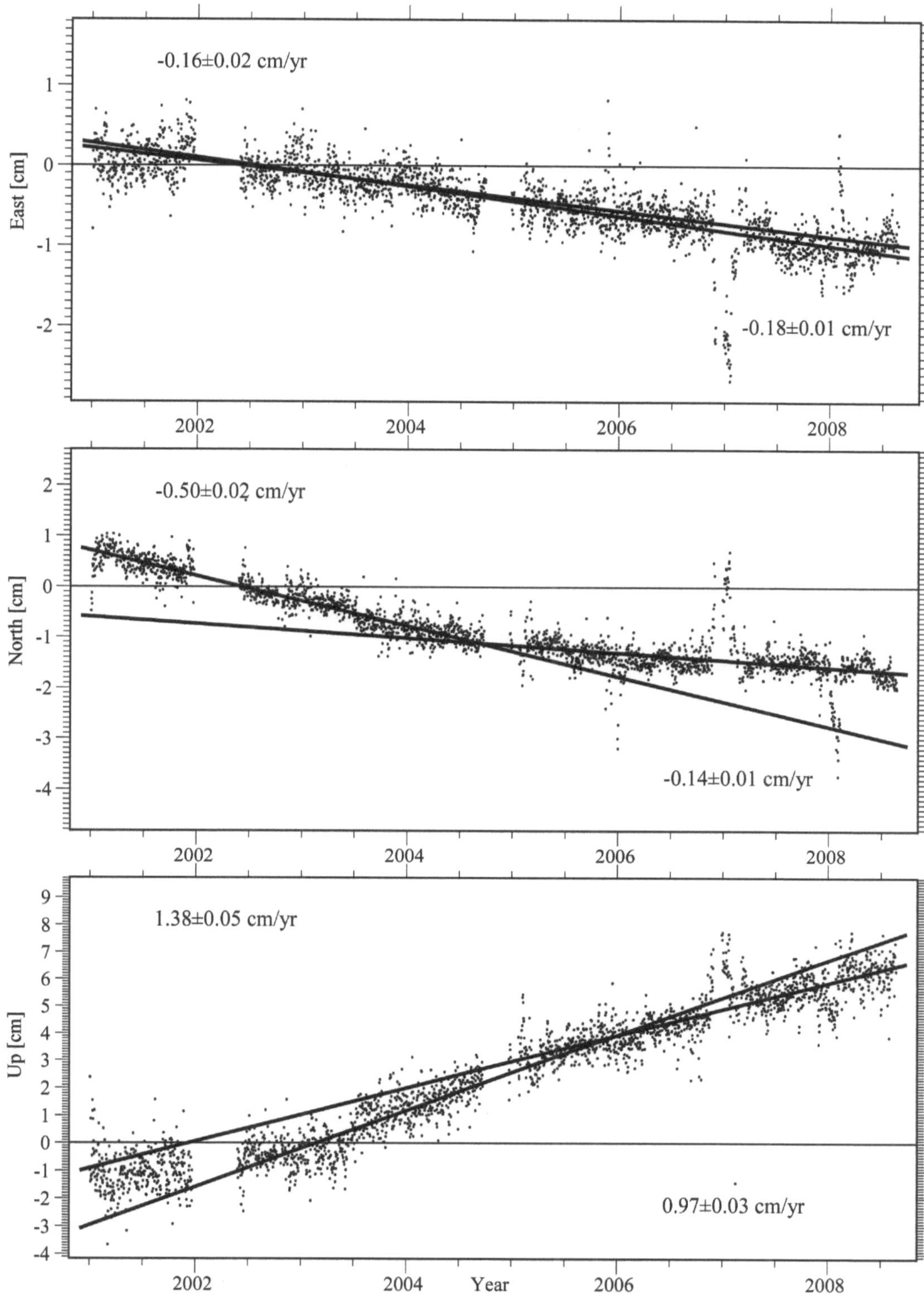

Fig. 2.11. Time series plot of displacement at SOHO. The horizontal displacements are relative to stable Eurasia but the uplift relative to ITRF 2005 reference frame. The uncertainty is the square of covariance from the weighted least square fit. The sloping lines indicate the displacement rates, and the most obvious change in rate appears in the north-south horizontal component. The rate changed in late August 2004, and this is suggested to represent the termination of the inflation episode. Annual variations in all components have been estimated and removed from the time series. In the vertical, we estimate a peak-to-peak amplitude of 1.2 cm. The time series of all three components shows that during a short time, an erratic behaviour of the calculated displacements takes place. This occurs in late 2006/early 2007 and in early 2008, and it is caused by accumulation of ice on different parts of the antenna.

pattern changed slightly in the 1999 episode, with some earthquakes occurring in the south. The majority of the earthquakes preceded the sill formation by half a year (Fig. 2.8A) and were spatially located in the northern part of Eyjafjallajökull (Fig. 2.7). This early cluster of earthquakes may be related to the feeder dike of the sill. Earthquake activity continued during the emplacement of the sill, which took about half a year.

Hooper *et al.* (2009) also applied the new time series InSAR algorithm to ERS and ENVISAT data acquired over Katla. No significant systematic motion was detected on the flanks of Katla from 1995 to 1998 and from 2000 to 2006, other than that expected from thinning of Vatnajökull and Mýrdalsjökull ice caps.

2.8. Discussion

Extensive crustal deformation research utilizing GPS geodesy and satellite radar interferometry provides constraints on magmatic unrest at the neighbouring subglacial volcanoes Katla and Eyjafjallajökull in 1994–2004. A decade long unrest period at these twin volcanoes began in 1994 with a sill intrusion under the Eyjafjallajökull volcano, followed by another sill intrusion in 1999. The onset of the second sill intrusion under Eyjafjallajökull occurred at a similar time as the 1999 jökulhlaup at Katla. Our measurements suggest that a modest inflow of magma towards shallow levels continued at Katla from 1999 until 2004. The active deformation sources are shown schematically in Fig. 2.12.

Crustal deformation measurements in the Mýrdalsjökull area have revealed movements that are caused by several processes. Interpretation of crustal deformation at subglacial volcanoes requires detailed consideration of the role of both load-induced deformation and deformation due to magmatic sources. Crustal deformation of the Mýrdalsjökull region appears to be affected by at least three common sources of deformation: (1) glacio-isostatic uplift due to thinning of the ice cap (Pinel *et al.*, 2007, 2009), (2) an annual cycle in ground movements due to variable ice load (Grapenthin *et al.*, 2006) and, in addition, (3) magmatic processes, as witnessed by the outward horizontal displacements from an area in the centre of the Katla caldera, as well as excessive uplift rate in 1999–2004. The time-dependent unrest in the Katla volcano is also manifested by elevated seismic activity. The dominant crustal deformation of the Katla volcano from the late 2004 and onwards can be attributed to rebound due to thinning of the Mýrdalsjökull ice cap. Ice thinning is well constrained for the period 1999–2004. Repeated radar altimeter measurements from an aircraft between 1999 and 2004 indicate a loss of over 3 km^3 of ice during that time, primarily from the margins of the ice cap (Pinel *et al.*, 2007). Beneath and at the edge of an ice cap, vertical uplift dominates as the crustal response to ice loss, with horizontal displacements an order of magnitude less (ratio <0.3) (Pinel *et al.*, 2007, 2009); this applies to both the immediate elastic response and to the final, relaxed state of the crust. For a point source approximation of a magma chamber, the ratio between vertical displacement and horizontal is > 0.5. For Katla, Pinel *et al.* (2007) showed that the observed ratio nears 1. Furthermore, the rate of horizontal displacement is 1–2 cm/yr away from the Katla caldera edge, implying that observed horizontal deformation is principally due to magma accumulation. However, most of the vertical component of displacement observed outside the ice cap is likely due to glacio-isostatic response.

Seismicity at Goðabunga is remarkably regular, persistent and spatially concentrated (Fig. 2.7). Precise hypocentral locations reveal a zone of seismicity at 1.5 km depth (Soosalu *et al.*, 2006a). The displacement field associated with the earthquakes appears to be very localized; it is not detected at a tilt station 8 km away, and apart from GOLA, none of the GPS benchmarks are affected. The seasonal trend in earthquake activity in the Goðabunga region has been interpreted by Einarsson and Brandsdóttir (2000) as a triggering effect due to increased groundwater pressure within the volcano during

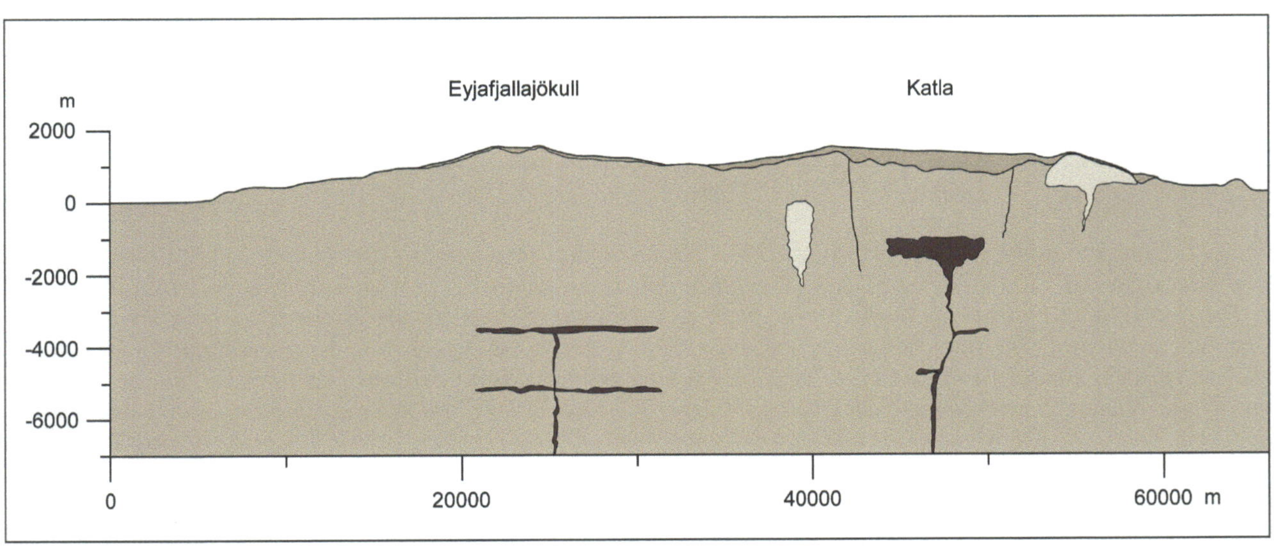

Fig. 2.12. Schematic cross-section from west to east across the Eyjafjallajökull and Katla volcanoes. Magmatic intrusions in 1994–2004 are drawn in red, rhyolitic domes and the cryptodome in yellow; glaciers are shown in light blue.

summertime melting of the overlying ice cap. More recently, Einarsson *et al.* (2005) and Soosalu *et al.* (2006a) have proposed that persistent earthquakes in the Goðabunga region are generated by an intruding cryptodome, that is, an ascending pocket of silicic or intermediate magma at shallow depth. This argument is based on the nature and style of the earthquake activity, the apparent displacement field, and the geographic proximity to the Katla caldera. Almost all outcrops of Katla rocks that are close to or immediately outside the caldera rim are silicic.

The inferred volume of magma accumulated in a shallow magma chamber at Katla during the 1999–2004 inflation period is approximately 0.01 km^3. This volume is miniscule compared to the erupted volume in 1918, when about 1 km^3 was erupted. The new magma was emplaced at shallow levels within the plumbing system of the Katla volcano. The high level of seismicity associated with inflation of the volcano during this period suggests that the volcano is close to failure and a new eruption or intrusion event may be initiated if magma flow towards shallow levels resumes. New material has been added to the shallow magma chamber, rather small volumes in this case, but this can thermally lubricate the pathways for deeper seated magma and give a fast track for the material to reach the surface. With the volcano in an agitated state, an eruption can take place without prolonged precursory signals. However, these effects decay with time as the magma solidifies unless new material is added.

Furthermore, the suspected cryptodome can cause explosive volcanism if it ascends above the ground and makes a dome, which can collapse and generate pyroclastic flows, or, if a basaltic magma intrudes into a rhyolitic cryptodome. In addition, basaltic eruptions that take place under the Mýrdalsjökull ice cap will be phreato-magmatic. This, together with a possible differentiation of the material in a shallow magma chamber, favours explosive activity, possibly reaching plinian magnitudes.

Acknowledgements

We thank Benedikt Bragason and the rescue team in Vík for assistance with field logistics. We thank all the people who participated in collection of geodetic data. Gabrielle Stockmann commented on earlier versions of this chapter. Comments from the reviewer, Heidi Soosalu, helped us to significantly improve the chapter. Also, the suggestions of the editor Anders Schomacker have been most helpful. The following institutions and research grants supported this study: the Icelandic Roads Authority; the Icelandic government; the University of Iceland Research Fund; the Icelandic research council (RANNÍS) and the European Union VOLUME (018471) project. The figures were produced using the GMT public domain software (Wessel and Smith, 1998).

References

Altamimi, Z., Collilieux, X., Legrand, J., Garayt, B., & Boucher, C., 2007. ITRF2005: A new release of the International Terrestrial Reference Frame based on time series of station positions and Earth Orientation Parameters. Journal of Geophysical Research 112, B09401.

Arnadottir, T., Geirsson, H., Hreinsdottir, S., Jonsson, S., LaFemina, P., Bennett, R., Decriem, J., Holland, A., Metzger, S., Sturkell, E., & Villemin, T., 2008. Capturing crustal deformation signals with a new high-rate continuous GPS network in Iceland. Eos, Transactions, American Geophysical Union 89(53). Fall Meet. Suppl., Abstract G43A-0650.

Berardino, P., Fornaro, G., Lanari, R., & Sansosti, E., 2002. A new algorithm for surface deformation monitoring based on small baseline differential SAR interferograms. IEEE Transactions on Geoscience and Remote Sensing 40, 2375–2383.

Björnsson, S. & Einarsson, P., 1981. Jarðskjálftar - "Jörðin skalf og pipraði af ótta", Náttúra Íslands (2. útgáfa). Reykjavík, Almenna bókafélagið, 121–155.

Björnsson, H., Pálsson, F., & Gudmundsson, M.T., 2000. Surface and bedrock topography of Myrdalsjökull, South Iceland: The Katla caldera, eruption sites and routes of jökulhlaups.. Jökull 49, 29–46.

Brandsdóttir, B. & Menke, W., 1990. Icequakes in Entujökull and Kötlujökull. Jökull 39, 96–98.

Dach, R., Hugentobler, U., Frides, P., & Meindl, M., 2007. Bernese GPS software version 5.0. Switzerland, Astronomical Institute, University of Bern.

Dahm, T. & Brandsdóttir, B., 1997. Moment tensors of microearthquakes from the Eyjafjallajökull volcano in south Iceland. Geophysical Journal International 130, 183–192.

Eggertsson, S., 1919. Ymislegt smávegis vidvíkjandi Kötlugosinu 1918. Eimreidin 25, 212–222.

Einarsson, P., 1987. Compilation of earthquake fault plane solutions in the North Atlantic and Arctic Oceans. In: Kasahara, K. (Ed), Recent plate movements and deformation, Geodynamics Series, vol. 20, Washington, DC, American Geophysical Union, 47–62.

Einarsson, P., 1991. Earthquakes and present-day tectonism in Iceland. Tectonophysics 189, 261–279.

Einarsson, P. & Brandsdóttir, B., 2000. Earthquakes in the Myrdalsjökull area, Iceland, 1978–1985: Seasonal correlation and relation to volcanoes. Jökull 49, 59–73.

Einarsson, P. & Saemundsson, K., 1987. Earthquake epicenters 1982–1985 and volcanic systems in Iceland (map). In: Sigfússon, Th. (Ed), Í Hlutarsins Eðli: Festschrift for Thorbjorn Sigurgeirsson. Reykjavík, Menningarsjóður.

Einarsson, P., Soosalu, H., Sturkell, E., Sigmundsson, F., & Geirsson, H., 2005. Virkni í Kötlueldstödinni og nágrenni hennar sídan 1999 og hugsanleg thróun atburdarásar (Activity of Katla and surrounding area since 1999 and potential scenarios). In: Guðmundsson, M.T., Gylfason, Á.G. (Eds), Haettumat vegna eldgosa og hlaupa frá vestanverdum Mýrdalsjökli og Eyjafjallajökli (Assessment of hazard due to eruptions and jökulhlaups from western Mýrdalsjökull and Eyjafjallajökull). Reykjavík, Iceland, University Publ., 151–158.

Ferretti, A., Prati, C., & Rocca, F., 2001. Permanent scatterers in SAR interferometry. IEEE Transactions on Geoscience and Remote Sensing 39, 8–20.

Geirsson, H., Árnadóttir, Th., Völksen, C., Jiang, W., Sturkell, E., Villemin, T., Einarsson, P., Sigmundsson, F., & Stefánsson, R., 2006. Current plate movements across the Mid-Atlantic Ridge determined from 5 years of continuous GPS measurements in Iceland. Journal of Geophysical Research 111, B09407.

Grapenthin, R., Sigmundsson, F., Geirsson, H., Arnadottir, T., & Pinel, V., 2006. Icelandic rhythmics: Annual modulation of land elevation and plate spreading by snow load. Geophysical Research Letters 33, L24305.

Guðmundsson, M.T. & Högnadóttir, Þ., 2005. Ísbrádnun og upptakarennsli jöulhlaupa vegna eldgos í Eyjafjallajökull og vestanverdum Mýrdalsjökuli. (The melting distribution and origin of the jökulhlaups that can be caused by eruptions from western Mýrdalsjökull and Eyjafjallajökull). In: Guðmundsson, M.T., Gylfason, Á.G. (Eds), Haettumat vegna eldgosa og hlaupa frá vestanverdum Mýrdalsjökli og Eyjafjallajökli (Assessment of hazard due to eruptions and jökulhlaups from western Mýrdalsjökull and Eyjafjallajökull). Reykjavík, Iceland, University Publ., 158–179.

Guðmundsson, M.T., Högnadóttir, Þ., Kristinsson, A.B., & Guðbjörnsson, S., 2007. Geothermal activity in the subglacial Katla caldera, Iceland, 1999–2005, studied with radar altimetry. Annals of Glaciology 45, 66–72.

Guðmundsson, Ó., Brandsdóttir, B., Menke, W., & Sigvaldason, G.E., 1994. The crustal magma chamber of the Katla volcano in South Iceland revealed by 2-D seismic undershooting. Geophysical Journal International 119, 277–296.

Hooper, A., 2008. A multi-temporal InSAR method incorporating both persistent scatterer and small baseline approaches. Geophysical Research Letters 35, L16302.

Hooper, A., Pedersen, R., & Sigmundsson, F., 2009. Constraints on magma intrusion at Eyjafjallajökull and Katla volcanoes in Iceland, from time series SAR interferometry. In: Bean, C.J., Braiden, A.K., Lokmer, I., Martini, F., O'Brien, G.S. (Eds), VOLUME project, EU PF6 (No. 018471). ISBN 978-1-905254-39-2. VOLUME Project Consortium, Dublin.

Hooper, A., Segall, P., & Zebker, H., 2007. Persistent scatterer InSAR for crustal deformation analysis, with application to Volcán Alcedo, Galápagos. Journal of Geophysical Research 112, B07407.

Hooper, A., Zebker, H., Segall, P., & Kampes, B., 2004. A new method for measuring deformation on volcanoes and other natural terrains using InSAR persistent scatterers. Geophysical Research Letters 31, L23611.

Jakobsdóttir, S.S., 2008. Seismicity in Iceland: 1994–2007. Jökull 58, 75–100.

Jakobsson, S.P., 1979. Petrology of recent basalts of the eastern volcanic zone, Iceland. Acta Naturalia Islandica 26, 1–103.

Jóhannesson, H., Jakobsson, S.P., & Saemundsson, K., 1990. Geological map of Iceland, sheet 6, South-Iceland (third edition). Reykjavík, Icelandic Museum of Natural History and Geodetic Survey.

Jóhannsson, G., 1919. Kötlugosid 1918. (The Katla eruption of 1918). Reykjavík, Bókaverslun Ársaels Árnasonar, 72.

Jónasson, K., 1994. Rhyolite volcanism in the Krafla central volcano, North-East Iceland. Bulletin of Volcanology 56, 516–528.

Jónsdóttir, K., Tryggvason, A., Roberts, R., Lund, B., Soosalu, H., & Bödvarsson, R., 2007. Habits of a glacier covered volcano: seismicity and a structure study of the Katla volcano, South Iceland. Annals of Glaciology 45, 169–177.

Jónsdóttir, K., Roberts, R., Tryggvason, A., Lund, B., Pojhola, V., Jakobsdóttir, S.S., & Bödvarsson, R., 2008. Local lp-events study in a glaciated volcanic environment in south Iceland. Geophysical Research Abstracts 10, EGU2008-A-08854, EGU General Assembly, Vienna.

Jónsson, J., 1998. Eyjafjöll drög að jarðfraeði. Rannsóknastofnunin Nedri Ás, Hveragerði, Ísland. Publ. 53, 1–111.

Jónsson, G. & Kristjánsson, L., 2000. Aeromagnetic measurements over Mýrdalsjökull and vicinity. Jökull 49, 47–58.

Kampes, B.M., 2005. Displacement parameter estimation using permanent scatterer interferometry. PhD Thesis. Delft University of Technology, Holland.

Lacasse, C., Sigurdsson, H., Carey, S.N., Jóhannesson, H., Thomas, L.E., & Rogers, N.W., 2007. Bimodal volcanism at the Katla subglacial caldera, Iceland: Insight into the geochemistry and petrogenesis of rhyolitic magmas. Bulletin of Volcanology 69, 373–399.

LaFemina, P.C., Dixon, T.H., Malservisi, R., Árnadóttir, T., Sturkell, E., Sigmundsson, F., & Einarsson, P., 2005. Geodetic GPS measurements in South Iceland: strain accumulation and partitioning in a propagating ridge system. Journal of Geophysical Research 110, B11405.

Larsen, G., 2000. Holocene eruptions within the Katla volcanic system, south Iceland: characteristics and environmental impact. Jökull 49, 1–28.

Larsen, G., Dugmore, A.J., & Newton, A.J., 1999. Geochemistry of historical-age silicic tephras in Iceland. The Holocene 9, 463–471.

Óladóttir, B.A., Larsen, G., Thordarson, Th., & Sigmarsson, O., 2005. The Katla volcano, S-Iceland: holocene tephra stratigraphy and eruption frequency. Jökull 55, 53–74.

Óladóttir, B.A., Sigmarsson, O., Larsen, G., & Thordarson, Th., 2008. Katla volcano, Iceland: Magma composition, dynamics and eruption frequency as depicted by the Holocene tephra layer record. Bulletin of Volcanology 70, 473–493.

Óskarsson, N., Sigvaldason, G.E., & Steinthórsson, S., 1982. A dynamic model of rift zone petrogenesis and the regional petrology of Iceland. Journal of Petrology 23, 28–74.

Pedersen, R. & Sigmundsson, F., 2004. InSAR based sill model links spatially offset areas of deformation and seismicity for the 1994 unrest episode at Eyjafjallajökull volcano, Iceland. Geophysical Research Letters 31, L14610.

Pedersen, R. & Sigmundsson, F., 2006. Temporal development of the 1999 intrusive episode in the Eyjafjallajökull volcano, Iceland, derived from InSAR images. Bulletin of Volcanology 68, 377–393.

Pinel, V., Sigmundsson, F., Sturkell, E., Geirsson, H., Einarsson, P., Guðmundsson, M.T., & Högnadóttir, T., 2007. Discriminating volcano deformation due to magma movements and variable surface loads: application to Katla subglacial volcano, Iceland. Geophysical Journal International 169, 325–338.

Pinel, V., Albino, F., Sigmundsson, F., Sturkell, E., Geirsson, H., Einarsson, P., & Guðmundsson, M.T., 2009. Consequences of local surface load variations for volcanoes monitoring: application to Katla subglacial volcano, Iceland. In: Bean, C.J., Braiden, A.K., Lokmer, I., Martini, F., O'Brien, G.S. (Eds), VOLUME project, EU PF6 (No. 018471). ISBN 978-1-905254-39-2. VOLUME Project Consortium, Dublin.

Poland, M.P. & Lu, Z., 2008. Radar interferometry observations of surface displacements during pre- and co-eruptive periods at Mount St. Helens, Washington, 1992–2005. In: Sherrod, D.R., Scott, W.E., Stauffer, P.H. (Eds), A volcano rekindled: the renewed eruption of Mount St. Helens, 2004–2006. Professional Paper 1750, United States Geological Survey, Menlo Park, CA, Chapter 18.

Rist, S., 1967. Jökulhlaups from the ice cover of Mýrdalsjökull on June 25, 1955 and January 20, 1956.. Jökull 17, 243–248.

Roberts, M.J., Magnússon, E., Bennett, R., Geirsson, H., Sturkell, E., Björnsson, H., Pálsson, F., & Rott, H., 2006. Meltwater dynamics beneath Skeiðararjökull from continuous GPS measurements and seismic observations. Eos, Transactions, American Geophysical Union 87(52). Fall Meet. Suppl., Abstract C31A-1232.

Schmidt, D.A. & Bürgmann, R., 2003. Time-dependent land uplift and subsidence in the Santa Clara valley, California, from a large interferometric synthetic aperture radar data set. Journal of Geophysical Research 108, 2416–2428.

Sigmundsson, F., 2006. Iceland geodynamics, crustal deformation and divergent plate tectonics. Chichester, Springer–Praxis Publishing Ltd, 209 pp.

Sigmundsson, F., Einarsson, P., Bilham, R., & Sturkell, E., 1995. Rift-transform kinematics in south Iceland: deformation from Global Positioning System measurements, 1986 to 1992. Journal of Geophysical Research 100, 6235–6248.

Sinton, J.M., Wilson, D.S., Christie, D.M., Hey, R.N., & Delaney, J.R., 1983. Petrological consequences of rift propagation on oceanic spreading ridges. Earth and Planetary Science Letters 62, 193–207.

Soosalu, H., Jónsdóttir, K., & Einarsson, P., 2006a. Seismicity crisis at the Katla volcano, Iceland – signs of a cryptodome? Journal of Volcanology and Geothermal Research 153, 177–186.

Soosalu, H., Lippitsch, R., & Einarsson, P., 2006b. Low-frequency earthquakes at the Torfajökull volcano, south Iceland. Journal of Volcanology and Geothermal Research 153, 187–199.

Sturkell, E., Einarsson, P., Roberts, M.J., Geirsson, H., Gudmundsson, M.T., Sigmundsson, F., Pinel, V., Guðmundsson, G.B., Ólafsson, H., & Stefánsson, R., 2008. Seismic and geodetic insights into magma accumulation at Katla subglacial volcano, Iceland: 1999 to 2005. Journal of Geophysical Research 113, B03212.

Sturkell, E., Einarsson, P., Sigmundsson, F., Geirsson, H., Ólafsson, H., Pedersen, R., De Zeeuwvan Dalfsen, E., Linde, A.T., Sacks, I.S., & Stefánsson, R., 2006. Volcano geodesy and magma dynamics in Iceland. Journal of Volcanology and Geothermal Research 150, 14–34.

Sturkell, E., Sigmundsson, F., & Einarsson, P., 2003. Recent unrest and magma movements at Eyjafjallajökull and Katla volcanoes, Iceland. Journal of Geophysical Research 108(B8), 2369.

Sveinsson, G., 1919. Kötlugosid 1918 og afleidingar thess (The Katla eruption of 1918 and its consequences). Reykjavík, Prentsmidja Gutenbergs, 61p.

Thórarinsson, S., 1975. Katla og annáll Kötlugosa. Árbók Ferdafélags Íslands. Ferdafélag Íslands, Reykjavík, 125–149.

Thordarson, T. & Larsen, G., 2007. Volcanism in Iceland in historical time: volcano types, eruption styles and eruptive history. Journal of Geodynamics 43, 118–152.

Thordarson, T., Miller, D.J., Larsen, G., Self, S., & Sigurdsson, H., 2001. New estimates of sulfur degassing and atmospheric mass-loading by the 934 AD Eldgjá; eruption, Iceland. Journal of Volcanology and Geothermal Research 108, 33–54.

Thoroddsen, Þ., 1925. Die Geschichte der isländischen Vulkane. Kongelige Danske Videnskabernes Selskab Skrifter, Naturvidenskabelig og Matematisk Afdeling, 8, Raekke, IX, Copenhagen, 458 pp.

Tryggvason, E., 1973. Seismicity, earthquake swarms and plate boundaries in the Iceland region. Bulletin of the Seismological Society of America 63, 1327–1348.

Vogfjörð, K. & Slunga, R., 2008. Imaging subsurface mass movement through relative earthquake locations in the Katla volcano, Iceland, and the possible location of a magma chamber. Reykjavík, Iceland, IAVCEI General Assembly.

Wessel, P. & Smith, W.H.F., 1998. New, improved version of generic mapping tools released. Eos, Transactions, American Geophysical Union 79(47), 579.

3

Katla: Tephrochronology and Eruption History

Guðrún Larsen

Institute of Earth Sciences, University of Iceland, Askja, Sturlugata 7, IS-101 Reykjavík, Iceland (glare@raunvis.hi.is)

3.1. Introduction

Eruptions below thick ice are nowhere more common than in Iceland where substantial parts of the volcanic zones are covered by glaciers. Explosive hydromagmatic basaltic eruptions within ice caps are, for Iceland, the characteristic volcanic activity (Fig. 3.1).

One of the most active volcanic systems in Iceland is the Katla volcanic system (KVS) on the Eastern Volcanic Zone (EVZ) in Iceland, partly covered by the 600 km^2 Mýrdalsjökull ice cap (Jakobsson, 1979). The central volcano comprises the massif below the ice cap, mostly built up during the last 0.7 million years (Jóhannesson *et al.*, 1990) with a 100 km^2 ice-filled caldera (Björnsson *et al.*, 2000). The Holocene volcanic activity has occurred both below the ice cap and on the ice-free flanks of the volcano but during the well-documented last millennium it has apparently been confined to the caldera as explosive basaltic eruptions (e.g. Thorarinsson, 1975). Effusive basaltic eruptions on fissures outside the ice appear to be of early Holocene age.

The fissure swarm of the KVS is 'embryonic' in the sense that it is poorly developed and defined by single volcanic fissure which extends some 60–70 km from the edge of Mýrdalsjökull ice cap almost to the Vatnajökull ice cap in the northeast (Larsen, 2000; Thordarson and Larsen, 2007). During the last glacial a short branch of the fissure swarm is thought to have extended southwards from the central volcano (Jóhannesson and Sæmundsson, 1998) but this part has been inactive during the Holocene.

Tephrochronology has played a major role in unravelling the eruption history of Katla. In prehistoric time, it is the only source of information on its explosive eruptions, type of activity, eruption frequency and geochemistry, and also a dating tool of effusive eruptions, jökulhlaups and major environmental changes in the Katla region. In historical time, written sources have supplemented the tephrochronological studies.

First and foremost the Holocene eruption history of the Katla volcano is characterized by hydromagmatic basaltic eruptions below the Mýrdalsjökull ice cap, probably mostly within the Katla caldera. The number of explosive basaltic eruptions that broke through the overlying ice and left a tephra layer outside Mýrdalsjökull is thought to exceed 300 (Óladóttir *et al.*, 2005; Larsen and Eiríksson, 2008). An unknown number of eruptions may never have become subaerial. Explosive silicic eruptions, also affected by the ice/meltwater environment, have occurred sporadically (Newton *et al.*, 2000; Larsen *et al.*, 2001). Effusive basaltic eruptions on the flanks of the central volcano and on the fissure swarm have produced lava flows ranging from a few km^2 to 780 km^2. Some of these effusive eruptions have been dated by tephrochronology. Most Katla eruptions are accompanied by jökulhlaups from below Mýrdalsjökull.

This paper provides a short review of the Holocene activity of the KVS and the environmental changes that have resulted from tephra fall, lava emission and jökulhlaup activity around the volcanic system.

3.2. Geological Setting

The KVS lies on the southern sector of the EVZ in South Iceland (Fig. 3.2). The 80 km long system comprises an ice-capped central volcano, most often referred to as Mýrdalsjökull after the ice cap, and an associated SW–NE trending fissure swarm (Jakobsson, 1979).

The Mýrdalsjökull hyaloclastite massif reaches an altitude of 1,380 m a.s.l. under the ice cover and encompasses an ice-filled caldera with an area of 100 km^2 and a depth of 700 m (Björnsson *et al.*, 2000). The eruption site(s) of all observed Katla eruptions lie within the caldera, and those that can be located through sightings or direct observations lie close to its SE boundary. The Mýrdalsjökull central volcano is sometimes been referred to as the Katla central volcano, but here *Katla* (or Katla volcano) refers to specific eruption sites and *Katla caldera* to the caldera as a whole.

Three outlet glaciers, Sólheimajökull, Kötlujökull and Entujökull, descend onto the lowlands to the south, east and northwest through deep breaches in the caldera walls. The breaches and the glaciers are also pathways for jökulhlaups resulting from eruptions within the Katla caldera, which drain onto Mýrdalssandur and Sólheimasandur outwash plains in the east and south, and into the Markarfjót river to the northwest.

The Mýrdalsjökull massif is surrounded by agricultural communities and their summer grazing fields on the east, south and west sides. The farms closest to the Katla eruption site of the last two centuries are located 13–14 km to the south and the village of Vík in the Mýrdalur district lies about 20 km south of Katla. The popular Þórsmörk recreational area, a former farming and

DEVELOPMENTS IN QUATERNARY SCIENCES
VOLUME 13 ISSN 1571-0866

Fig. 3.1. The eruption cloud of Katla in October 1918. Photo: Kjartan Guðmundsson.

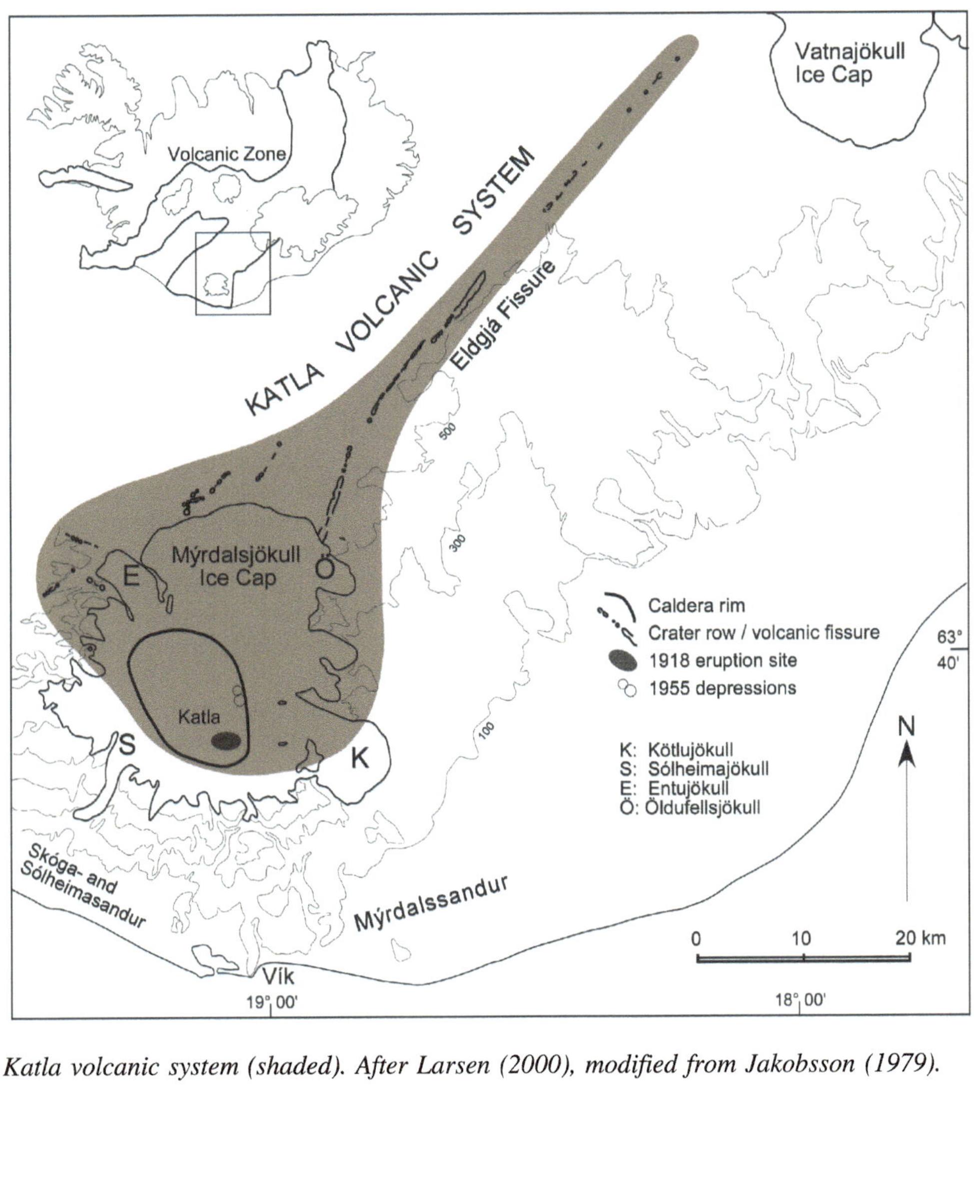

Fig. 3.2. The Katla volcanic system (shaded). After Larsen (2000), modified from Jakobsson (1979).

summer grazing area, extends onto the west slopes of the massif. During the past centuries, the main traffic routes between southern and eastern Iceland have passed on the south and north side of the massif. Highway 1 runs along the southern margin of the massif and across three flood plains formed by glacial rivers and volcanogenic jökulhlaups.

3.3. Holocene Eruptions in the Katla Volcanic System

Knowledge of Holocene eruptive activity in the KVS is based on two data sets. The shorter record consists of written accounts of eruptions and accompanying tephra fall, earth fires and jökulhlaups, compiled since the settlement of Iceland in the AD 870s. The longer record comprises the products of its eruptions, mainly tephra layers and lava flows, and to some extent jökulhlaup deposits. Historical eruption *dates* are presented in the respective years AD. The *age* of prehistoric eruptions and jökulhlaups is in calibrated years (expressed as 'years old' or 'years ago') unless otherwise specified.

3.3.1. Eruptions and Jökulhlaups: The Written Accounts

Eruptions since the tenth century in the KVS have been mentioned in document. Katla eruptions are often referred to as Kötluhlaup, probably because the jökulhlaups caused more damage than the tephra fall did. Jökulhlaup on Mýrdalssandur is mentioned for the first time in the late twelfth century (Bishop sagas, 1858). 'Darkness' was reported in connection with the 1262 eruption in Mýrdalsjökull and the first direct reference to tephra fall was in ~1357 (Storm, 1888). Earth fire mentioned in the *Book of Settlement* refers to lava flows. Detailed descriptions of eruptions and/or jökulhlaups exist since the Katla eruption of 1625 (e.g. Safn til sögu Íslands IV). An official report on the 1918 eruption contains day-to-day description of the course of events (Sveinsson, 1919) and collections of eyewitness accounts add valuable details (Jóhannsson, 1919; Eggertsson, 1919). Thorarinsson (1975) briefly summarized this information in his paper on Katla and the Annal of Katla eruptions.

Altogether 15 eruptions that can be attributed to the KVS are mentioned in documents since the settlement in the 870s, including the earth fire now known as the Eldgjá eruption. The date of 12 Katla eruptions is reported to the year and of additional two to within a few years (Table 3.1). The time of the year is known for these eruptions since 1580, all of which began during the period May–November, implying that Katla eruptions are more likely to begin during the spring–fall season than in wintertime.

The duration of documented Katla eruptions since 1625 has varied from 2 weeks to over 5 months

Table 3.1. Historical eruptions of the Katla system (Larsen, 2000).

Eruption site	Eruption year/century	Date	Length days	Preceding years
Katla	(1955*)	June 25th		(37)
Katla	1918*	October 12th	24	58
Katla	1860*	May 8th	20	37
Katla	1823*	June 26th	28	68
Katla	1755*	October 17th	~120	34
Katla	1721*	May 11th	>100	61
Katla	1660*	November 3rd	>60	35
Katla	1625*	September 2nd	13	13
Katla	1612*	October 12th		32
Katla	1580*	August 11th		80
Katla	~1500†			
Katla	15th century			
Katla	~1440†			(24)
Katla	1416*			(59)
Katla	~1357‡			(95)
Katla	1262*			17
Katla	1245*			(66)
Katla	~1179*			
Katla	12th century			
Eldgjá	934/938§			
Katla	~920**			(16)
Katla	9th century			

Note: Tephra layers, thought to represent the above eruptions have been identified in soil sections around the Mýrdalsjökull central volcano. They have been chemically analysed to verify their origin, with the exception of K-1823 and the two twelfth century layers. Three eruptions, previously labelled K-1311, K~1000 and K-x, have been dropped from the list as the tephra layers representing them belonged to other eruptions on the Katla system. The latter two are part of the Eldgjá tephra. Dates as summarized by (*) Thorarinsson (1975), (†) Larsen and Thorarinsson (1984), (‡) Einarsson *et al.* (1980), (§) Hammer *et al.* (1980) and Zielinski *et al.* (1995), (**) Hafliðason *et al.* (1992).

(Table 3.1). The average repose period since AD 1500 is 47 years with maximum deviation of 34 years. Of those, the shortest repose period is 13 years, between the eruptions in AD 1612 and 1625. The longest is, supposedly, about 95 years between the eruptions in AD 1262 and ~1357.

The documented Katla eruptions are described as being accompanied by often widespread tephra fall, lightning in the eruption cloud, thunder and enormous jökulhlaups consisting of meltwater, ice and volcanic debris (e.g. Safn til sögu Íslands IV; Sveinsson, 1919; Jóhannsson, 1919). Although the opening stages of Katla eruptions are subglacial, they apparently melt their way through the overlying ≥400 m thick ice cover in a matter of hours, if precursory earthquakes felt in nearby areas are taken to mark the beginning of the eruptions.

A sustained eruption column of vapour, gases and tephra usually develops during the first hours of subaerial activity at Katla. On the first day of the 1918 event the eruption cloud reached an elevation of 14 km above sea level as measured from Reykjavík, about 160 km west of the volcano (Eggertsson, 1919). At closer range it was seen as a fast rising cloud with a convoluted upwind margin, and the basal part was described as black, becoming whitish in the upper reaches (Sveinsson, 1919; Jóhannsson, 1919). Tephra fall commenced at a distance of 30 km within an hour of the first cloud sighting.

Tephra fall can occur at any time during the eruptions but explosive activity is usually most intense, and tephra production greatest, during the first days (Safn til sögu Íslands IV, 1907–1915; Sveinsson, 1919). Katla tephra is deposited both as lobate fans and as thin veils around the volcano. The reported tephra fall area in the 1918 eruption exceeded 50,000 km^2 on land (Fig. 3.3). Tephra from the first day of Katla eruptions has reached the Faroe and Shetland Islands and in the 1625 eruption tephra reached the European mainland (Thorarinsson, 1981 and references therein). Airborne tephra and water-transported volcanic debris are the only historically recorded volcanic products of Katla eruptions.

Basaltic Katla tephra is coal-black to brownish black and is often described as 'sand' or 'ash and pumice' in contemporary documents. In the 1625 eruption, friable fist-size clasts reached as far as Álftaver, >30 km southeast of Katla (the 1918 eruption site, Safn til sögu Íslands IV, 1907–1915). Reported thicknesses at distances of 25–30 km in the 1755 eruption were 1–2 ft or 30–60 cm in open areas and 3–4 ft or 90–120 cm closer to the source (Ólafsson, 1978), which correlates well with maximum measured thickness of the Katla 1755 tephra, 120 cm, about 15 km east of Katla (1918 eruption site).

Floods on the Skóga- and Sólheimasandur plains are described in the *Book of Settlement* (1967) and refer to a volcanogenic jökulhlaup in the early tenth century, perhaps the same jökulhlaup that left deposits in sections east of the Sólheimasandur plain (Dugmore, 1987 and unpublished data). Since AD 1179 all documented Katla jökulhlaups have escaped from the caldera through the Kötlujökull pass onto Mýrdalssandur with the exception of a minor 'jökulhlaup' from below Sólheimajökull during the 1860 eruption (Safn til sögu Íslands IV, 1907–1915; Hákonarson, 1860). Jökulhlaups accompanying Katla eruptions are a mixture of meltwater, ice blocks and volcanic debris. Huge icebergs are carried by the largest jökulhlaups (Fig. 3.4) and left on the flood plain or discharged into the sea. Flood routes after 1625 are described in contemporary literature (Safn til sögu Íslands IV, 1907–1915). All documented jökulhlaups since 1660 have raised the sandur surface and moved the shore outwards.

The historical jökulhlaups have, reportedly, emerged from the Kötlujökull glacier either shortly before or shortly after an eruption plume becomes visible. They discharge in several outbursts onto Mýrdalssandur, those of the first day usually being most voluminous. The first 15 km of their route lies below, within or on top of the Kötlujökull glacier, and they may emerge out from under

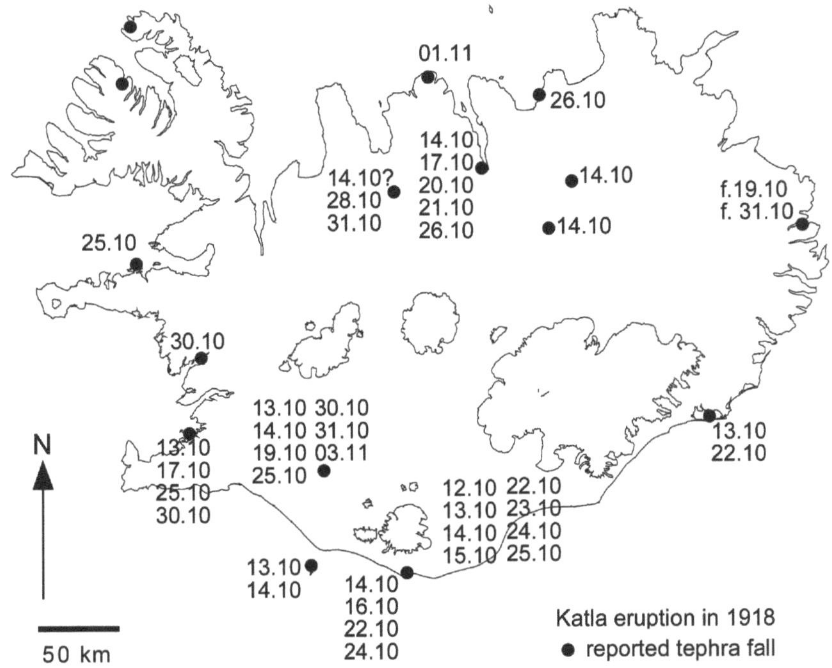

Fig. 3.3. Localities and dates of reported tephra fall in the 1918 Katla eruption. The numerous localities in the area east of Mýrdalsjökull are grouped together. Exact dates of tephra fall in the Northwest peninsula are not known (Guðrún Larsen and Sigurður Ásbjörnsson, unpublished data).

Fig. 3.4. Icebergs left on the Mýrdalssandur outwash plain by the 1918 jökulhlaup. Note the two men on top of the lower iceberg. Photo: Kjartan Guðmundsson.

its snout or break out well above the ice margin (Safn til sögu Íslands IV, 1907–1915). The locations of the outlets determine the routes across Mýrdalssandur, some of which are shown in Fig. 3.5. The major jökulhlaups may have outlets in three or more locations but the minor ones may discharge through a single outlet.

Earth fire, a traditional description of flowing lava, is said to have overflowed two settlements, one of them densely populated (*Book of Settlement*) in the area now covered by the Álftaver lava from S-Eldgjá. Part of the settlers fled to Mýrdalur in the west. Others remained despite the damage, including the settler Hrafn Hafnarlykill, whose son was nicknamed Aurgoði or 'lord of the mud'. This occurred during or shortly after the period of Norse settlement in Iceland, traditionally the period AD 870–930.

Damages to property, livestock and people resulting from tephra fall, earth fire, jökulhlaups and lightning are recorded in contemporary documents. The worst recorded damage by tephra fall was in 1755 when 50 farms were abandoned, some of them permanently. Two people were killed by lightning in the 1755 eruption. The worst damage to property by lava flows was in the Eldgjá eruption ~934 when settlers had to abandon farms in at least two settlements in Álftaver and nearby areas (*Book of Settlement*). The greatest damage to property by jökulhlaups is most difficult to assess. Shortly before 1179, jökulhlaup destroyed several farms and a few churches in the easternmost part of Mýrdalur (Bishop sagas, 1858). A fjord in the same area, mentioned

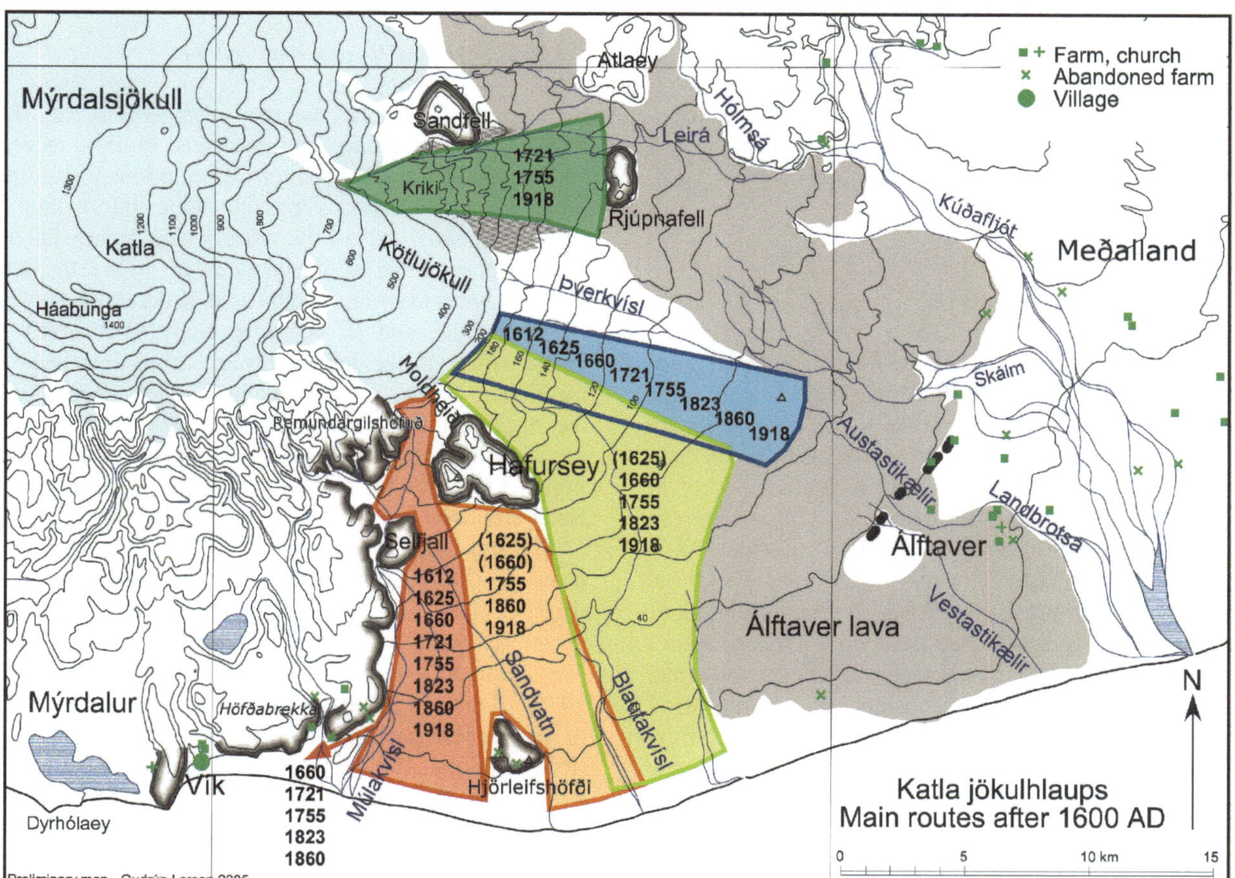

Fig. 3.5. Generalized main flood routes of Katla jökulhlaups after AD 1600, based on descriptions in contemporary documents. Eruption years in brackets indicate no direct reference to that particular flood route. Adapted from Larsen et al. (2009).

in the *Book of Settlement*, had probably been filled in before AD 1100. The greatest loss of livestock in a jökulhlaup may, however, have been in 1918 when several hundred sheep were lost (Sveinsson, 1919).

3.3.2. Tephra Stratigraphy and Explosive Volcanism

The thick soils around Mýrdalsjökull contain a 8,000–9,000 year record of tephra layers from the KVS and the other volcanic systems of the EVZ. Preservation conditions are excellent on the east, south and west side of Mýrdalsjökull but poor north of the ice cap. Numerous soil sections with tephra layers have been measured around Mýrdalsjökul to obtain information on the type, frequency and magnitude of explosive eruptions on the KVS (e.g. Thorarinsson, 1958; Larsen, 1978, 1981, 1996, 2000; Einarsson *et al.*, 1980; Larsen *et al.*, 2001; Óladóttir *et al.*, 2005) as well as changes in magma composition with time (Óladóttir *et al.*, 2007, 2008). A second goal is to construct a tephrochronology that can be used to date other events on the KVS such as lava emission and jökulhlaups as well as glacier fluctuations, ecological changes and early settlements around Mýrdalsjökull (e.g. Larsen, 1978, 1979, 2000; Haraldsson, 1981; Sveinbjarnardóttir, 1982; Larsen and Thorarinsson, 1984; Buckland *et al.*, 1986; Dugmore, 1987, 1989; Dugmore and Sugden, 1991; Dugmore *et al.*, 2000; Casely and Dugmore, 2004; Smith, 2004; Larsen *et al.*, 2005; Smith and Haraldsson, 2005; Smith and Dugmore, 2006; Smith *et al.*, submitted for publication).

Studies of individual tephra layers erupted within the KVS have been carried out in order to map their dispersal and thickness (Thorarinsson, 1969, 1971, 1980, 1981; Larsen, 1978, 1979, 1996, 2000; Einarsson *et al.*, 1980; Larsen *et al.*, 2001) as well as to characterize the tephra (Einarsson *et al.*, 1980; Guðmundsdóttir, 1998; Larsen *et al.*, 2001; Óladóttir, 2003).

3.3.2.1. Tephra characteristics

Basaltic Katla tephra, emitted in hydromagmatic eruptions within the Mýrdalsjökull ice cap, is the most common volcanic product of the KVS. As previously mentioned, it is coal-black to brownish black in the field and it consists mostly of highly fragmented, poorly to moderately vesiculated glass with grain sizes in the ash and lapilli range (Einarsson *et al.*, 1980; Larsen, 2000; Óladóttir *et al.*, 2005). Crystals are scarce. The lithic component, when present, consists of minor amounts of greyish white, felsic rock fragments, brown hyaloclastites or greyish basalts. Most Katla tephra layers show distinct bedding due to intermittent deposition and shifting wind strength and wind directions during the eruption. A fine-grained lower part and a coarser upper part characterize some of the layers (e.g. K$\sim$1357, Einarsson *et al.*, 1980), implying that the first erupted tephra is more highly fragmented than that of later stages, probably as a result of abundant meltwater at the eruption site during the early stages of the eruptions. The opposite has also been observed (e.g. K-1755, Guðmundsdóttir, 1998), indicating less favourable water-to-magma mass ratio in the early stages, possibly as a result of rapid escape of meltwater from the eruption site or high initial eruption rates.

The basaltic tephra emitted from the partly subglacial and partly subaerial 75-km-long Eldgjá fissure in the AD $\sim$934 eruption ranges from purely phreatomagmatic Katla-type tephra to purely magmatic tephra covering the range from fine ash to bombs (Larsen, 2000, Thordarson and Larsen, unpublished data). The formation and deposition of the Eldgjá tephra was more complex than that of 'normal' Katla eruptions as manifested by alternating beds of phreatomagmatic and magmatic tephra and much higher portion of lithics than in typical Katla tephra (Larsen *et al.*, 2004).

Silicic tephra has erupted sporadically from within the Mýrdalsjökull central volcano during the last 8,000 years (Newton, 1999; Larsen *et al.*, 2001; Óladóttir *et al.*, 2008). Silicic Katla tephra layers (SILK layers) appear in the soil around Mýrdalsjökull as yellowish or greenish grey layers of ash and lapilli, sometimes containing elongate, highly vesiculated needle-like grains of translucent glass and/or black, poorly vesiculated equant grains. All SILK layers identified so far are of prehistoric age (Table 3.2) and those which have been mapped have thickness axes pointing towards the Katla caldera. Whether the eruptions sites lie along the caldera fracture or within the caldera has not been resolved. Minor silicic components are present in several predominantly basaltic prehistoric and historical tephra layers but are not considered as a separate category.

3.3.2.2. Tephra dispersal and volumes

Numerous soil sections covering the uppermost part of the KVS tephra record have been measured in detail to verify and supplement the descriptions of eruptions in historical time. Many descriptions provide solid information on tephra dispersal and on the areas where tephra fall was most intense, often with estimates of tephra thickness. Tephra layers from all documented Katla eruptions have been identified and all but three analysed to verify the origin. Five Katla tephra layers from eruptions not mentioned in any documents have been added to the record (Thorarinsson, 1975; Larsen, 1978, 2000) bringing the record to 21 eruptions including the Eldgjá eruption. Figure 3.6 shows the directions of the thickness axes in 15 Katla eruptions and in the Eldgjá eruption.

Basaltic tephra from several major Katla eruptions has been mapped to estimate the magnitude of the eruptions (Figs. 3.7 and 3.8). The volumes, calculated as uncompacted or freshly fallen tephra, range from about 0.01 to over 1 km^3 for the historical tephra layers (Thorarinsson, 1975; Larsen, 2000). The uncompacted volumes of the largest prehistoric eruptions may exceed 2 km^3 (Róbertsdóttir, 1992). These are, however, minimum volumes because an unknown portion of the volcanic products in Katla eruptions is either left at the eruption site or transported by jökulhlaups onto the flood plains and into the sea.

The Eldgjá tephra (Fig. 3.9) is the largest basaltic tephra layer originating from the KVS, about 4.5 km^3

Table 3.2. Radiocarbon dates and age estimates for 12 SILK tephra layers.

SILK tephra	^{14}C age BP	Lab. No.	Reference
SILK YN	1676±12	GU-7091*	Dugmore *et al.* (2000)
SILK UN	2660±50	SSR-2805	Larsen *et al.* (2001)
SILK MN	2975±12	GU-7021*	Shore *et al.* (1995)
SILK LN	3139±40	GU-7019*	Larsen *et al.* (2001)
SILK N4	c. 3600		Larsen *et al.* (2001)
SILK N2	c. 4200		Larsen *et al.* (2001)
SILK N1	c. 4900		Larsen *et al.* (2001)
SILK A1	c. 5000		Larsen *et al.* (2001)
SILK A5	c. 6100		Larsen *et al.* (2001)
SILK A7	c. 6200		Larsen *et al.* (2001)
	6400±80	U-4604	Larsen *et al.* (2001)
SILK A8	c. 6400		Larsen *et al.* (2001)
SILK A9	c. 6600		Larsen *et al.* (2001)

*Average of several radiocarbon dates including those indicated by the Lab. No.

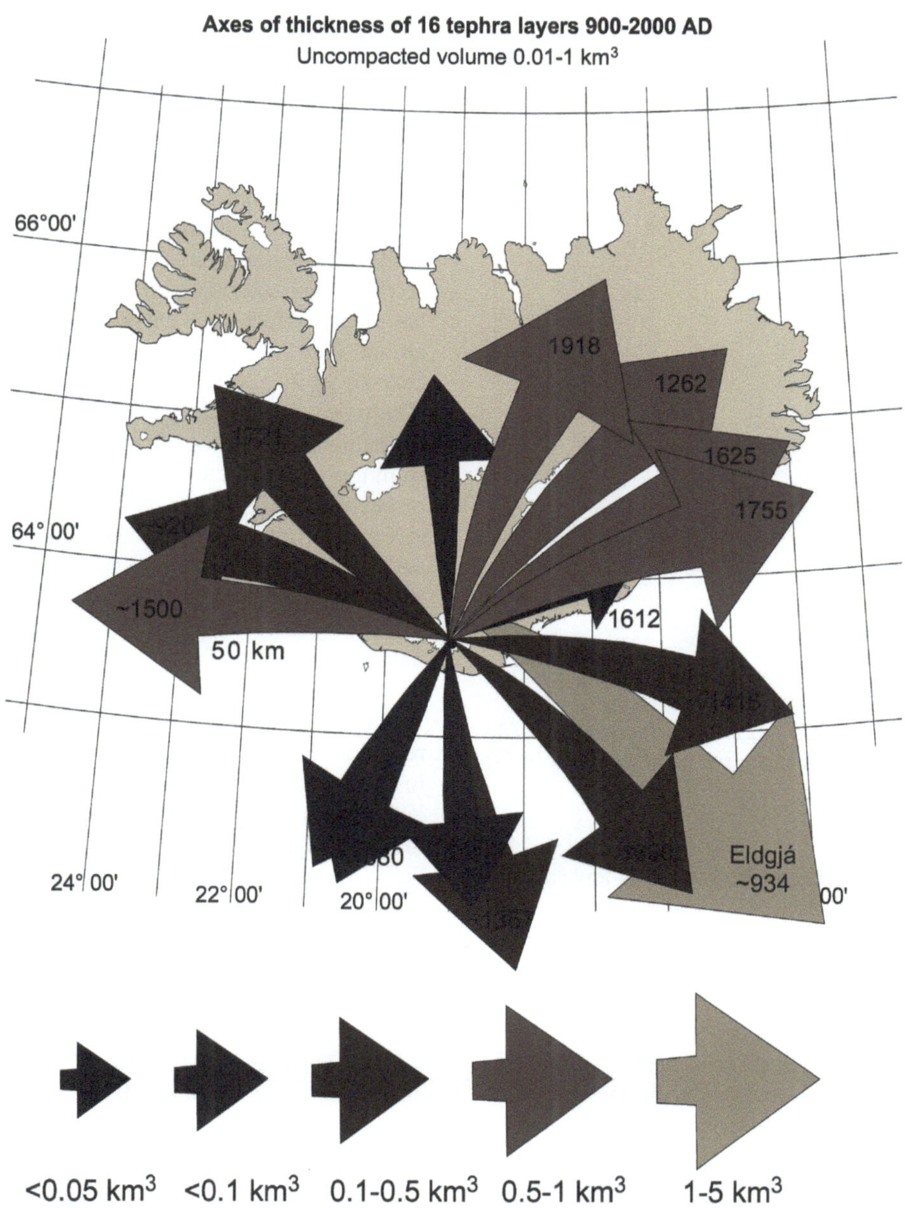

Fig. 3.6. Main directions of tephra dispersal in historical eruptions in the Katla volcanic system. Note the infrequent westwards dispersal. Adapted from Larsen et al. (2009).

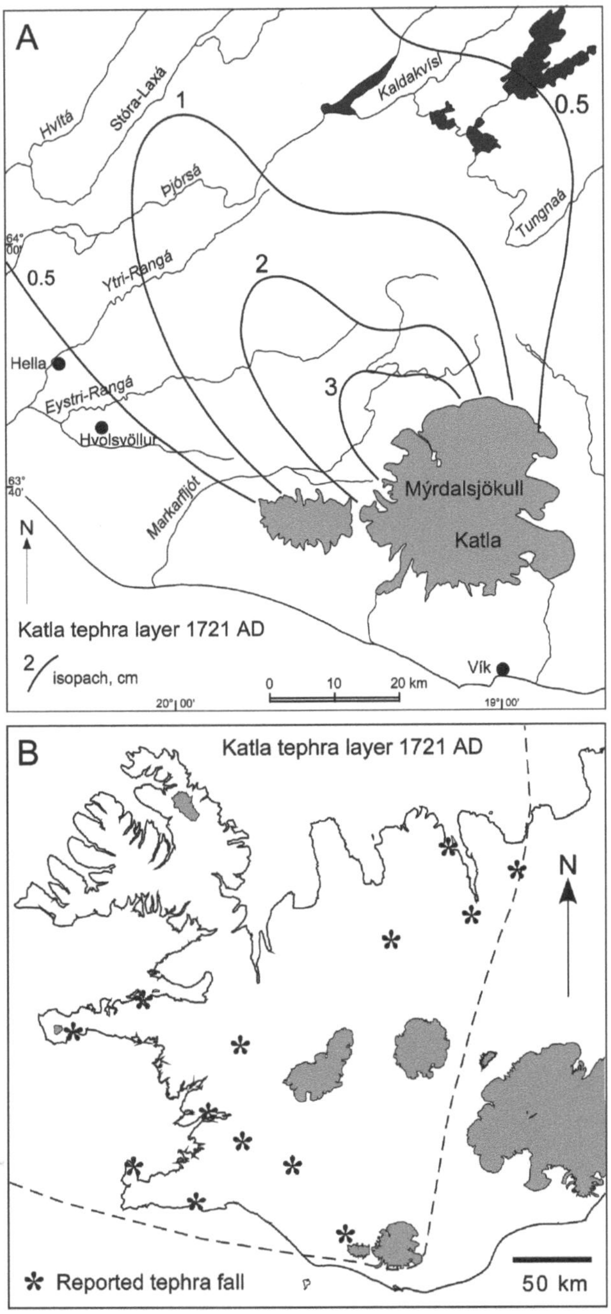

Fig. 3.7. Isopach map of the 1721 Katla tephra layer (A) and reported tephra fall (B). Adapted from Larsen (1978) and Thorarinsson (1955).

deposited on land (calculated as freshly fallen tephra), and one of the four largest tephra layers erupted in historical time. A tephra layer that had previously been assigned to an eruption around AD 1000 (Thorarinsson, 1967, 1975) was found to be a part of the Eldgjá tephra (Larsen, 1996) and the AD 1000 eruption has been abandoned.

Basaltic Katla eruptions last days, weeks or months, and tephra fall can occur at any time during the eruptions. Hence the dispersal of tephra differs from that of shorter-lived plinian and subplinian eruptions. Many Katla tephra layers are lobate and/or form thin 'veils' around the volcano (Figs. 3.7 and 3.8). In the larger Katla eruptions tephra fall may reach most of Iceland (Fig. 3.3) but is usually minor in distal areas. Table 3.3 lists some of the Katla eruptions and the areas covered within specific isopachs (Larsen *et al.*, 2009).

The largest historical Katla tephra layers are of the order 1 km^3 of uncompacted tephra. (The Eldgjá eruption, tephra deposits and lava flow, is treated separately below.) The largest prehistoric Katla tephra layers, the K-E and K-N of Róbertsdóttir (1992) are significantly larger or $>2\,km^3$ as freshly fallen.

The SILK tephra layers mapped so far have volumes in the range of 0.05–0.3 km^3, calculated as uncompacted tephra (Larsen *et al.*, 2001). Most of the SILK layers are lobate, with two or three well-defined main lobes, the largest layer being a notable exception (Fig. 3.10). Some of the lobes are narrow (Fig. 3.11), indicating short-lived events. This implies that at least some of the eruptions consisted of several relatively short-lived explosive events separated by quiet periods of unknown length. The location of the vent areas can be assessed by extrapolating the thickness axes (Fig. 3.12) which indicate intra-caldera vents.

3.3.2.3. Frequency of explosive eruptions as revealed by tephra stratigraphy

Large near-continuous soil sections have been measured in several locations around Mýrdalsjökull. For best result, selected parts of nearby sections, containing maximum number of tephra layers, are combined into composite soil sections representative of the area. Three such sections have been selected to exemplify the tephra stratigraphy around the KVS (Fig. 3.13): Einhyrningsflatir to the west (Larsen *et al.*, 2005), Klifurárgil-Álftagróf to the south (Dugmore, 1987; Larsen, unpublished data) and Atlaey-Rjúpnafell to the east (Óladóttir *et al.*, 2005). Distances from the centre of the Katla caldera are 20, 15 and 25 km, respectively. All sections are much closer to the KVS than to the other major tephra producers, i.e. Hekla, Veiðivötn and Grímsvötn volcanic systems. Key tephra layers, easily recognizable in the field on their macroscopic characteristics (colour, grain size, grain type, bedding) link the sections (Table 3.4). The majority of the key layers have been radiocarbon dated (Table 3.4).

In each section, 90–210 tephra layers have been recorded. Macroscopic characteristics of all tephra layers in each section were described in order to identify tephra layers, both from the KVS and other nearby systems. The majority of the tephra layers in the eastern section have been geochemically analysed (Óladóttir *et al.*, 2008), as well as selected layers in other two sections, but there identification of Katla tephra layers is mostly based on field characteristics.

The western section (Larsen *et al.*, 2005) contains ~170 tephra layers deposited in about 8,400 years (about 7,500 ^{14}C years). At least 90 have the macroscopic characteristics of basaltic Katla tephra and another 10 have those of silicic Katla tephra. At least 15 layers have the characteristics of Hekla, Veiðivötn, Eyjafjallajökull and Torfajökull tephra. A large part of the remaining thin deposits is black ash likely to originate in the KVS. The lowermost part of the section is resting on a very thick tephra deposit from one of the subaerial volcanic

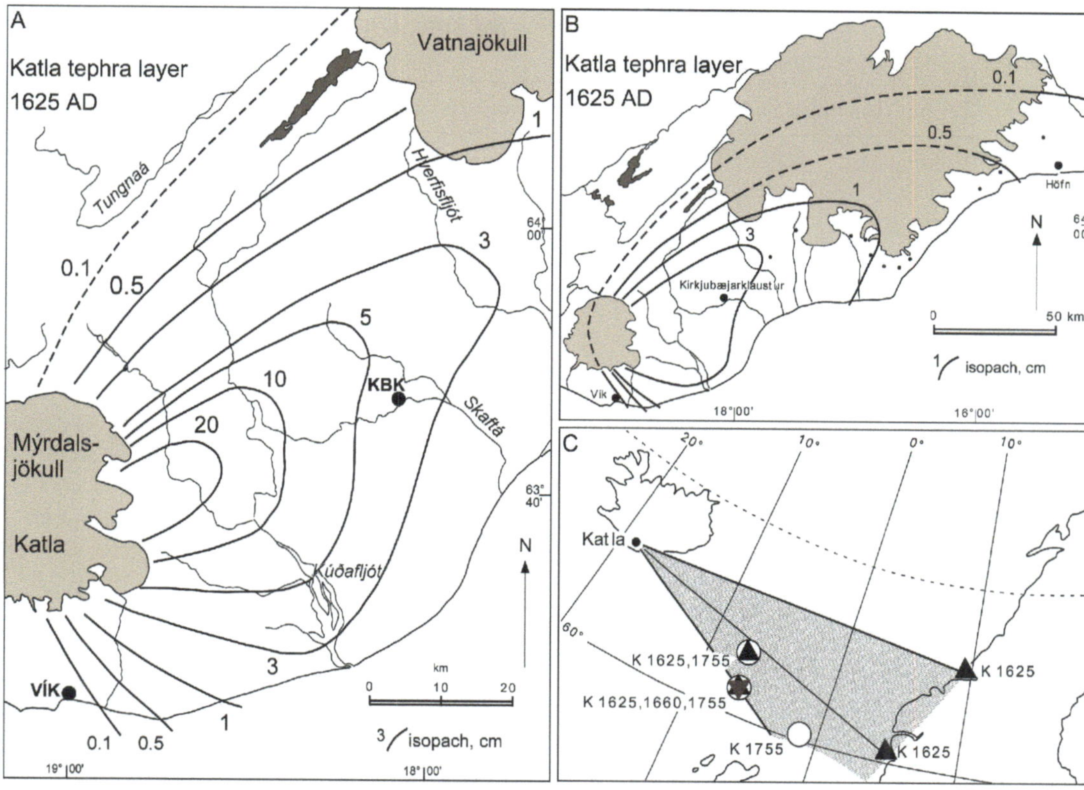

Fig. 3.8. Isopach map of the 1625 Katla tephra layer (A and B) and reported tephra fall outside Iceland (C). After Thorarinsson (1981) and Larsen (2000).

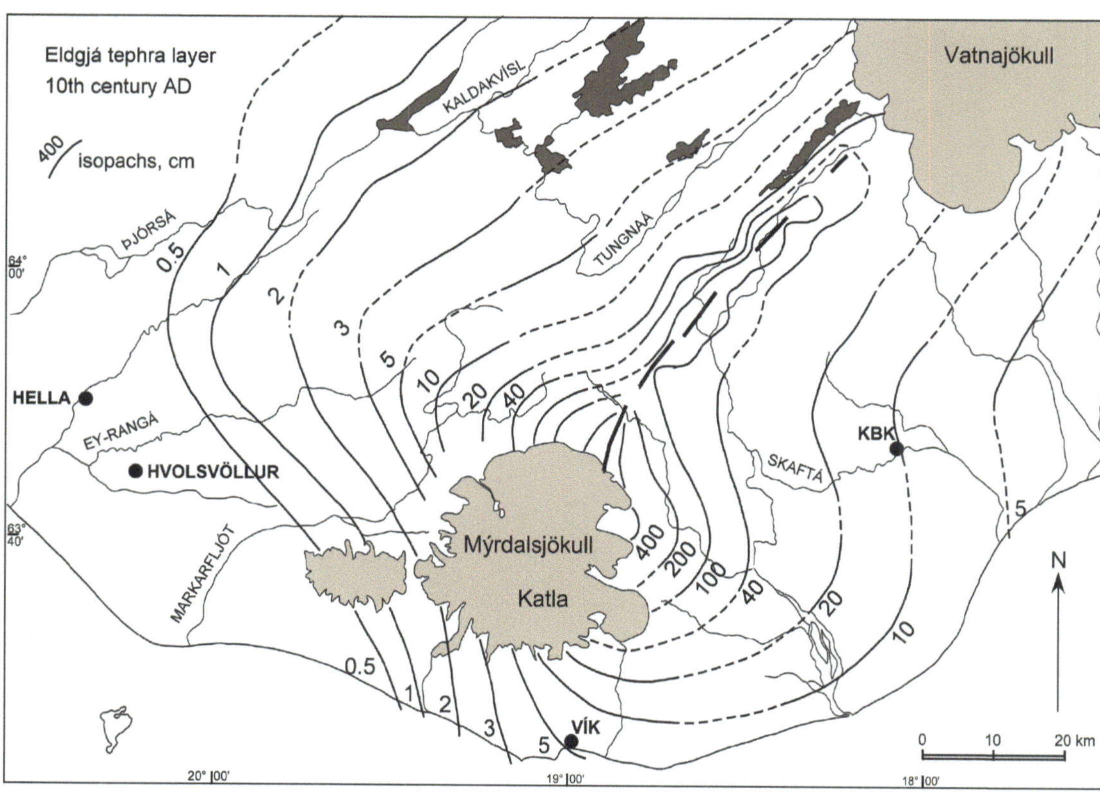

Fig. 3.9. Isopach map of the Eldgjá tephra layer. After Larsen (2000).

Table 3.3. Volume of basaltic tephra deposited on land in selected Katla eruptions, together with maximum distance to and area within the 1 and 10 cm isopachs, respectively.

KVS tephra layer	Tephra (uncompacted)	1 cm isopach		10 cm isopach	
	Volume on land (km³)	Maximum distance (km)	Tephra fall area on land (km²)	Maximum distance (km)	Tephra fall area on land (km²)
K-1755	0.8	120	~6000	35	~900
K-1721	0.33	80	~2800	†	†
K-1660	0.26	~65	~2100	20	~225
K-1625	0.5	130	~5500	40	~570
K-1612	0.04	35	~300	†	†
K-1500	0.5	170	~6800	†	†
K~1357	~0.2	*	*	20	~130
K-1262	0.48	~90	~4550	20	~160
E~934	~4.5	*	*	45	~4600
K~920	0.27	70	~2200	30	~155

*Isopach extends out into the sea.
†Isopach does not extend into vegetated areas.

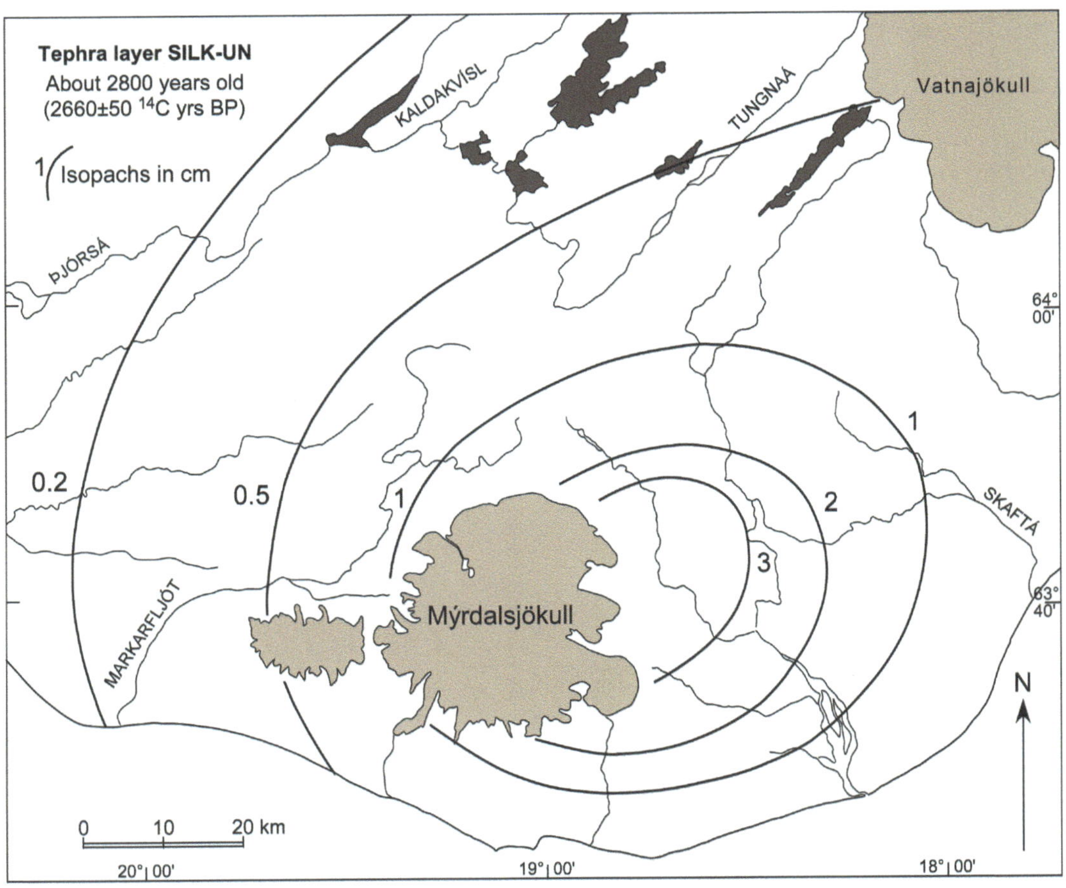

Fig. 3.10. Isopach map of the silicic Katla tephra layer SILK-UN. Modified from Larsen et al. (2001).

fissures on the flanks of Mýrdalsjökull that may have delayed soil formation in the area. Numerous jökulhlaup deposits are present in soil sections in the valleys west of Mýrdalsjökull (e.g. Sigurðsson, 1988; Smith, 2004; Smith and Haraldsson, 2005; Larsen *et al.*, 2005). Volcanogenic jökulhlaups are treated separately in this volume (see Russell *et al.*).

The south section is modified (re-measured) after Dugmore (1987) and covers about 7,700 years (about 6,800 ^{14}C years). About 90 tephra layers are present, of which at least 60 have the characteristics of basaltic Katla tephra and another 6 those of silicic Katla tephra. At least 14 have the characteristics of Hekla, Veiðivötn and Grímsvötn tephra, but the origin of the remaining

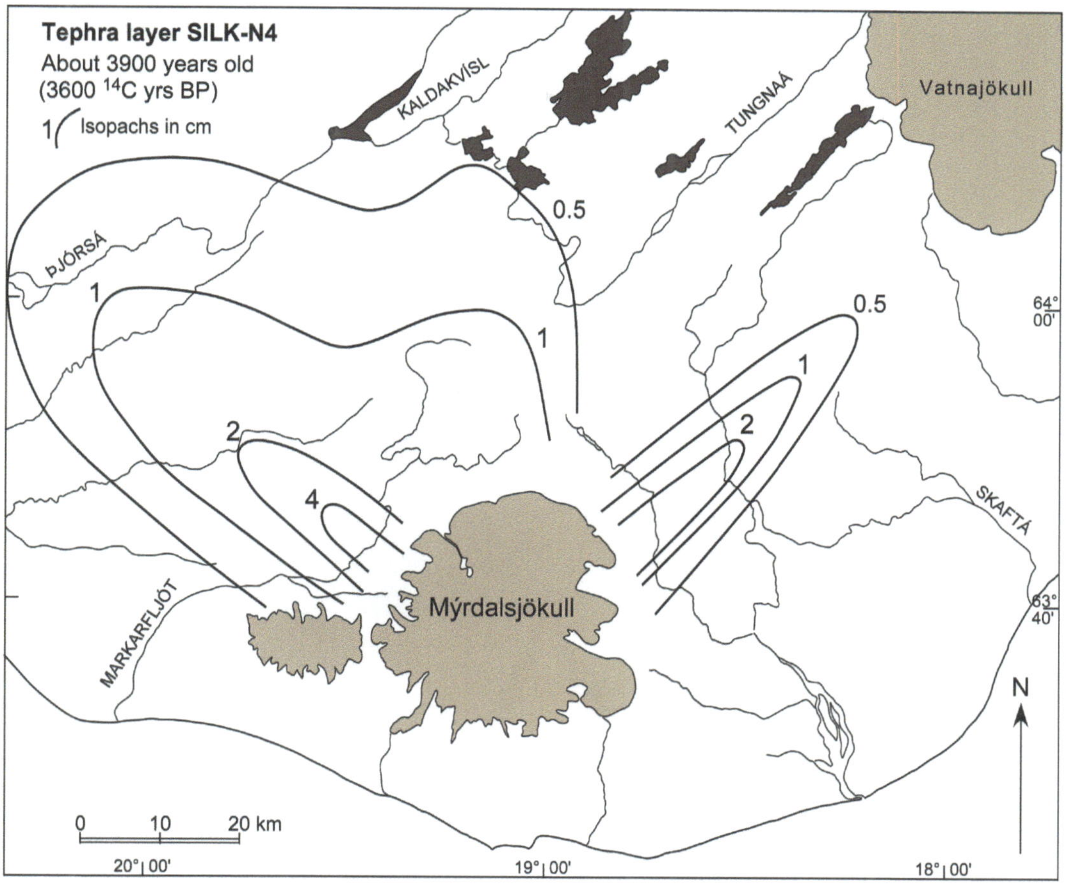

Fig. 3.11. Isopach map of the silicic Katla tephra layer SILK-N4. Modified from Larsen et al. (2001).

tephra layers cannot be determined from macroscopic characteristics. Many of the soil sections in the south contain jökulhlaup deposits or water-transported tephra discharged from Sólheimajökull (Dugmore, 1987, 1989; Russell *et al.*, this volume).

The east section contains ~210 tephra layers deposited in about 8,400 years (about 7,500 ^{14}C years). The majority of the tephra layers have been geochemically analysed (Óladóttir *et al.*, 2005, 2008), and the remaining layers attributed to source according to macroscopic characteristics. In all, 182 tephra layers were attributed to KVS on grounds of field characteristics, including 10 SILK layers. Of the 172 basaltic ones 111 were analysed for major elements and Katla origin was confirmed for all the layers. Ten of the 20 historical Katla tephra layers, as well as the Eldgjá tephra, are preserved in the soils east of Mýrdalsjökull. Assuming that the modern weather pattern (wind directions and wind strength) has remained similar during the last 8,400 years, the number of basaltic Katla eruptions can be doubled and a realistic value seems to be 300–350 eruptions (Óladóttir *et al.*, 2005). Altogether 16 or 17 SILK layers have been identified in the soils around Katla (Newton *et al.*, 2000; Larsen and Eiríksson, 2008) and a few others are anticipated. The pre-Holocene Skógar-Vedde tephra is not included in this tally.

The eruption frequency has varied through the last 8,400 years (Óladóttir *et al.*, 2005). Figure 3.14 shows 'tephra layer frequency' in the east section per millennium, i.e. the Katla tephra layers arranged in 1,000 year bins. There are two distinct peaks: between 2,000 and 4,000 years and between 7,000 and 8,000 years ago. Average tephra layer frequency is 20.5 layers or 41 eruptions per millennium, assuming that half of the basaltic Katla eruptions are present in the east section. Historical time (the last 1,100 years) has the lowest recorded tephra layer frequency – and number of eruptions – only half of the average of the last ~8,500 years. It is worth noting that the largest known prehistoric Katla eruptions, K-E and K-N, occurred during the peak between 2,000 and 4,000 years ago (Róbertsdóttir, 1992).

3.3.2.4. A note on pre-Holocene tephra from the KVS

Only one tephra layer in Icelandic soils is known to originate from the Katla system during the last glacial termination. In Iceland it is known as the Skógar tephra (Norðdahl and Hafliðason, 1992) or as layer F (Sæmundsson, 1991) and in the North Atlantic outside Iceland it is known as the Vedde ash (Mangerud *et al.*, 1984), recently revised in the Greenland ice cores to an age of 12,170 ± 114 years (before AD 2000, Rasmussen *et al.*, 2006). This tephra consists of two main glass types: a major component of acid composition and a minor component of basaltic composition together with sporadic occurrences of intermediate compositions (Norðdahl and Hafliðason, 1992; Lacasse *et al.*, 1995). Water-transported pumice heaps deposited directly on the bedrock, known as Pumice Hills, on the south slopes of

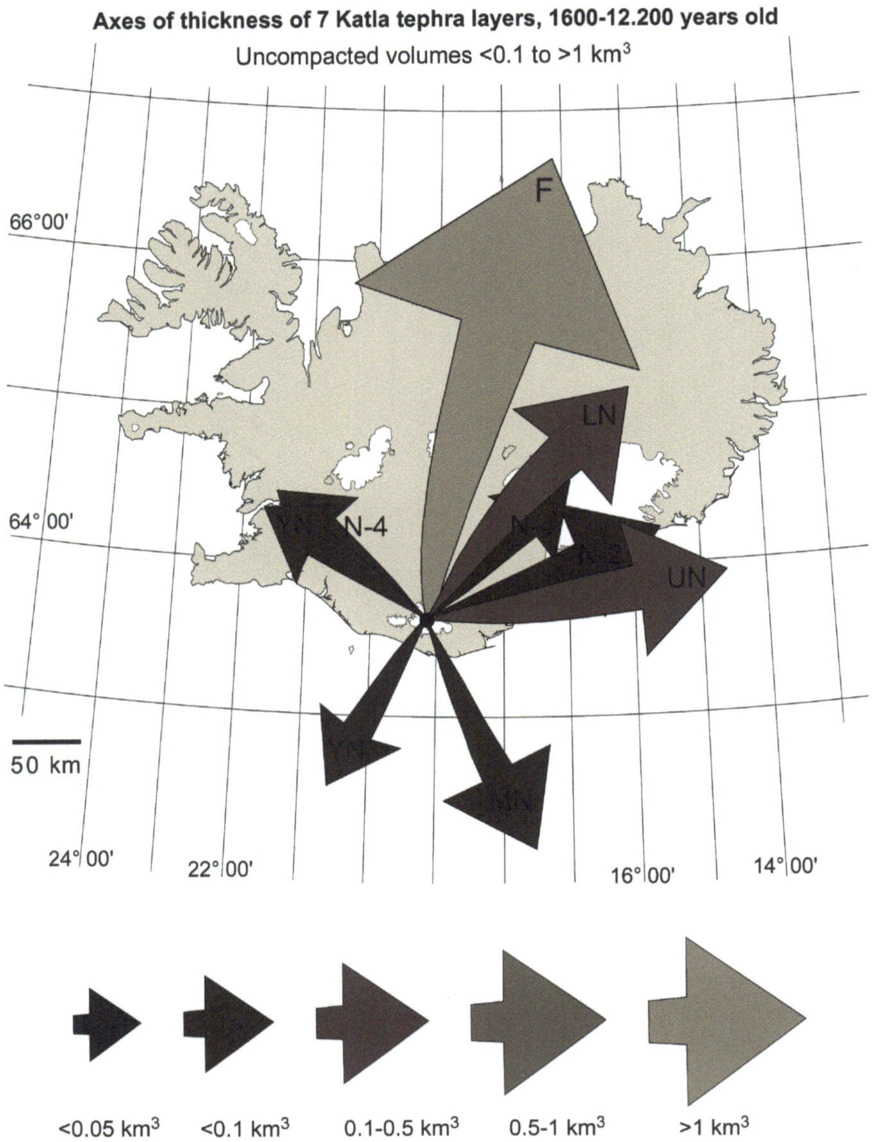

Fig. 3.12. Main directions of tephra dispersal in silicic Katla eruptions. For dates see Table 3.2 and main text. Modified from Larsen and Eiríksson (2008).

Mýrdalsjökull are also thought to belong to this eruption, as well as the acid nunataks around the Katla caldera (Lacasse *et al.*, 1995). The origin of the acid component of the Skógar-Vedde tephra is, accordingly, almost certainly within the Katla caldera. Two closely spaced eruptions on the KVS could explain the basaltic component or it could be part of the same eruption on volcanic fissure extending outside the caldera and onto the fissure swarm. However, no craters from this time are preserved outside the ice cap.

In Iceland, most of the Skógar-Vedde tephra was deposited on the last Icelandic ice sheet. Occurrences and tephrochronological applications of the Vedde ash outside Iceland have been widely treated (e.g. Mangerud *et al.*, 1984; Wastegård *et al.*, 1998, 2000; Hafliðason *et al.*, 2000) and are beyond the scope of this paper. When this large tephra layer was deposited on the ice sheet the tephra on the ablation areas was *washed off* the ice into the proglacial environment and carried by streams and rivers into lakes and into the sea. The tephra deposited in the accumulation area was temporarily buried and then gradually moved to the ablation area, to be *washed out* of the melting ice into the proglacial areas after a long residence time in the ice.

Substantial amounts of the Skógar-Vedde tephra were deposited on the outlet glaciers and the ice-free areas in North Iceland between Eyjafjörður in the west and Axarfjörður in the east, where thick deposits have been found in sediments formed in ice-dammed lakes (Norðdahl, 1983; Norðdahl and Hafliðason, 1992) as well as in lows and hollows in proglacial areas (Sæmundsson, 1991). The tephra deposits thin westwards to Skagi (Björck *et al.*, 1992) as well as eastwards (Norðdahl, personal communication, 2008). This implies that the thickness axis of the airborne tephra lies towards NNE (Fig. 3.12). The western and eastern limits of the airborne Skógar-Vedde tephra in Iceland are not known. The Torfadalsvatn occurrence may be a primary airfall, according to the reconstruction of the last Icelandic ice sheet about 12,100 years ago (about 10,300 ^{14}C years BP, Norðdahl and Pétursson, 2005). Sporadic occurrences reported from inland lakes such as Hestvatn (Geirsdóttir

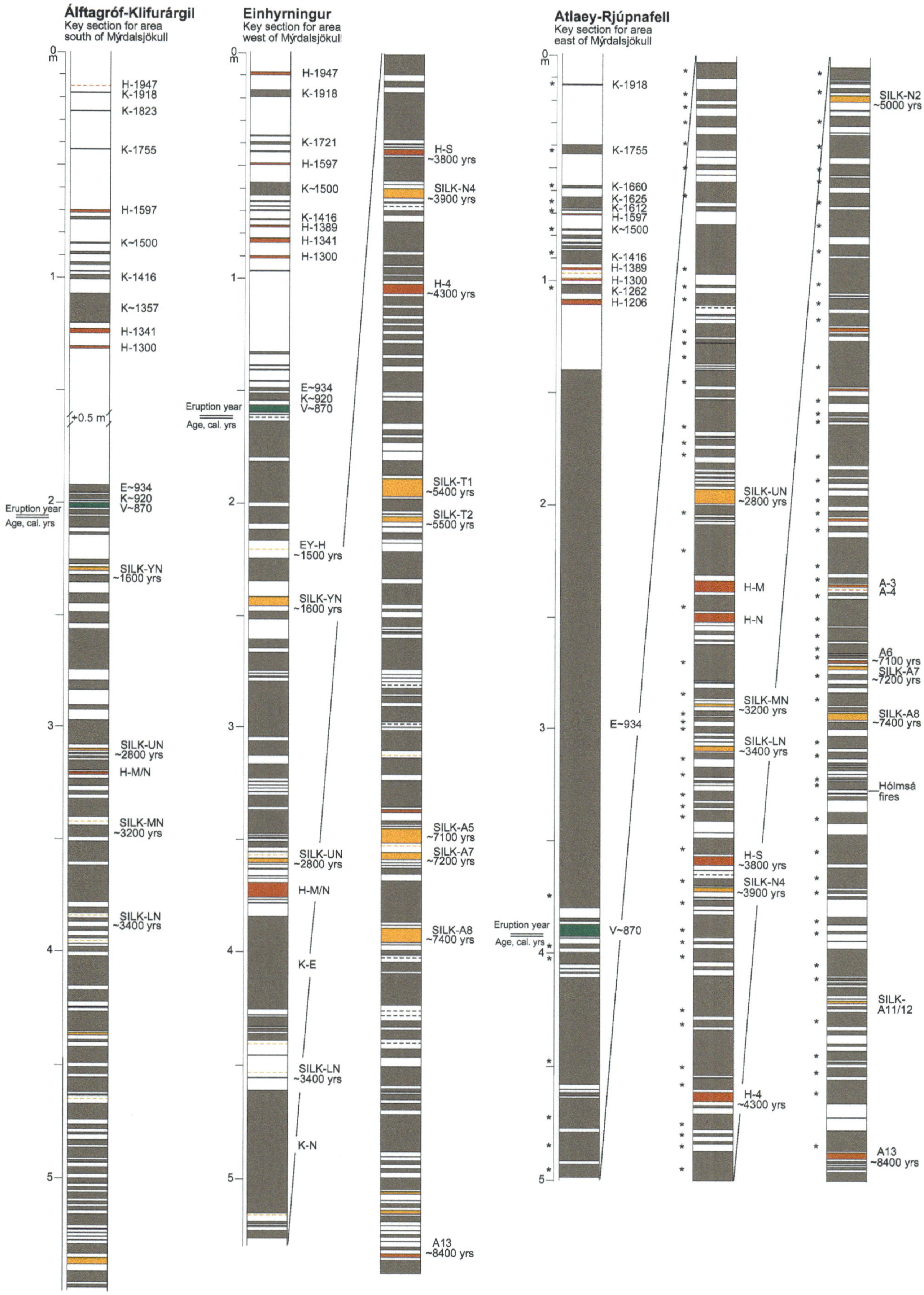

Fig. 3.13. Key sections for tephra layers (composite soil sections) south, west and east of Mýrdalsjökull. Age is in calibrated years before the present time but rounded to the nearest hundred years. Key tephra layers and their respective radiocarbon dates are listed in Table 3.4. Origin of tephra layers is indicated by colour. Grey: black tephra layers, most of which having the macroscopic characteristics of basaltic tephra from the Katla system, those verified by chemical analysis are marked by asterisk. Yellow: Katla system, silicic tephra. Red: Hekla system. Green: Veiðivötn system. Brown: Other systems, silicic tephra. Adapted from Dugmore (1989), Larsen et al. (2005) and Óladóttir et al. (2005).

Table 3.4. Key tephra layers in the soil around Mýrdalsjökull.

Key tephra layers	^{14}C age BP	Calibrated age range AD/BC (2 sigma)	Rounded calibrated age	Reference
EY-H	1540±50	AD 415–616	~1500	Smith and Haraldsson (2005)
SILK YN	1676±12	AD 338–418	~1600	Dugmore *et al.* (2000)
SILK UN	2660±50	917–777 BC	~2800	Larsen *et al.* (2001)
SILK MN	2975±12	1261–1128 BC	~3200	Shore *et al.* (1995)
SILK LN	3139±40	1512–1367 BC	~3400	Larsen *et al.* (2001)
Hekla-S	3515±55	1977–1692 BC	~3800	Larsen *et al.* (2001)
SILK N4	c. 3600		~3900	Larsen *et al.* (2001)
Hekla-4	3826±12	2309–2203 BC	~4300	Dugmore *et al.* (1995)
Layer A6	c. 6100		~7100	Larsen *et al.* (2001)
SILK A7	c. 6200		~7200	Larsen *et al.* (2001)
SILK A8	c. 6400		~7400	Larsen *et al.* (2001)
Layer A13	7505±42	6445–6330 BC	~8400	Dugmore (1987, 1989)

Note: Dates are calibrated using Calib 5.0.2. In column 4 and in the main text all ages have been rounded to the nearest hundred years and are referred to as 'years ago' or 'years old' (from the present time).

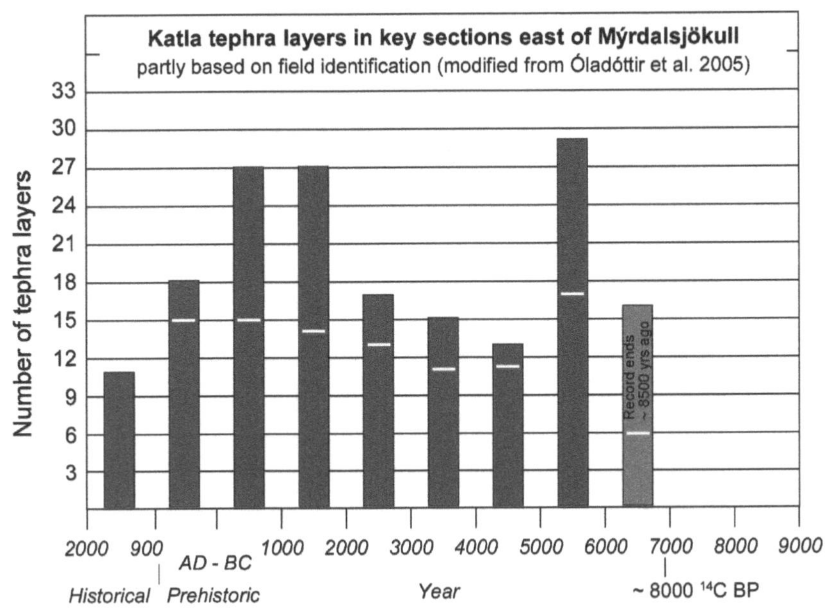

Fig. 3.14. Frequency of basaltic Katla tephra layers per 1,000 years in the key section east of the Katla volcano. White bars indicate the number of Katla tephra layers verified by chemical characteristics, those above are identified by field characteristics. Modified from Óladóttir et al. (2005).

et al., 1998) could, however, be later wash-out from the melting ice sheet.

3.3.3. Lava Flows and Effusive Volcanism

In contrast to the high frequency of explosive eruptions within the Mýrdalsjökull ice cap recorded in the tephra stratigraphy, volcanic activity on the ice-free areas of the KVS is infrequent. Jakobsson (1979) mapped 34 individual lava units originating on the KVS, all basaltic. Later work has shown that in many cases several units can be attributed to a single eruption (e.g. Larsen, 1979; Jóhannesson *et al.*, 1990) which reduces the number to about 10 eruptions outside the Mýrdalsjökull ice cap. It is possible that some lava flows emerging out from under the ice on Fimmvörduháls, between Mýrdalsjökull and Eyjafjallajökull, also belong to the KVS (Torfason and Jónsson, 2005).

Most of the lava flows are small, from a few km^2 to some tens of km^2, with notable exceptions which are described below as 'fires'. The former occurred on short fissures on the western and northern periphery of the Mýrdalsjökull massif and are thought to be more than 4,500 years old (Jakobsson, 1979; Jóhannesson *et al.*, 1990). In the two 'fires', fissures also opened up on the NE trending fissure swarm and they are the largest and most hazardous Holocene events of the KVS. In all, lava flows from the KVS cover about 890 km^2.

3.3.3.1. Major events: The 'Fires'

The older 'fires' (Hólmsá fires) are about 7,700 years old (about 6,800 ^{14}C years BP, Larsen, 2000). The lava flows, now partly covered by the products of the younger fires, can be shown to fit into the regional tephra stratigraphy of S-Iceland at a specific stratigraphical level (Fig. 3.13). The lavas flowed along depressions and river channels down the Álftaversafréttur area at least as far as Atlaey (Fig. 3.15) and possibly all the way to the coast of that time. Their southwards extension is hidden below an extensive cover of younger lava. The volume of lava is tentatively estimated to be ca. 5 km^3. The length of the eruptive fissure is not known, but a minimum length of 8–10 km outside the present ice margin is inferred by the paths taken by the lava.

The younger 'fires' (Eldgjá eruption) were dated by tephrochronology to the early tenth century (Larsen, 1979), and by ice-core chronology to around AD 934–938 (Hammer *et al.*, 1980; Zielinski *et al.*, 1995). The ca. 75 km long fissure extends from the Katla caldera beneath the Mýrdalsjökull ice cap in the southwest, through the mountainous terrain northeast of the ice cap to Eldgjá proper and continues intermittently to Stakafell mountain (Fig. 3.15). It is the longest known eruptive fissure to be active in historical time. The eruption produced a widespread basaltic tephra layer, composed of several distinct units and two major lava fields (Robson, 1957; Miller, 1989; Larsen, 1996; Thordarson *et al.*, 2001).

The Eldgjá eruption is of particular importance and deserves some deliberation because of its magnitude and because it is the cause of the most extensive environmental changes brought about by volcanic activity in Iceland during the last 11 centuries (Larsen, 2000). It was also the greatest volcanic pollution event from a flood lava eruption of the last 11 centuries, responsible for SO_2 emissions of about 220 Mt, corresponding to about 440 Mt of sulphuric aerosols (Thordarson *et al.*, 2001).

About 60 km of the Eldgjá fissure were subaerial. Voluminous lava flows emanated from the fissure segments and were accompanied by vigorous degassing and explosive activity that resulted in rheomorphic spatter

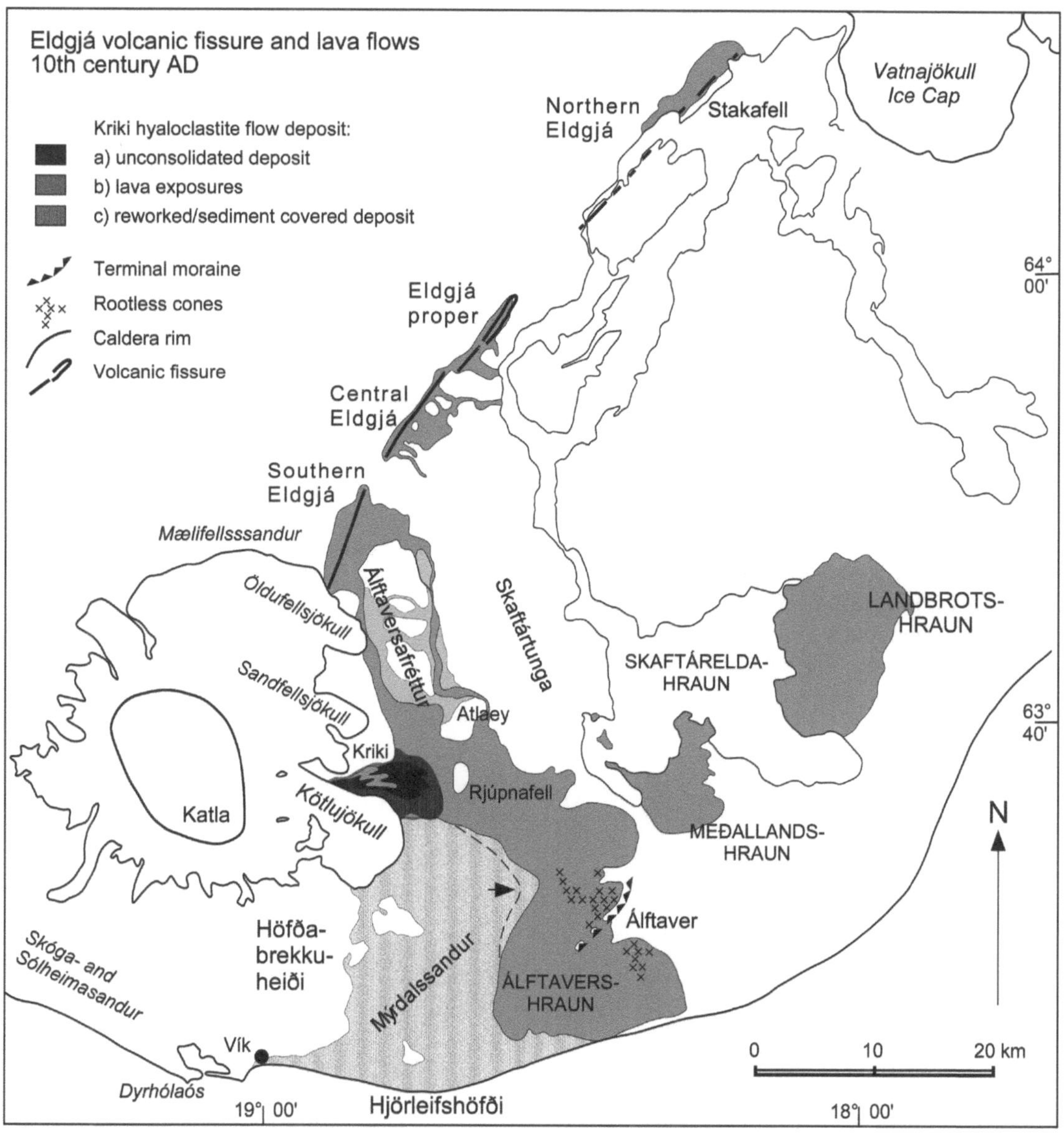

Fig. 3.15. The subaerial part of the Eldgjá fissure and the lava flows of the AD ~934 Eldgjá fires (dark shading). Arrow indicates buried margin. The subglacial part of the Eldgjá fissure extended into the Katla caldera. Also shown is the Kriki hyaloclastite flow. Exposed parts of the lava flows from the Hólmsá fires are indicated by light shading. The outlines of the 1783–1785 Skaftár fires lava flows are shown for comparison. After Larsen (2000).

deposit (Robson, 1957; Miller, 1989). The lavas were channelled along river gorges and valleys down to the lowland of Landbrot, Meðalland and Álftaver, where the lava flowed into the sea. Their combined area is close to 780 km^2 and the volume is about 18.6 km^3, making the Eldgjá lavas the largest flood lava of the last 11 centuries (Thordarson *et al.*, 2001). Lava productivity was highest on the $\geq$8-km-long S-Eldgjá segment where the 345 km^2 Álftaver lava emanated, whereas all other subaerial segments contributed to the remaining 435 km^2 lava field.

About 15 km of the Eldgjá fissure opened up below Mýrdalsjökull where hydromagmatic explosive activity was dominant throughout the eruption. The subglacial part of the fissure is likely to have been discontinuous but tephra dispersal indicates a main source west of Öldufellsjökull and another within the Katla caldera (Larsen, 2000). At least 30 discrete fall units have been identified, with grain sizes ranging from fine ash to coarse lapilli with highly variable proportions of juvenile clasts and lithic fragments, reflecting temporal and spatial variations in magma-to-water mass ratio. The tephra layer covers only about 20,000 km^2 on land but a large offshore dispersal is anticipated (Fig. 3.9). The tephra volume deposited on land is estimated to be about 4.5 km^3 (corresponding to 0.9 km^3 DRE) and total volume may exceed 6 km^3 (1.3 km^3 DRE).

A hyaloclastite flow may have accompanied the Eldgjá eruption (Larsen, 2000). The Kriki hyaloclastite flow deposits emerge out from under the present north margin of Kötlujökull at 600 m a.s.l. and can be followed for 6–7 km until they disappear at ca 300 m a.s.l. below younger sediments from glacial rivers and jökulhlaups. The main body consists of three subunits or facies: (1) a lowermost irregularly jointed lava and pillow lava with a highly irregular surface; (2) hyaloclastite breccia consisting of poorly consolidated scoriaceous tephra in the ash to bomb size ranges, pillows fragments and small isolated pillows; and (3) poorly consolidated layered hyaloclastite tuff with occasional cross-bedded or slumped layers. Vertical or subvertical protrusions of pillow basalt extend upwards and thin horizontal lava sheets are also intercalated with the tephra. Narrow (20–50 cm) dykes emerge from a protrusion and extend to the top of the flow deposit. The Kriki hyaloclastite flow is very similar to the standard hyaloclastite unit defined and described by Berg (1985) and Berg and Sigvaldason (1991), with the exception that the lowermost facies, regularly jointed lava, has not been observed. The volume of the Kriki deposits outside the glacier margin is about 0.5 km^3 (Larsen, 2000).

Jökulhlaups caused by activity on the subglacial part of the Eldgjá fissure emerged from the Öldufellsjökull and N-Sandfellsjökull outlet glaciers at the east margin of Mýrdalsjökull, from Sléttjökull at the north margin and possibly also Sólheimajökull in the south (Larsen, 2000; Kjær *et al.*, 2004).

3.3.4. Evolution of the Magma System in 8,400 Years Revealed by Tephra Layers

The basaltic magma erupted from the KVS is Fe–Ti basalt with small but significant temporal variations in major elements. Katla tephra layers spanning 8,400 years have been analysed to elucidate the magma system below the central volcano during that time (Óladóttir *et al.*, 2005, 2008). In accordance with changes in major element compositions, the activity at Katla can be divided into eight different evolutionary periods. Compositional changes over the past 8,400 years suggest a cyclic behaviour in the magmatism resulting from changes and evolution of plumbing systems beneath the volcano. A simple system is replaced by a more complex sill and dyke system that finally develops into a shallow crustal magma chamber (Fig. 3.16). The activity has passed through two such cycles with corresponding changes in the eruption frequency. When the plumbing system evolves from a simple system to a sill and dyke complex, the eruption frequency increases. When the sill and dyke complex is replaced by a magma chamber it drops again. With the relatively low eruption frequency in historical time the Katla volcano is now in a period of a simple plumbing system according to this model. Katla has also produced small-volume SILK layers in prehistoric times,

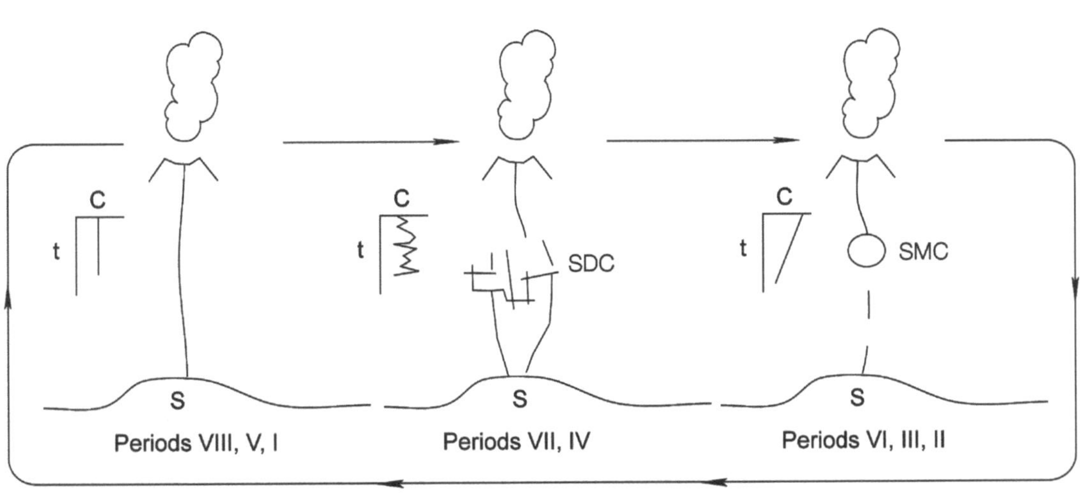

Fig. 3.16. Cartoon showing evolution of the plumbing system below the Mýrdalsjökull central volcano as suggested by Óladóttir et al. (2008). Roman numbers indicate the eight evolutionary periods.

which were primarily erupted during high-frequency eruption periods. The production of silicic melts is most likely due to increased influx of basaltic magma causing partial crustal melting (Óladóttir *et al.*, 2008).

3.4. Jökulhlaups Associated with Eruptions in the KVS

Jökulhlaups generated by eruptions within the Katla caldera escape through gaps in the wall of the ice-filled caldera, through which the outlet glaciers Kötlujökull, Sólheimajökull and Entujökull emerge. The Mýrdalssandur flood plain east of Mýrdalsjökull and the Skógasandur flood plain south of the ice cap are largely formed by volcanogenic jökulhlaups. The Markarfljót area and Landeyjar district to west have also been extensively modified by Katla jökulhlaups. All known historical jökulhlaups have drained towards east onto Mýrdalssandur.

Katla eruptions begin as subglacial eruptions but commonly melt their way through the overlying ice and become subaerial in 1–4 h (Safn til sögu Íslands IV). Large volumes of ice are rapidly melted in the process and drain from the eruption sites in jökulhlaups that are a mixture of meltwater, ice and volcanic debris. The volcanic material is mostly highly fragmented basaltic ash and lapilli (Einarsson, 1975), which allow rapid heat transfer to the meltwater/ice environment during the subglacial phase of eruptions and generation of voluminous jökulhlaups (e.g. Björnsson, 2003; Guðmundsson, 2003). These jökulhlaups have been defined as debris flows (Jónsson, 1982; Maizels, 1993), as water floods (Karlsson, 1994; Tómasson, 1996) and as alternating between mud flows and water flows (Björnsson, 1993) on their 35–40 km route from the eruption site to the coast. Katla jökulhlaups are multiple events. Jökulhlaups can occur throughout an eruption and may continue for some time afterwards. Those of the first day are usually the largest and transport the greatest ice and sediment load onto the flood plain and into the sea. Temporary forelands of sediment and ice are formed, extending kilometres into the sea.

In the largest of the jökulhlaups discharging onto Mýrdalssandur the first phase of the flood breaks through the ice inside the glacier terminus and flows supraglacially until it drains onto the flood plain at the periphery of Kötlujökull. When subglacial channels have been established, the main phase of the jökulhlaup in most cases emerges at the southern corner of Kötlujökull, breaching the ice margin and flooding the western part of Mýrdalssandur. The 1918 jökulhlaup, as calculated from cross sections of its main channels, may have peaked at 300,000 m^3/s and the volume of meltwater has been estimated to be as much as 8 km^3 (Tómasson, 1996). Volume estimates of water-transported volcanic debris, based on topographic changes on Mýrdalssandur and offshore deposition in 1918, vary between 0.7 and 1.6 km^3, respectively (Larsen and Ásbjörnsson, 1995; Tómasson, 1996). An area of up to 700 km^2 is flooded in the largest jökulhlaups (Fig. 3.17).

Sedimentary structures indicate a velocity of up to 55 km/h (15 m/s) in channels (Maizels, 1993). The velocity of the leading edge of the 1918 jökulhlaup on Mýrdalssandur plain was, however, close to 20 km/h or 6 m/s according to eyewitness accounts (Jóhannsson, 1919), in good accordance with translatory wave theory (Elíasson *et al.*, 2006). Travel time of future jökulhlaups from the Kötlujökull glacier margin down to Highway 1, which crosses the 30 km wide flood plain, could be as short as 30–40 min.

Jökulhlaups emerging from Entujökull in the west drain into the Markarfljót basin. Deposits of at least 10 prehistoric floods have been detected and dated by tephrochronology (Fig. 3.18; Haraldsson, 1981; Sigurðsson, 1988; Smith, 2004; Smith and Haraldsson, 2005; Larsen *et al.*, 2005; Smith and Dugmore, 2006). The youngest jökulhlaup from Entujökull occurred about 1,200 years ago and devastated a mature birch wood on the Markarfljót sandur plain (Knudsen and Eggertsson, 2005). The two largest jökulhlaups took place about 3,500 and 4,400 years ago. Cross sections of their main channel, as determined by flood marks on the slopes of the basin, indicate peak discharge of 200,000–250,000 m^3/s for these floods, respectively (Gröndal *et al.*, 2005). Over 1,000 km^2, mostly in the Landeyjar districts, could be inundated in future floods of this magnitude (Fig. 3.19) and travel time from Entujökull down to Highway 1 could be as short as 4 h (Hólm and Kjaran, 2005).

The last major jökulhlaup(s) to emerge from Sólheimajökull onto Skógasandur and Sólheimasandur took place in the AD tenth century and has already been mentioned in Section 3.3.1. Apparently it advanced in two branches that were channelled through gorges on both sides of the glacier (Fig. 3.20) and most likely, also through an outlet at the glacier snout. Flood deposit from the eastern branch has been identified on the slopes east of the flood plain and dated by tephrochronology to early tenth century according to its position just above the AD ~870 settlement tephra, but so far it has not been possible to assess whether it belongs to the ~920 Katla eruption or the ~934 Eldgjá eruption (Larsen, 2000; Larsen and Dugmore, unpublished data). At least four flood deposits left by prehistoric jökulhlaups younger than 3,800 years (3,480 ± 60 ^{14}C years BP) have been identified (Dugmore, 1987, 1989) – the largest being the Skógasandur hlaup 1,200–1,300 years ago (1,280 ± 100 ^{14}C years BP; Larsen, 1978).

3.5. Environmental Changes

3.5.1. *Tephrochronology as a Tool to Date Glacier Fluctuations*

Marginal fluctuations of Entujökull, Sólheimajökull and Kötlujökull are recorded by their moraines, some of which have been dated by tephrochronology. The best record comes from the work on Sólheimajökull (Fig. 3.21) where consecutive end and lateral moraines since the mid-Holocene have been mapped and dated (Dugmore,

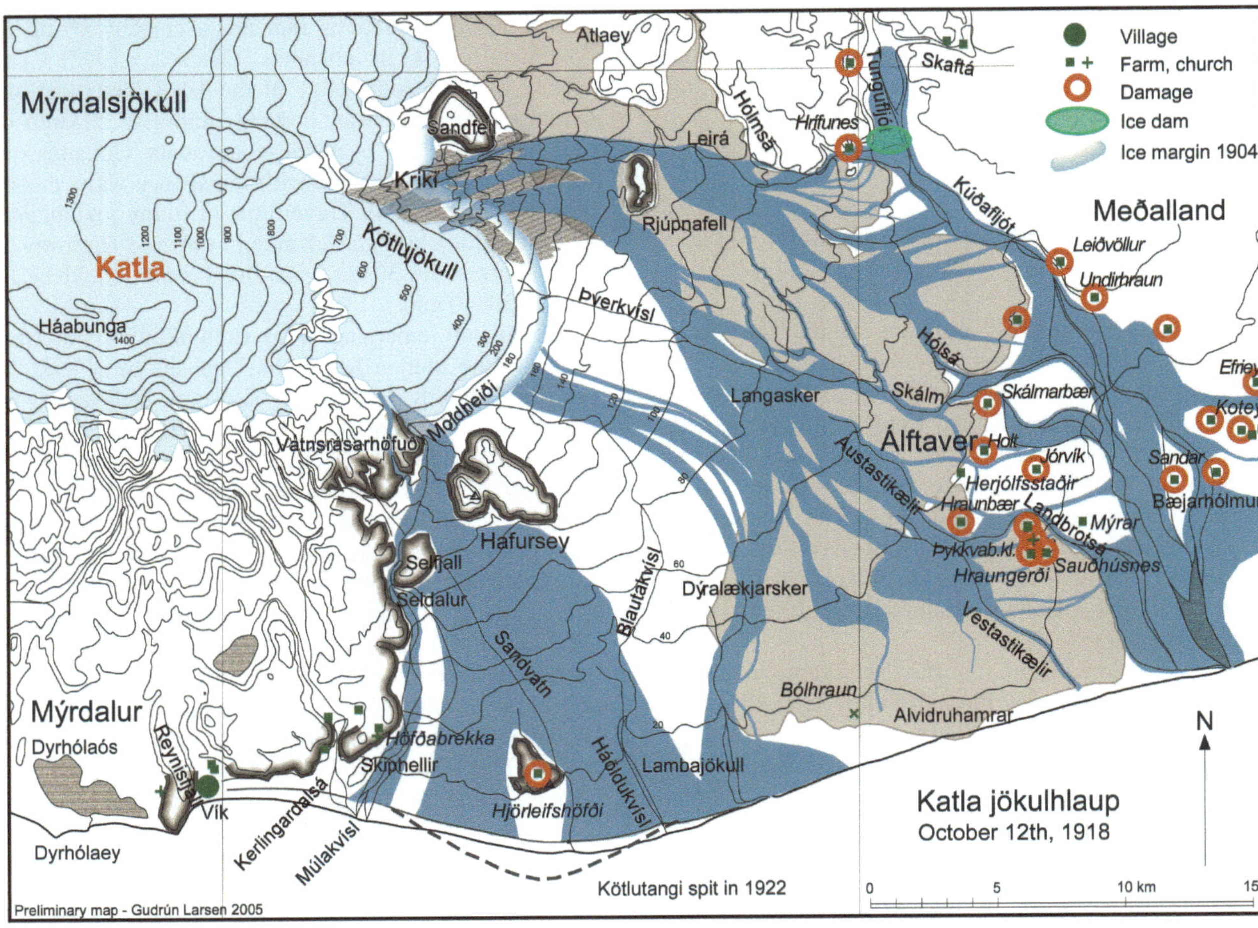

Fig. 3.17. Flood routes of the jökulhlaup on the first day of the 1918 Katla eruption, October 12th. Blue line in front of Kötlujökull follows the early twentieth century moraines. Adapted from Larsen et al. (2009).

1987, 1989; Dugmore and Sugden, 1991; Ingólfsson *et al.*, this volume). The outermost of these moraines also marks the maximum extent of Sólheimajökull since the last glaciation, the mid-Holocene Drangagil stage, and its gradual retreat to the present position.

Due to extensive cover of the tenth century Eldgjá lava and post-Eldgjá jökulhlaup deposits on Mýrdalssandur the only end moraines of Kötlujökull that have not been destroyed are those of the twentieth century. However, basal till and underlying small soil-patches are preserved on an older lava flow up to 5.5 km in front of the present margin of Kötlujökull (Schomacker *et al.*, 2003). A dated tephra layer, SILK-YN (1,676 $\pm$ 12 ^{14}C year BP, Table 3.2), identified in the soil together with a radiocarbon date of a twig below the moraine (1,610 $\pm$ 40 ^{14}C year BP) indicates that Kötlujökull extended across this lava 1,500–1,600 years ago, but its termination is not known.

In the Entujökull foreland at least four ice positions, which are prerequisites for the observed paths of tephrochronologically dated jökulhlaups from Entujökull, have been determined and assigned approximate dates (Smith, 2004). All lie outside the Little Ice Age (LIA) maximum. The outermost of these ice positions, the Emstrur stage, appears to be somewhat younger than 3,800 years (about 3,500 ^{14}C years BP) and may correspond to the Hólsárgil stage (about 3,100 ^{14}C years BP) at Sólheimajökull (Dugmore, 1987, 1989; Dugmore and Sugden, 1991). The Mófell stage (about 1,600 ^{14}C years BP) at Entujökull is of similar age as the basal till in front of Kötlujökull reported by Schomacker *et al.* (2003). The Botn stage (about 1,200 ^{14}C years BP) at Entujökull is of similar age as the Yztagil stage (1,200–1,400 ^{14}C years BP) at Sólheimajökull and tenth century ice positions have been identified in both areas (Dugmore, 1987, 1989; Smith, 2004).

Other glaciers from the Mýrdalsjökull ice cap that have been dated by tephrochronology and/or other methods do not show fluctuations corresponding to those described above and had maximum advance during the LIA (e.g. Casely and Dugmore, 2004). Dugmore and Sugden (1991) suggest catchment changes and ice-divide migration as possible explanation for the Sólheimajökull fluctuations. An altered topography within the Katla caldera as a consequence of the huge tenth century Eldgjá eruption has been suggested to explain changed flood routes and some of the changes observed at Sólheimajökull (Larsen, 2000). Intra-caldera changes over time may best explain the sometimes synchronous pre-LIA movements of the three outlet glaciers coming from the Katla caldera.

Independent tephrochronological corroboration that the Mýrdalsjökull ice cap has existed for the last 8,400 years comes from the sulphur content in the volcanic glass from Katla, which indicates that phreatomagmatic explosive activity has dominated throughout this period

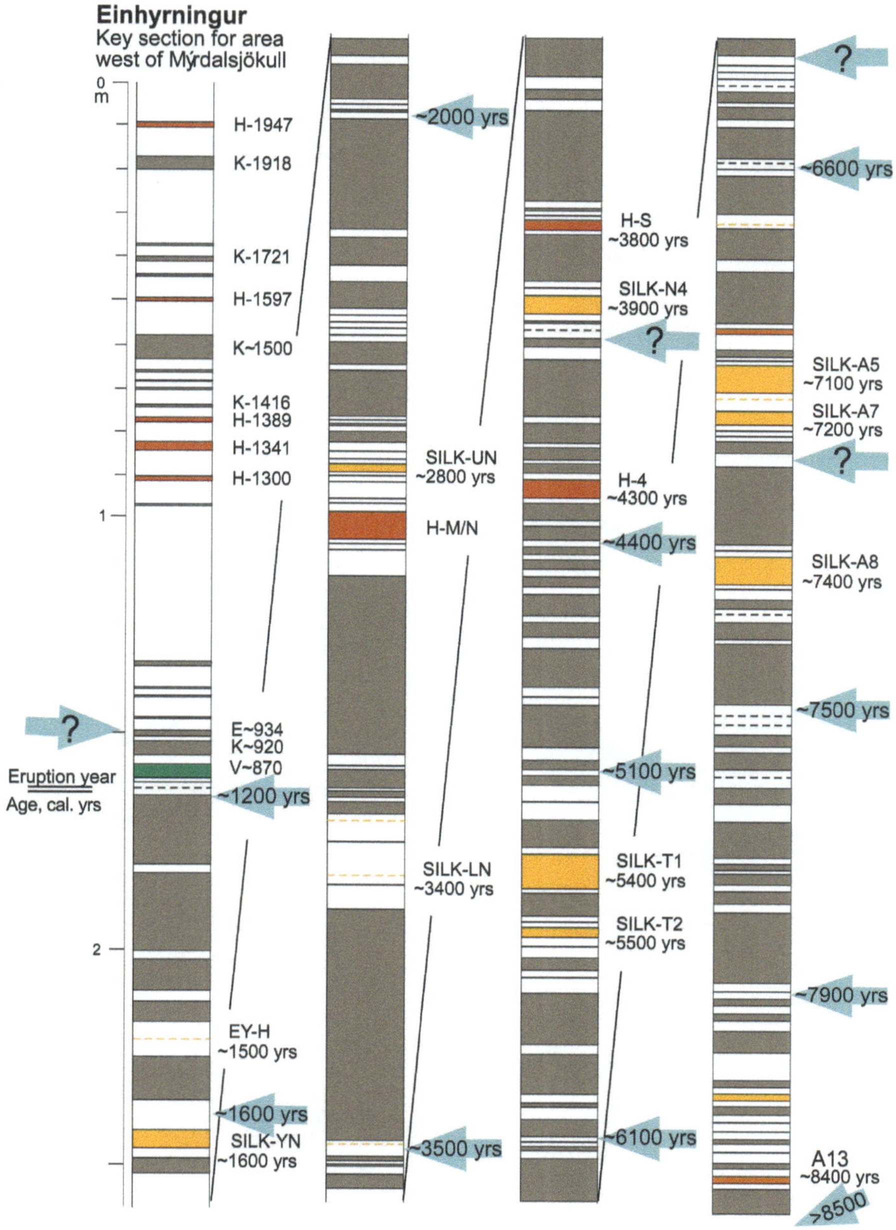

Fig. 3.18. Chronology of jökulhlaups flowing westwards from Mýrdalsjökull. Arrows indicate where deposits or erosional features associated with the jökulhlaups fit into the tephra stratigraphy. Legend as in Fig. 3.13. After Larsen et al. (2005).

(Óladóttir *et al.*, 2007). This implies that the Mýrdalsjökull massif has been capped by ice at least since the early Holocene and, perhaps, that it never was totally ice-free.

3.5.2. Damage and Environmental Changes Resulting from Eruptions in the KVS

Environmental changes brought about by eruptions in the KVS can be divided into four categories, i.e. those resulting from (1) tephra fall, (2) lava emanation, (3) jökulhlaups and hyaloclastite flows and (4) volcanic gas emissions. Such changes have also been referred to as damages during the 11 centuries since Iceland was settled. The first three will be briefly treated below but the reader is referred to Thordarson *et al.* (2001, 2003 and references therein) and to Stothers (1998) for evaluation of the effects of volcanic gas emissions of large eruptions on the KVS and in general.

Volcanic eruptions in the Katla system have caused more changes to its immediate (<30 km distance) surroundings than any other volcano since the settlement in the ninth century, with the possible exception of the Öræfajökull volcano (cf. Thorarinsson, 1958). Lava emission, jökulhlaups and tephra fall have radically changed an area of some 4,000 km^2 during this period. Earlier Holocene changes are less well known and more difficult to estimate.

3.5.2.1. Environmental changes east of Mýrdalsjökull

Repeated tephra fall in Holocene Katla eruptions has contributed to, and is the prime reason for, the thick soil

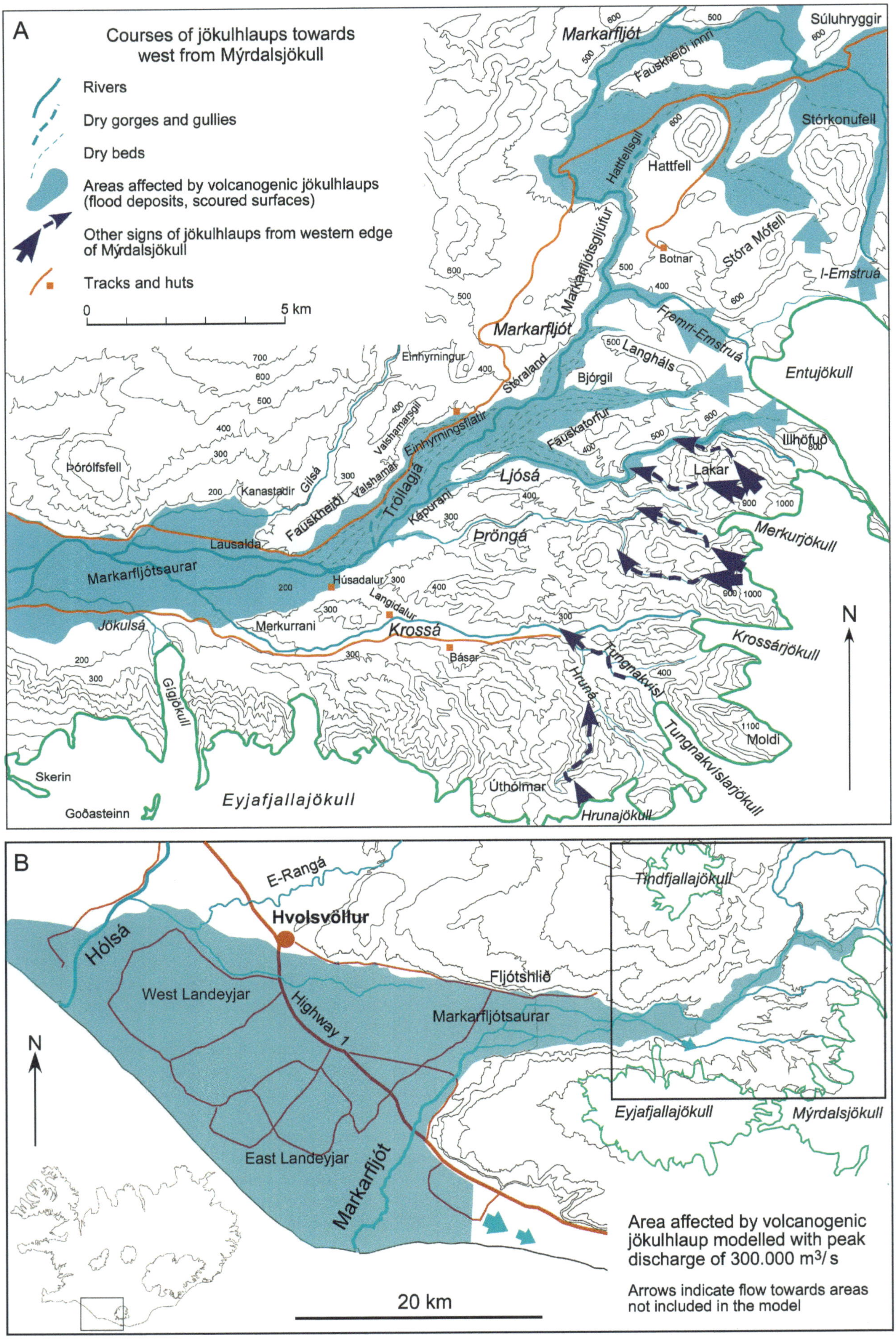

Fig. 3.19. (A) Flood routes towards west from Mýrdalsjökull. After Smith et al. (submitted for publication). (B) Flooded area in a jökulhlaup with peak discharge of 300,000 m^3/s according to the model of Hólm and Kjaran (2005).

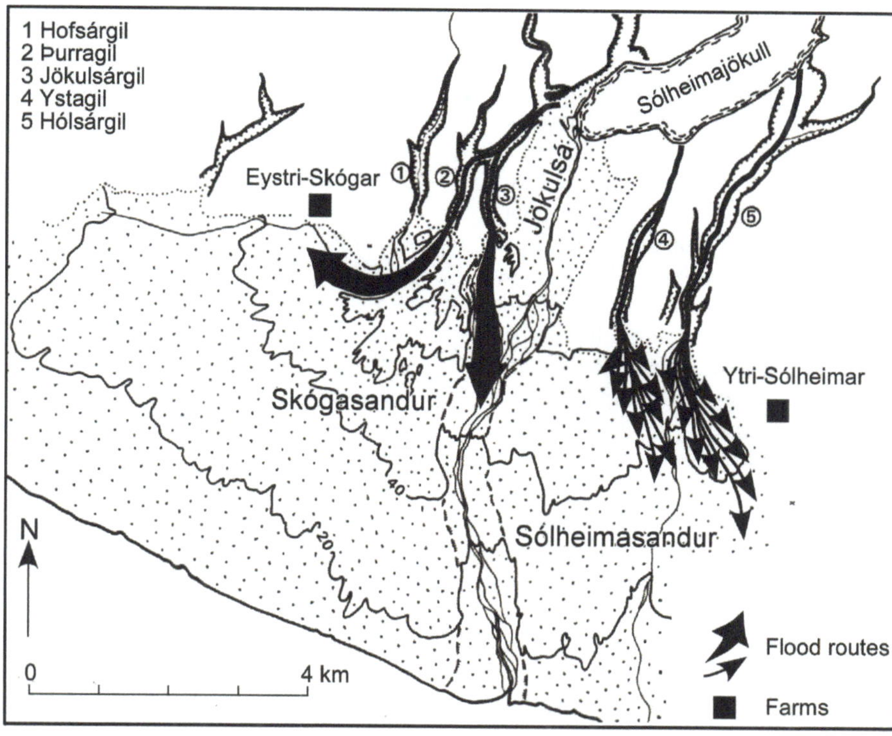

Fig. 3.20. Flood routes of historical jökulhlaup(s) on Skóga and Sólheimasandur, possibly in AD ~920 or ~934. Also shown are the two farms threatened by the flood described in the Book of Settlement. *The sediment fan west of Jökulsá was deposited by the Skógasandur hlaup 1,200–1,300 years ago. Modified from Larsen (1978).*

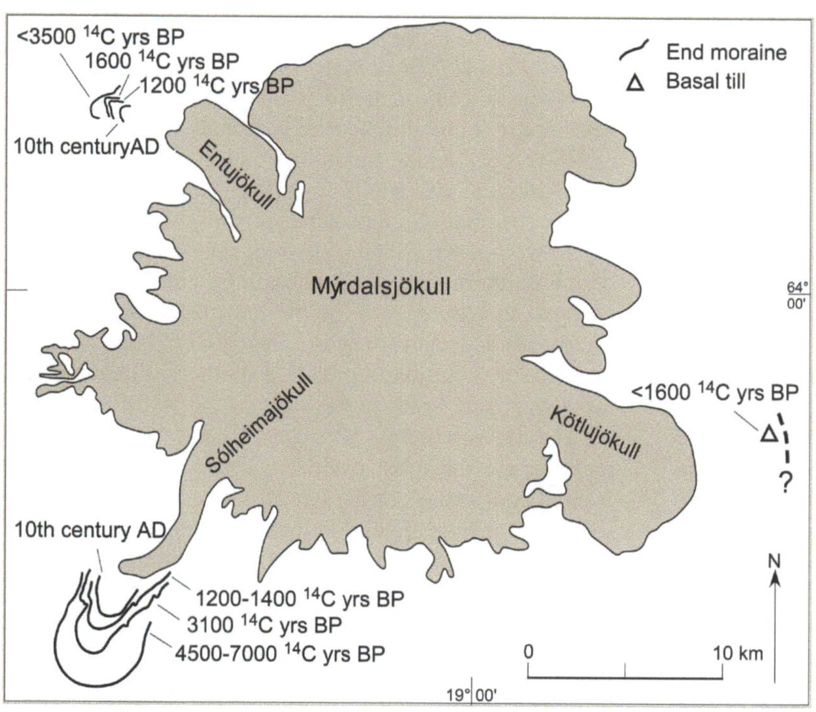

Fig. 3.21. Mid- and late Holocene fluctuations of Sólheimajökull and Entujökull dated by tephrochronology, after Dugmore (1989) and Smith (2004). Note that here the age is in ^{14}C years BP. Also shown is a locality 3.5–5.5 km east of Kötlujökull where basal till overlies organic material and SILK tephra dated to 1,610±40 and 1,675±12 ^{14}C years BP, respectively (Schomacker et al., 2003).

in the areas around Mýrdalsjökull. Apart from that, tephra fall from Katla volcano is a *relatively* minor contributor to environmental change east of Mýrdalsjökull. The Eldgjá tephra is, however, exceptional because of its volume and dispersal. Of the 20,000 km^2 affected by the tephra fall on land some 2,600 km^2 were covered by over 20 cm thick tephra and severely damaged. Of these, roughly 600 km^2 were buried below more than 100 cm thick deposits and permanently laid waste.

The Eldgjá lava fields changed the topography, hydrology and utilization potential of large areas east of the Mýrdalsjökull massif. About 800 km^2 of land were covered by new lava that raised the topography, blocked waterways and permanently changed the run-off pattern of an area extending from the Mýrdalssandur in the west to Landbrot in the east. Among these changes was the obstruction of jökulhlaup routes to the east by the Álftaver lava. A considerable part of the 800 km^2 overrun by the lavas was vegetated.

Volcanogenic jökulhlaups have affected even larger area than the lavas. During the last millennium, all major jökulhlaups have emerged through the Kötlujökull pass and discharged onto Mýrdalssandur. The large ones had peak discharges of about 300,000 m^3/s. The record of jökulhlaups in historical Katla eruptions is mostly based on documents and is fairly complete. The jökulhlaup event that accompanied the Eldgjá eruption, with floods emanating from Öldufellsjökull, N-Sandfellsjökull and Kötlujökull, was probably the largest in historical time. A hyaloclastic flow from below Kötlujökull at Kriki apparently dates from the Eldgjá event. Very little evidence of prehistoric east-going jökulhlaups is accessible because of the extensive cover of younger lava flows in the Álftaver and Meðalland districts.

The extensive cover of the Eldgjá lavas makes reconstruction of the pre-Eldgjá environment difficult. The >7,700-year-old Hólmsá lavas flowed down along the east side of the Mýrdalsjökull massif and disappear below the Eldgjá lavas near Atlaey, and their southwards extension is therefore unknown. Their source vents are not known, possibly overrun by the Eldgjá lavas or located west of Eldgjá, at Mælifellssandur. The Hólmsá lavas were channelized along topographical lows, possibly eroded by ice or glacial rivers at the end of the last glaciation. Smaller lava flows from a slightly younger event followed their course and affected rivers locally. All these lava flows became vegetated and accumulated soil still preserved to various extent. The erosional history of this soil is, however, only partly understood.

The pre-Eldgjá environment can be reconstructed up to a point (Fig. 3.22). The area north of Sandfell-Atlaey had thick soil cover and vegetation on the hyaloclastic bedrock and old lavas. Spring-fed rivers and smaller brooks flowed between thick banks of soil. The dominating feature south of Sandfell-Atlaey was apparently an outwash fan in front of Kötlujökull (the general shape of the Álftaver lava field indicates that the topography was a wide fan with an apex at Kötlujökull, which the lava swayed around). Neither the Laki lava nor the Eldgjá lavas in Meðalland and Álftaver existed at that time, leaving drainage of glacial water and jökulhlaups open as far eastwards as the >4,000-year-old Botnahraun lava. A large moraine, still visible in the Álftaver district, extended across the eastern part of the fan with a lake or fen nested behind its crest. The main rivers, Hólmsá and Tungufljót, may have retained separate courses to the south coast or converged at some point. The edge of the Kötlujökull glacier lay several kilometres to the east and – assuming that the Kriki hyaloclastite flow had not yet been deposited – may have adjoined (or coalesced with) the S-Sandfellsjökull. The shore lay farther inland, with Hjörleifshöfði and possibly also Mýrnahöfði as headlands.

From the *Book of Settlement* we know that the area now known as Álftaver was vegetated, as well as the area to the north of Álftaver, because two settlements were established there in the ninth century. We also know from the *Book of Settlement* that an inlet or lagoon, Kerlingarfjörður, lay west of Hjörleifshöfði and that the area north of it was vegetated.

The changes caused by the Eldgjá eruption were most radical in areas within 30 km east of the Mýrdalsjökull massif that were affected by heavy tephra fall, lava flows, a hyaloclastite flow and jökulhlaups. The lava from S-Eldgjá followed river valleys and gorges to the low areas, forcing rivers out of their beds. In Álftaver, the lava banked against and was deflected westwards by the moraine (Fig. 3.15), filling in the lake or fen in the process, then turning southwards to the coast of that time, possibly extending it seawards. The lava fronts on the sandur east of Kötlujökull obstructed previous routes of meltwater and jökulhlaups to the east. Hyaloclastite flows at Kriki may have blocked previous meltwater routes and changed the topography below Kötlujökull (and consequently the pre-eruption meltwater paths). A tephra blanket more than 1 m thick suffocated the existing vegetation and filled in gullies and depressions. Extensive soil erosion followed, which may have resulted in complete denudation locally, e.g. in the westernmost part of Skaftártunga.

The extent and effects of jökulhlaups accompanying the Eldgjá event cannot, however, be realistically estimated because the Eldgjá lavas and post-Eldgjá jökulhlaups in the Mýrdalssandur area have covered most of their tracks. The nickname Aurgoði (Lord of the mud) of a second-generation Norse settler in that area implies that water-transported sediments (aur = mud) occurred within his estate. Broadly speaking, areas along the eastern periphery of the Mýrdalsjökull ice cap may have been affected to a greater or lesser extent by the Eldgjá jökulhlaups. The inlet or lagoon at the Hjörleifshöfði headland had been filled in before the *Book of Settlement* was written. Its earliest version was most likely composed around AD 1100 (Rafnsson, 1974, 2001).

When volcanic activity was resumed after the +200 year repose following the Eldgjá eruption, new conditions had developed on the sandur plain in front of Kötlujökull. The Álftaver lava was now a high area that formed a barrier obstructing meltwater flow to the east. Hyaloclastite flow(s) at Kriki had, possibly, raised the topography on the northern side of Kötlujökull by tens of metres and changed the topography below the glacier significantly. The combined effect was to direct jökulhlaups escaping through the Kötlujökull pass in a southerly direction, into the lower lying areas to the west of the lava fields.

The environmental changes caused by the Eldgjá eruption are the most extensive since the settlement of Iceland and perhaps of the late Holocene in Iceland. They still have lasting effects. Perhaps the most important of these are the restriction of Katla jökulhlaups to the Kötlujökull outlet glacier, relieving the current Markarfljót and Landeyjar area of catastrophic floods – at least for the time being – and the development of the current Mýrdalssandur flood plain as we know it today.

3.5.2.2. Environmental changes west and south of Mýrdalsjökull

Tephra layers from Katla are the prime reason for the thick soil west of Mýrdalsjökull. Here, however, the Katla tephra has had relatively large environmental impact. This is partly because the two largest known Holocene Katla layers, K-E and K-N, have west-trending thickness axes (Róbertsdóttir, 1992) and buried vegetation in large areas

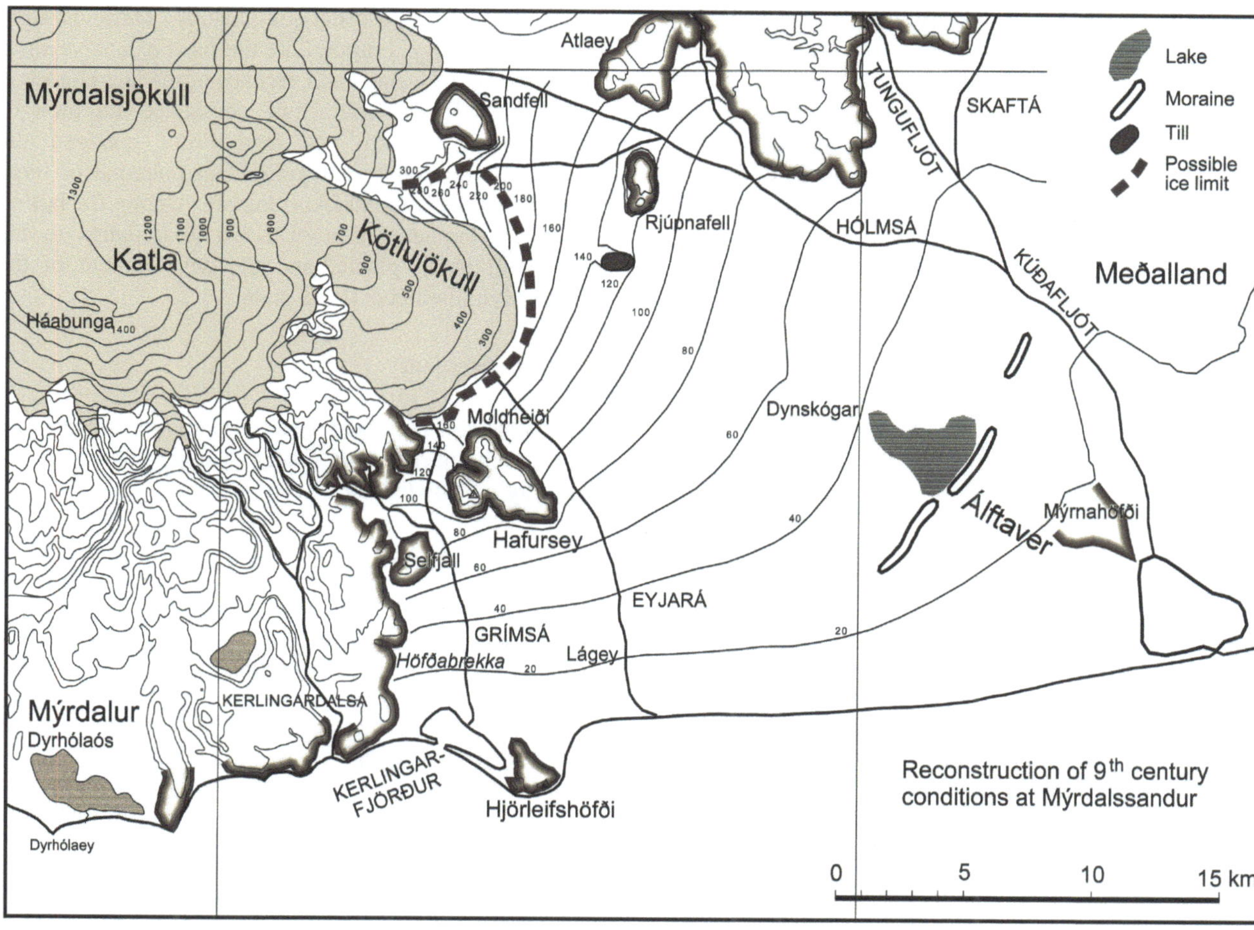

Fig. 3.22. Reconstruction of conditions on the Mýrdalssandur flood plain before the Eldgjá eruption and the emplacement of the Eldgjá lava flows in the early tenth century and, possibly, the emplacement of the Kriki hyaloclastite flow. The reconstruction is based on the assumption that the description of the shoreline west of Hjörleifshöfði, as deduced from the Book of Settlement, *is largely correct. The old shore must then have been close to where the 20 m a.s.l. contour lies today. Accordingly, thickness of 20 m has been 'lifted' off the sandur plain and, for simplicity, also in the area covered by the Eldgjá lava although thickness of 20 m is likely to be an underestimate. Kerlingarfjörður is 'modelled' after the Dyrhólaós lagoon (Imsland and Larsen, 1993). The lake behind the Álftaver moraine is placed where the rootless cones in the Eldgjá lava are most numerous. Broken thick line suggests the shape of the Kötlujökull glacier before the Kriki deposits were emplaced.*

to the west of Mýrdalsjökull below 50–100 cm thick primary deposits about 3,000 and 3,500 years ago (Fig. 3.13). Katla tephra is easily eroded, and locally the remobilized tephra formed still thicker deposits. Furthermore, two fairly thick tephra layers were deposited in the area shortly before the settlement in the ninth century. Only a thin soil layer (3–5 cm, Fig. 3.13) had formed on top of the these tephra layers when people arrived in Iceland with their livestock. Several farms/settlements in the proximal areas (within 25–30 km of Katla) are known from the first centuries of human presence but all were abandoned relatively early (Sveinbjarnardóttir, 1982). The thin layer of soil was easily punctured and it may have been further weakened by clearing of trees and scrubs. Once the vegetation cover was punctured the underlying tephra layers were easily eroded and as a result large areas were partly or totally denuded.

The lava flows on the west and northwest side of Mýrdalsjökull cover 60–70 km^2. Their number is uncertain but six to seven are anticipated. The lava flows were channelled along valleys and river gorges towards the lowlands but disappear below sediments in the Markarfljót valley.

At least 10 jökulhlaups, the largest having peak discharge of about 250,000 m^3/s, emanated westwards from Entujökull between 7,900 and 1,200 years ago. The west-going jökulhlaups carved the lower Markarfljótsgljúfur and the dry gorge Tröllagjá into the Holocene lava pile and modified the upper Markarfljótsgljúfur (Figs. 3.18 and 3.19). Their effects on the lowlands to the south are assumed to be extensive but have not been fully recorded. However, the interval between these jökulhlaups was long enough to allow the areas to revegetate and, in some cases at least, to develop mature birchwoods. Since the tenth century Eldgjá eruption no jökulhlaups have emerged through the Entujökul pass to the west. This, and the confinement of Markarfjót river within systems of dykes, has allowed the Markarfljótsaurar alluvial plain to revegetate and become popular leisure area.

South of Mýrdalsjökull, the largest environmental impact resulting from a Katla eruption is probably the

Skógasandur jökulhlaup 1,200–1,300 years ago, emplaced during the Yztagil stage of Sólheimajökull (Dugmore, 1987, 1989). This jökulhlaup deposited a large fan of volcanic debris, mostly black ash and lapilli, in front of the gullies Þurragil (Dry gully) and Jökulsárgil (Fig. 3.20). The historical jökulhlaup on the Sólheima and Skógasandur mentioned in the *Book of Settlement* may also have caused considerable environmental changes, cf. the comment in the *Book of Settlement*: 'These floods created the Sólheimasandur plain' (Í þeim vatnagangi varð Sólheimasandur).

3.6. Summary

Holocene volcanism within the KVS falls into three categories:

1. Explosive hydromagmatic basaltic eruptions on short volcanic fissures below the Mýrdalsjökull ice cap are the most common events of the Katla system. These eruptions are accompanied by tephra fall and jökulhlaups and appear to be the typical Katla eruptions as far back as the soil section record goes. The number of Holocene basaltic eruptions is unknown but may exceed 300.
2. Explosive silicic eruptions from vents below the ice cap, apparently within the caldera. They are accompanied by tephra fall and probably by jökulhlaups as well. At least six eruptions are known from soil sections in the surrounding area and several others are anticipated. These are the second most common eruptions in the Katla system.
3. Predominantly effusive basaltic eruptions within the fissure swarm and along the margin of the central volcano. Most of these eruptions probably had an explosive hydromagmatic component in cases where the fissures extended below the ice cap. The longest fissures are up to 75 km long. Huge lava flows have accompanied some of these eruptions, which are the least common type of activity in the Katla system.

The eruption frequency of the KVS has varied throughout the Holocene with two distinct peaks 2,000–4,000 and 7,000–8,000 years ago. The eruption frequency of the last millennium is the lowest in the record. A cyclic behaviour is observed in evolution of plumbing systems beneath the central volcano. A simple system is replaced by a more complex sill and dyke system that finally develops into a shallow crustal magma chamber.

Jökulhlaups, emerging from the Kötlujökull glacier onto the Mýrdalssandur flood plain east of the Mýrdalsjökull ice cap, have accompanied Katla eruptions since the late twelfth century. During the tenth century Eldgjá eruption jökulhlaups may have emerged from several other outlet glaciers along the east and south rim of the ice cap. Prior to the Eldgjá eruption large jökulhlaups emerged from the Entujökull outlet glacier in the west onto the Markarfljót and Landeyjar flood plains. Peak discharge in the largest floods was up to 300,000 m^3/s.

Extensive environmental changes have resulted from the volcanism in the KVS during the Holocene caused by tephra fall, lava flows and jökulhlaups. Those brought about by the Eldgjá eruption are the most extensive since the settlement of Iceland and perhaps of the late Holocene in Iceland. Heavy tephra fall devastated an area of 600 km^2, and 800 km^2 of lava flows changed the topography, hydrology and utilization potential of large areas east of the Mýrdalsjökull massif. Among the lasting effects are the restriction of Katla jökulhlaups to the Kötlujökull outlet glacier and the development of the current Mýrdalssandur flood plain.

Acknowledgement

The manuscript benefited from constructive reviews by Anders Schomacker and Stefan Wastegård which are gratefully acknowledged.

References

Berg, S., 1985. Structure, depositional environment and mode of emplacement of basaltic hyaloclastites and related lavas and sedimentary rocks: Plio-Pleistocene of the Eastern Volcanic Rift Zone, Nordic Volcanological Institute Report 8502, Southern Iceland, 91 pp.

Berg, S. & Sigvaldason, G., 1991. Pleistocene mass-flow deposits of basaltic hyaloclastite on a shallow submarine shelf, South Iceland. Bulletin of Volcanology 53, 597–611.

Bishop sagas (Biskupa sögur), 1858. Hið íslenska bókmenntafélag. Kaupmannahöfn, 1858–1878, 952 pp.

Björck, S., Ingólfsson, Ó., Hafliðason, H., Hallsdóttir, M., & Anderson, N.J., 1992. Lake Torfadalsvatn, a high resolution record of the North Atlantic ash zone 1 and the last glacial-interglacial environment changes in Iceland. Boreas 21, 15–22.

Björnsson, H., 1993. Ýmis sjónarmið um eðli Kötluhlaupa. Kötlustefna 27.-29.mars 1993. Rannsóknir á eldvirkni undir Mýrdalsjökli, 11–13. Science Institute Research Report RH-3-93, University of Iceland, Reykjavík.

Björnsson, H., 2003. Subglacial lakes and jökulhlaups in Iceland. Global and Planetary Change 35, 255–271.

Björnsson, H., Pálsson, F., & Guðmundsson, M.T., 2000. Surface and bedrock topography of Mýrdalsjökull, Iceland: the Katla caldera, recent eruption sites and routes of jökulhlaups. Jökull 49, 29–46.

Book of Settlement, 1967. Íslensk fornrit I. Landnámabók. Hið íslenzka fornritafélag, Reykjavík, 525 pp.

Buckland, P.C., Perry, D.W., Gislason, G.M., & Dugmore, A.J., 1986. The pre-Landnám fauna of Iceland: a paleontological contribution. Boreas 15, 173–184.

Casely, A.F. & Dugmore, A.J., 2004. Climate change and "anomalous" glacier fluctuations: the southwest outlets of Mýrdalsjökull, Iceland. Boreas 33, 108–122.

Dugmore, A.J., 1987. Holocene glacier fluctuations around Eyjafjallajökull, South Iceland. Ph.D. Thesis, University of Aberdeen.

Dugmore, A.J., 1989. Tephrochronological studies of Holocene glacier fluctuations in south Iceland. In: Oerlemans, H. (Ed), Glacier fluctuations and climatic change. Dordrecht, Reidel, 37–57.

Dugmore, A.J. & Sugden, D., 1991. Do the anomalous fluctuations of Sólheimajökull reflect ice-divide migration. Boreas 20, 105–113.

Dugmore, A.J., Larsen, G., & Newton, A.J., 1995. 7 tephra isochrones in Scotland. The Holocene 5, 257–266.

Dugmore, A.J., Newton, A.J., Larsen, G., & Cook, G.T., 2000. Tephrochronology, environmental change and the Norse settlement of Iceland. Environmental Archaeology 5, 21–34.

Eggertsson, S., 1919. Ýmislegt smávegis viðvíkjandi Kötlugosinu 1918. Eimreiðin 1919, 212–222.

Einarsson, S., 1975. Vikur á Mýrdalssandi-forkönnun. Volcanic Material Committee Report GN 7503, 4 pp.

Einarsson, E.H., Larsen, G., & Thorarinsson, S., 1980. The Sólheimar tephra layer and the Katla eruption of ~1357. Acta Naturalia Islandica 28, 1–24.

Elíasson, J., Larsen, G., Guðmundsson, M.T., & Sigmundsson, F., 2006. Probabilistic model for eruptions and associated flood events in the Katla caldera, Iceland. Computational Geosciences 10, 179–200.

Geirsdóttir, Á., Harðardóttir, J., & Sveinbjörnsdóttir, Á., 1998. Umhverfisbreytingar á Íslandi frá síðjökultíma og fram á nútíma í ljósi setmyndunar í stöðuvötnum. Vorráðstefna Jarðfræðafélags Íslands 1998. Ágrip erinda og veggspjalda 9–10.

Gröndal, G.O., Larsen, G., & Elefsen, S., 2005. Stærðir forsögulegra hamfaraflóða í Markarfljóti-mæling á farvegum neðan Einhyrningsflata. In: Guðmundsson, M.T., Gylfason, Á.G. (Eds), Hættumat vegna eldgosa og hlaupa frá vestanverðum Mýrdalsjökli og Eyjafjallajökli. Reykjavík, Ríkislögreglustjórinn, Háskólaútgáfan, 99–104.

Guðmundsdóttir, E.R., 1998. Gjóska úr Kötlugosi 1755. Ásýnd, kornastærð og kornalögun. B.Sc. Thesis, University of Iceland, Reykjavík, 64 pp.

Guðmundsson, M.T., 2003. Melting of ice by magma–ice–water interactions during subglacial eruptions as an indicator of heat transfer in subaqueous eruptions. In: White, J.D.L., Smellie, J.L., Clague, D. (Eds), Explosive subaqueous volcanism, Geophysical Monograph 140, Washington, D.C., American Geophysical Union, 61–72.

Hafliðason, H., Larsen, G., & Ólafsson, G., 1992. The recent sedimentation history of Thingvallavatn, Iceland. Oikos 64, 80–95.

Hafliðason, H., Eiríksson, J., & Kreveld, S.V., 2000. The tephrochronology of Iceland and the North Atlantic region during the Middle and Late Quaternary: a review. Journal of Quaternary Science 15, 3–22.

Hákonarson, M., 1860. Katla. Íslendingur 1(8), 61–62 og (9), 67.

Hammer, C.U., Clausen, H.B., & Dansgaard, W., 1980. Greenland ice sheet evidence of postglacial volcanism and its climatic impact. Nature 288, 230–235.

Haraldsson, H., 1981. The Markarfljót sandur area, southern Iceland: Sedimentological, petrographical and stratigraphical studies. Ph.D. Thesis, STRIAE 15, Uppsala University, Uppsala, 65 pp.

Hólm, S.L. & Kjaran, S.P., 2005. Reiknilíkan fyrir útbreiðslu hlaupa úr Entujökli. In: Guðmundsson, M.T., Gylfason, Á.G. (Eds), Hættumat vegna eldgosa og hlaupa frá vestanverðum Mýrdalsjökli og Eyjafjallajökli. Reykjavík, Ríkislögreglustjórinn, Háskólaútgáfan, 197–210.

Imsland, P. & Larsen, G., 1993. Kerlingarfjörður-hvar var hann, hvernig var hann og hver urðu örlög hans. Kötlustefna 27.-29.mars 1993. Rannsóknir á eldvirkni undir Mýrdalsjökli, 23. Science Institute Research Report RH-3-93, University of Iceland, Reykjavík.

Jakobsson, S.P., 1979. Petrology of recent basalts of the Eastern Volcanic Zone, Iceland. Acta Naturalia Islandica 26, 1–103.

Jóhannesson, H., Jakobsson, S.P., & Sæmundsson, K., 1990. Geological map of Iceland, sheet 6, South-Iceland (3rd ed.). Reykjavík, Icelandic Museum of Natural History and Iceland Geodetic Survey.

Jóhannesson, H. & Sæmundsson, K., 1998. Geological map of Iceland. Tectonics. 1,500.000. Reykjavík, Icelandic Institute of Natural History.

Jóhannsson, G., 1919. Kötlugosið 1918. Reykjavík, Bókaverslun Ársæls Árnasonar, 72 pp.

Jónsson, J., 1982. Notes on Katla volcanoglacial debris flows. Jökull 32, 61–68.

Karlsson, Th., 1994. Kötluhlaup 1918-vangaveltur um eðli hlaupsins og hámarksrennsli. Kötlustefna 26. febrúar 1994. Jarðfræðafélag Íslands, 10–12.

Kjær, K.H., Sultan, L., Krüger, J., & Schomacker, A., 2004. Architecture and sedimentation of outwash fans in front of the Mýrdalsjökull ice cap, Iceland. Sedimentary Geology 172, 139–163.

Knudsen, Ó. & Eggertsson, Ó., 2005. Jökulhlaupaset við Þverá í Fljótshlíð. In: Guðmundsson, M.T., Gylfason, Á.G. (Eds), Hættumat vegna eldgosa og hlaupa frá vestanverðum Mýrdalsjökli og Eyjafjallajökli. Reykjavík, Ríkislögreglustjórinn, Háskólaútgáfan, 113–121.

Lacasse, C., Sigurðsson, H., Jóhannesson, H., Paterne, M., & Carey, S., 1995. Source of Ash Zone 1 in the North Atlantic. Bulletin of Volcanology 57, 18–32.

Larsen, G., 1978. Gjóskulög í nágrenni Kötlu. B.Sc. Hons. Thesis, University of Iceland, 57 pp.

Larsen, G., 1979. Um aldur Eldgjárhrauna. Náttúrufræðingurinn 49, 1–26.

Larsen, G., 1981. Tephrochronology by microprobe analysis. In: Self, S., Sparks, R.S.J. (Eds), Tephra Studies. Dordrecht, D. Reidel Publishing Company, 95–102.

Larsen, G., 1996. Gjóskutímatal og gjóskulög frá tíma norræns landnáms á Íslandi. In: Grímsdóttir, G.Á. (Ed), Landnám á Íslandi. Ráðstefnurit V, Vísindafélag Íslendinga, 81–106.

Larsen, G., 2000. Holocene eruptions within the Katla volcanic system, south Iceland: characteristics and environmental impact. Jökull 49, 1–28.

Larsen, G. & Thorarinsson, S., 1984. Kumlateigur í Hrífunesi í Skaftártungu IV. Árbók hins íslenska fornleifafélags 31–47.

Larsen, G. & Ásbjörnsson, S., 1995. Volume of tephra and rock debris deposited by the 1918 jökulhlaups on western Mýrdalssandur, South Iceland. Abstracts, International Glaciological Society, 20–25 August 1995, Reykjavík.

Larsen, G. & Eiríksson, J., 2008. Late Quaternary terrestrial tephrochronology of Iceland – frequency of explosive eruptions, type and volume of tephra deposits. Journal of Quaternary Science 23, 109–120.

Larsen, G., Newton, A.J., Dugmore, A.J., & Vilmundardóttir, E.G., 2001. Geochemistry, dispersal, volumes and chronology of Holocene silicic tephras from the Katla volcanic system, Iceland. Journal of Quaternary Science 16, 119–132.

Larsen, G., Thordarson, Th., & Miller, J.D., 2004. Nature, style and magnitude of explosive activity in the 934-40 Eldgjá flood lava eruption in S-Iceland. Abstracts, IAVCEI General Assembly, 15.-19. nóv. 2004 (www.free.cl/iavcei2004), Pucon, Chile.

Larsen, G., Smith, K., Newton, A., & Knudsen, O., 2005. Jökulhlaup til vesturs frá Mýrdalsjökli: Ummerki um forsögulega hlaup niður Markarfljót. In: Guðmundsson, M.T., Gylfason, Á.G. (Eds), Hættumat vegna eldgosa og hlaupa frá vestanverðum Mýrdalsjökli og Eyjafjallajökli. Reykjavík, Ríkislögreglustjórinn, Háskólaútgáfan, 75–98.

Larsen, G., Guðmundsson, M.T., & Sigmarsson, O., 2009. Katla. In: Sólnes, J. et al. (Eds), Náttúruvá á Íslandi-Tekist á við náttúruöflin í 1100 ár. Eldgosavá. Viðlagatrygging Íslands, Reykjavík.

Maizels, J., 1993. Lithofacies variations within sandur deposits: the role of runoff regime, flow dynamics and sediment supply characteristics. Sedimentary Geology 85, 299–325.

Mangerud, J., Lie, S.E., Furnes, H., Kristianssen, I.L., & Lømo, L., 1984. A Younger Dryas ash bed in western Norway and its possible correlation with tephra in cores from the Norwegian Sea and the North Atlantic. Quaternary Research 21, 85–104.

Miller, J., 1989. The 10th century eruption of Eldgjá, southern Iceland. Nordic Volcanological Institute Report 8903, 30 pp.

Newton, A.J., 1999. Ocean-transported pumice in the North Atlantic. Ph.D. Thesis, University of Edinburgh, 394 pp.

Newton, A.J., Dugmore, A.J., & Larsen, G., 2000. Katla – a source of widespread ocean-transported pumice. Iceland 2000. Modern Processes and Past Environments. Abstracts, 75. April 27–29, 2000. Keele University, Keele, England.

Norðdahl, H., 1983. Late Quaternary stratigraphy of Fnjóskadalur, central North Iceland, a study of sediments, ice-lake strandlines, glacial isostasy and ice-free areas. LUNDQUA Thesis 12, Lund University, 78 pp.

Norðdahl, H. & Hafliðason, H., 1992. The Skógar tephra, a Younger Dryas marker in North Iceland. Boreas 21, 23–41.

Norðdahl, H. & Pétursson H.G., 2005. Relative sea-level changes in Iceland, new aspects of the Weichselian deglaciation of Iceland. In: Caseldine, C., Russell, A., Harðardóttir, J., Knudsen, Ó. (Eds). Iceland – Modern processes and past environments, Developments in Quaternary Science 5, 25–78.

Óladóttir, B.A., 2003. Gjóskulagið E2 úr Kötlu, kornalögun og kornastærð. B.Sc. Thesis, University of Iceland, Reykjavík, 90 pp.

Óladóttir, B.A., Larsen, G., Thordarson, Th., & Sigmarsson, O., 2005. The Katla volcano S-Iceland, Holocene tephra stratigraphy and eruption frequency. Jökull 55, 53–74.

Óladóttir, B.A., Thordarson, Th., Larsen, G., & Sigmarsson, O., 2007. Survival of the Mýrdalsjökull ice cap through the Holocene thermal maximum: evidence from sulphur contents in Katla tephra layers (Iceland) from the last ~8400 years. Annals of Glaciology 45, 183–188.

Óladóttir, B.A., Sigmarsson, O., Larsen, G., & Thordarson, Th., 2008. Katla volcano, Iceland, magma composition, dynamics and eruption frequency as recorded by tephra layers. Bulletin of Volcanology 70, 475–493.

Ólafsson, E., 1978. Ferðabók Eggerts Ólafssonar og Bjarna Pálssonar um ferdir þeirra á Íslandi árin 1752–1757. Reykjavík, Örn og Örlygur.

Rafnsson, S., 1974. Studier i Landnámabók: kritiska bidrag till den isländska fristatstidens historia. Lund, Bibliotheca Historica Lundensis XXXI.

Rafnsson, S., 2001. Sögugerð Landnámabókar. Um íslenska sagnaritun á 12. og 13. öld. Ritsafn Sagnfræðistofnunar 35. Reykjavík, University of Iceland.

Rasmussen, S.O., Andersen, K.K., Svensson, A.M., Steffensen, J.P., Vinther, B.M., Clausen, H.B., Siggaard-Andersen, M.L., Johnsen, S.J., Larsen, L.B., Dahl-Jensen, D., Bigler, M., Röthlisberger, R., Fischer, H., Goto-Azuma, K., Hansson, M.E., & Ruth, U., 2006. A new Greenland ice core chronology for the last glacial termination. Journal of Geophysical research - Atmospheres 111, D06102.

Róbertsdóttir, B.G., 1992. Forsöguleg gjóskulög frá Kötlu. Vórráðstefna Jarðfræðafélags Íslands 1992, Ágrip erinda og veggspjalda, 8–9.

Robson, G.R., 1957. The volcanic geology of Vestur-Skaftafellssýsla, Iceland. Ph.D. thesis, University of Durham, 259 pp.

Safn til sögu Íslands IV. Kaupmannahöfn og Reykjavík, 1907–1915, 186–294.

Schomacker, A., Krüger, J., & Larsen, G., 2003. An extensive late Holocene glacier advance of Kötlujökull, Central South Iceland. Quaternary Science Reviews 22, 1427–1434.

Shore, J.S., Cook, G.T., & Dugmore, A.J., 1995. The ^{14}C content of modern vegetation samples from the flanks of the Katla volcano, Southern Iceland. Radiocarbon 37, 525–530.

Sigurðsson, F., 1988. Fold og vötn að Fjallabaki. Árbók Ferðafélags Íslands 1988, 181–202.

Smith, K.T., 2004. Holocene jökulhlaups, glacier fluctuations and palaeoenvironment, Mýrdalsjökull, South Iceland. Ph.D. Thesis, Institute of Geography, School of Geosciences, University of Edinburgh, UK.

Smith, K.T. & Haraldsson, H., 2005. A late Holocene jökulhlaup, Markarfljót, Iceland: nature and impacts. Jökull 55, 75–86.

Smith, K.T. & Dugmore, A.J., 2006. Jökulhlaups circa Landnám: mid to late first millennium AD floods in south Iceland and their implications for landscapes of settlement, South Iceland. Geografiska Annaler 88A, 165–176.

Smith, K.T., Larsen, G., Dugmore, A.J., Newton, A.J., & Knudsen, Ó., Holocene jökulhlaups from western Mýrdalsjökull, Iceland: a chronology. Quaternary Science Reviews, submitted for publication.

Storm, G., 1888. Islandske Annaler indtil 1578. Christiania, Det norske historiske Kildeskriftfond, 667 pp.

Stothers, R.B., 1998. Far Reach of the tenth Century Eldgjá eruption, Iceland. Climate Change 39, 715–726.

Sveinbjarnardóttir, G., 1982. Byggðaleyfar vid Einhyrningsflatir í Fljótshlíð. In: Thórarinsdóttir, H., et al. (Eds), Eldur er í norðri. Reykjavík, Sögufélag, 67–77.

Sveinsson, G., 1919. Kötlugosið 1918 og afleiðingar þess. Reykjavík, Prentsmiðjan Gutenberg, 61 pp.

Sæmundsson, K., 1991. Jarðfræði Kröflukerfisins. In: Mývatns, N., Garðarson, A., Einarsson, Á. (Eds), Hið íslenska náttúrufræðifélag, Reykjavík, 25–95.

Thorarinsson, S., 1955. Öskufall svo sporrækt varð og Kötlugosið 1721. Náttúrufræðingurinn 25, 87–98.

Thorarinsson, S., 1958. The Öræfajökull eruption of 1362. Acta Naturalia Islandica II 2, 1–100.

Thorarinsson, S., 1967. The eruptions of Hekla in historical times. In: Einarsson, T., Kjartansson, G., Thorarinsson, S. (Eds), The eruption of Hekla 1947–48, I. Reykjavík, Societas Scientiarum Islandica, 1–177.

Thorarinsson, S., 1969. A short review of tephrochronological studies in Iceland. Paris, The VIIIth INQUA Congress, 953–957.

Thorarinsson, S., 1971. Damage caused by tephra fall in some big Icelandic eruptions and its relation to the thickness of the tephra layers. Athens, Acta of the International Scientific Congress of the volcano of Thera, 213–236.

Thorarinsson, S., 1975. Katla og annáll Kötlugosa. Árbók Ferðafélags Íslands 1975, 125–149.

Thorarinsson, S., 1980. Langleiðir gjósku úr thremur Kötlugosum. Jökull 30, 65–73.

Thorarinsson, S., 1981. Greetings from Iceland. Geografiska Annaler 63A, 109–118.

Thordarson, T. & Larsen, G., 2007. Volcanism in Iceland in historical time, Volcano types, eruptions styles and eruptive history. Journal of Geodynamics 43, 118–152.

Thordarson, T., Miller, D.J., Larsen, G., Self, S., & Sigurdsson, H., 2001. New estimates of sulfur degassing and atmospheric mass-loading by the 934 AD Eldgjá eruption, Iceland. Journal of Volcanology and Geothermal Research 108, 33–54.

Thordarson, T., Self, S., Miller, D.J., Larsen, G., & Vilmundardóttir, E.G., 2003. Sulphur release from flood lava eruptions in the Veidivötn, Grímsvötn and Katla volcanic systems, Iceland. In: Oppenheimer, C., Pyle, D.M., Barclay, J. (Eds), Volcanic degassing, Geological Society of London, Special Publications 213, 103–121.

Tómasson, H., 1996. The Jökulhlaup from Katla in 1918. Annals of Glaciology 22, 249–254.

Torfason, H. & Jónsson, H.B., 2005. Jarðfræði við norðvestanverðan Mýrdalsjökul. In: Guðmundsson, M.T., Gylfason, Á.G. (Eds), Hættumat vegna eldgosa og hlaupa frá vestanverðum Mýrdalsjökli og Eyjafjallajökli. Reykjavík, Ríkislögreglustjórinn, Háskólaútgáfan, 45–73.

Wastegård, S., Björck, S., Possnert, G., & Wohlfarth, B., 1998. Evidence of the occurrence of Vedde Ash in Sweden, radiocarbon age estimates. Journal of Quaternary Science 13, 271–274.

Wastegård, S., Wohlfarth, B., Subetto, D.A., & Sapelko, T.V., 2000. Extending the known distribution of the Younger Dryas Vedde Ash into northwestern Russia. Journal of Quaternary Science 15, 581–586.

Zielinski, G.A., Germani, M.S., Larsen, G., Baille, M.G.L., Whitlow, S., Twickler, M.S., & Taylor, K., 1995. Evidence of the Eldgjá (Iceland) eruption in the GISP2 Greenland ice core: relationship to eruption processes and climatic conditions in the tenth century. The Holocene 5, 129–140.

4

Deglaciation and Holocene Glacial History of Iceland

Ólafur Ingólfsson*, Hreggviður Norðdahl and Anders Schomacker

University of Iceland, Institute of Earth Sciences, Askja, Sturlugata 7, IS-101 Reykjavík, Iceland
**Correspondence and requests for materials should be addressed to Ólafur Ingólfsson (e-mail: oi@hi.is)*

4.1. Introduction and Terminology

This paper gives a brief overview of the deglaciation and environmental history of Iceland since the Last Glacial Maximum (LGM) with focus on the glacial history of the Mýrdalsjökull area. It is primarily based on reviews by Ingólfsson (1991), Norðdahl (1991), Gudmundsson (1997), Norðdahl and Pétursson (2005), and Norðdahl *et al.* (2008).

In our chronology, ^{14}C dates on marine shells are reservoir corrected by 365 ± 20 ^{14}C years [$\Delta R = 24 \pm 15$], the apparent age for living marine organisms around Iceland (Håkansson, 1983). Weighted mean age and standard deviation of dated samples are calculated with the Radiocarbon Calibration Program (CALIB) Rev. 5.0.1 (Stuiver and Reimer, 1993). Terrestrial material is calibrated with the data set *Intcal04* (Reimer *et al.*, 2004) and marine material with the data set *Marine04* (Hughen *et al.*, 2004). In this paper, we report ages as calibrated kilo-years before present (cal. kyr BP = AD 1950) and all essential dates are listed in Table 4.1 with references to their original publication. This paper follows the chronostratigraphical terminology for Norden as proposed by Mangerud *et al.* (1974).

4.2. The Icelandic Ice Sheet at LGM (>25.0 cal. kyr BP)

Geomorphological data like glacial striae, extent of glacial deposits, and glacially sculptured landscapes, as well as glacial diamictons and terminal moraines on the shelf, suggest that Iceland was covered by an extensive ice sheet during the LGM (Fig. 4.1). Tills and striae on the island of Grímsey, 40 km off North Iceland, suggest that the LGM ice sheet reached and overrode the island (Einarsson, 1967). A 100 km long and up to 50 m high ridge at 200–350 m below sea level on the shelf and about 130 km west of the Breiðafjörður inlet in West Iceland has been suggested to be a submarine end moraine marking the LGM extent of the Icelandic ice sheet (Ólafsdóttir, 1975). Based on submarine geomorphology and degree of weathering on top of mountains, Hjort *et al.* (1985) suggested that the ice sheet extended 10–20 km onto the shelf off the Hornstrandir area, Northwest Iceland. Geophysical and bathymetrical data, showing position and extent of apparent end and terminal moraines as well as lateral extent and thickness of glacigenic and non-glacigenic sediments, have been used for confining the LGM ice sheet on the shelf (Egloff and Johnson, 1979). Sediment cores have provided samples to date and constrain the deglaciation and the post-glacial environmental history of the Iceland shelf (Syvitski *et al.*, 1999; Eiríksson *et al.*, 2000; Andrews *et al.*, 2000).

On land, glacial striae and erratics on top of high mountains, relative height of table-mountains, lip-elevation of corries, frost shattering on top of high mountains, and orientation of glacial striae and linear landforms have been used to infer the temporary thickness and flow direction of the ice sheet. There are contrasting opinions about the LGM thickness of the Icelandic ice sheet, ranging from 1,000–1,500 m to more than 2,000 m above the centre of Iceland (Rundgren and Ingólfsson, 1999) and it is still unclear whether some elements of the Icelandic flora and fauna may have survived the Weichselian glaciation in ice-free coastal nunataks. The LGM configuration of the Icelandic ice sheet (Fig. 4.1) is based on a synthesis by Norðdahl and Pétursson (2005) and a three-dimensional thermomechanical modelling by Hubbard *et al.* (2006), adopted by Norðdahl *et al.* (2008). Calculations by Hubbard *et al.* (2006) suggest that the Icelandic LGM ice sheet, with a surface area of about 330×10^3 km^2 (projected baseline area of about 220×10^3 km^2), may have contained more than 300×10^3 km^3 of ice with a substantial marine component, i.e. glaciers grounded below relative sea level (RSL) at that time (Fig. 4.1). The model of Hubbard *et al.* (2006) suggests the possibility that the Mýrdalsjökull massif with the Katla caldera acted as an effective barrier to southward drainage of the interior part of the ice sheet. There is, however, no credible reason to believe that the Mýrdalsjökull massif was an ice-free enclave at LGM, as suggested by Steindórsson (1963).

4.3. Bølling (15.4–13.9 cal. kyr BP) Deglaciation and Relative Sea Level

Relatively warm Atlantic water reached the shelf areas off North Iceland between 18.7 and 16.1 cal. kyr BP (Table 4.1;1–2) and the ice sheet retreated from the shelf off West Iceland at 14.9 cal. kyr ka BP (Table 4.1;3; Syvitski *et al.*,

DEVELOPMENTS IN QUATERNARY SCIENCES
VOLUME 13 ISSN 1571-0866

Table 4.1. List of the most important ^{14}C datings referred to in this chapter.

	Lab. No.	Sample	Age (^{14}C BP)	Calibrated age range (BP 95.4%, $\pm 2\sigma$)	Age (kcal BP)	Calibration dataset	References
1	CAMS-46527	96-1234GCC, 259 cm	15,720±70	18,560–18,809	18.7	Marine04	Andrews *et al.* (2000)
2	AAR-3383	HM107-05, 393.0—394.3 cm	13,980±90	15,703–16,569	16.1	Marine04	Eiríksson *et al.* (2000)
3	AA-12896	193030-006LCF, 1235 cm	13,105±85	14,480–15,263	14.9	Marine04	Syvitski *et al.* (1999)
4	Ua-21222	Stóri-Sandhóll	12,975±105				Ingólfsson and Norðdahl (2001)
	Ua-12021	Stóri-Sandhóll	12,880±85				Ingólfsson and Norðdahl (2001)
	Mean	Stóri-Sandhóll (ML at 150 m a.s.l.)	12,918±67	14,206–14,956	14.6	Marine04	
5	T-4468	Hvalvík (ML at >50 m a.s.l.)	13,020±90	14,273–15,125	14.7	Marine04	Pétursson (1986)
6	Lu-2195	Ásbakkar	12,870±110				Ingólfsson (1988)
	U-0641	Melar	12,290±160				Ingólfsson (1985)
	Lu-2192	Melaleiti	12,460±120				Ingólfsson (1985)
	Lu-2193	Melaleiti	12,830±110				Ingólfsson (1985)
	Lu-2377	Ásbakkar 3	12,310±110				Ingólfsson (1988)
	Lu-2379	Ás	12,380±110				Ingólfsson (1988)
	Mean	Ás- and Melabakkar Clifs (RSL at ~0 m a.s.l.)	12,548±48	13,829–14,101	14.0	Marine04	
7	Lu-3342	Melar	11,530±100				Ásbjörnsdóttir and Norðdahl (1995)
	U-2019	Ekruhorn	11,620±240				Kjartansson (1966)
	Lu-2778	Tjaldanes 1	11,640±110				Ásbjörnsdóttir and Norðdahl (1995)
	Lu-2776	Ballará 1	11,720±80				Ásbjörnsdóttir and Norðdahl (1995)
	Mean	Inner parts of Breiðafjörður (RSL at metre a.s.l.)	11,643±54	12,973–13,224	13.1	Marine04	
8	Lu-2056	Súluá	11,330±80				Ingólfsson (1985)
	Lu-2338	Heynes	11,430±140				Ingólfsson (1985)
	Mean	Arctic mollusc species	11,355±70	12,812–13,000	12.9	Marine04	
9	Mean	The Fossvogur sediments	11,435±35	12,869–13,036	13.0	Marine04	Sveinbjörnsdóttir *et al.* (1993)

(Continued)

10	Lu-3118	Helgafellsmelar	10,580±90				Ingólfsson *et al.* (1995)
	U-2898	Varmá, Helgafellsmelar	10,780±110				Ingólfsson *et al.* (1995)
	Mean	Helgafellsmelar (RSL at ~55 m a.s.l.)	10,661±70	11,634–12,310	12.0	Marine04	
11	T-4467	Kópasker (RSL at ~ 17 m a.s.l.)	10,570±80	11,351–12,064	11.7	Marine04	Pétursson (1986)
12	Lu-2401	Stóra-Laxá at Hrepphólar 1	10,110±140				Hjartarson & Ingólfsson (1989)
	Lu-2402	Stóra-Laxá at Hrepphólar 2	9,960±160				Hjartarson & Ingólfsson (1989)
	Lu-2403	Þjórsá 1	10,360±90				Hjartarson & Ingólfsson (1989)
	Lu-2404	Þjórsá 2, near Minnahof	10,220±90				Hjartarson & Ingólfsson (1989)
	AAR-1241	Búðafoss	10,290±140				Geirsdóttir *et al.* (1997)
	AAR-1242	Búðafoss	10,120±150				Geirsdóttir *et al.* (1997)
	AAR-1243	Búðafoss	10,130±170				Geirsdóttir *et al.* (1997)
	Mean	The inner Búði moraine	10,214±47	11,113–11,254	11.2	Marine04	
13	Lu-3802	Breiðdalsá	9,980±110				Norðdahl and Einarsson (2001)
	Lu-3800	Fargrihvammur	10,170±110				Norðdahl and Einarsson (2001)
	Lu-3668	Teigarhorn	10,480±120				Norðdahl & Einarsson (2001)
	Mean	Glacier advance in East Iceland	10,195±66	11,046–11,312	11.2	Marine04	
14	AAR-0809	Selfljót	9,77 0 ±150				New date
	AAR-0810	Selfljót	9,820±150				New date
	Mean	Fljótsdalshérað RSL ~30 m a.s.l.	9,795±107	10,427–11,052	10.7	Marine04	
15	Lu-2673	Fell 2	9,980±150			Marine04	Norðdahl & Hjort (1987)
	Lu-2675	Skógaeyri 1	10,050±90			Marine04	Norðdahl & Hjort (1987)
	Lu-3289	Vs-1	10,140±130			Marine04	Sæmundsson (1995)
	Lu-2674	Fell 3a	10,230±90			Marine04	Norðdahl & Hjort (1987)
	AAR-0886	VN-3	10,270±150			Marine04	Sæmundsson (1995)
	Mean	Glacier advance in Vopnafjörður, East Iceland	10,137±51	10,935–11,221	11.1		Andrews *et al.* (2000)
16	Estimate	Glacier advance in Eyjafjörður, North Iceland	10,140±40	11,053±11,212	11.1	Marine04	Norðdahl and Pétursson (2005)
17	W-482	Þjórsárbrú	8,065±400				Kjartansson (1966)
	W-913	Þjórsárbrú	8,170±300				Kjartansson (1966)
	Lu-2601	Árnes	7,800±60				Hjartarson (1988)
	Mean	The Þjórsárhraun lava flow	7,820±58	8,429–8,931	8.7	IntCal04	

(Continued)

Table 4.1. (Continued)

	Lab. No.	Sample	Age (^{14}C BP)	Calibrated age range (BP 95.4%, $\pm 2\sigma$)	Age (kcal BP)	Calibration dataset	References
18	TUa-298	Heggjadalsetra 1	10,325 ± 115				Bard *et al.* (1994)
	TUa-299	Heggjadalsetra 2	10,360 ± 80				Bard *et al.* (1994)
	GifA-91244	Torvlömyra, 640–642 cm	10,430 ± 240				Bard *et al.* (1994)
	GifA-92307	Torvlömyra, 623–626 cm	10,040 ± 260				Bard *et al.* (1994)
	GifA-91235	Torvlömyra, 623–626 cm	9,950 ± 160				Bard *et al.* (1994)
	Mean	The age of the Vedde Ash in West Norway	10,287 ± 57	11,824–12,372	12.1	IntCal04	
19	Lu-3345	Seltjörn (Saksunarvatn Tephra)	9,030 ± 110	9,777–10,491	10.1	IntCal04	Ingólfsson *et al.* (1995)
20	Lu-2371	Árdalsá 2 (RSL > 50 m a.s.l.)	12,510 ± 140	13,614–14,434	14.0	IntCal04	
21	Lu-2372	Ásbakkar – Ásgil 2	12,080 ± 120				Ingólfsson (1988)
	Lu-2373	Ásbakkar – Ásgil 3	11,910 ± 140				Ingólfsson (1988)
	Lu-2376	Ásbakkar 2	11,830 ± 100				Ingólfsson (1988)
	Lu-2196	Ásbakkar – Ásgil 3	11,980 ± 130				Ingólfsson (1985)
	Mean	Ás- and Melabakkar Clifs (RSL > 25 m a.s.l.)	11,939 ± 60	13,226–13,484	13.4	Marine04	
22	Lu-3346	Seltjörn (RSL > –2.5 m a.s.l.)	9,280 ± 470	9,312–12,029	10.7	IntCal04	Andrews *et al.* (2000)
23	AA-21798	Fúlavík	6,475 ± 95				Richardson (1997)
	AA-21799	Hraunnes	6,215 ± 95				Richardson (1997)
	Mean	Main raised shoreline at ~4 m a.s.l. (NE-Iceland)	6,345 ± 67	7,158–7,426	7.3	IntCal04	
24	Lu-3667	Stokkseyri	3,140 ± 70				Símonarson & Leifsdóttir (2002)
	AAR-4649	Litla-Hraun	3,145 ± 35				Símonarson & Leifsdóttir (2002)
	Mean	Flói (RSL at > 3 m a.s.l.)	3,144 ± 34	2,779–3,015	2.9	Marine04	
25	AAR-4650	Litla-Hraun (RSL at > 2 m a.s.l.)	2,625 ± 40	2,137–2,373	2.3	Marine04	Símonarson & Leifsdóttir (2002)
26	SUERC-2785		4,432 ± 32				Kirkbride *et al.* (2006)
	SUERC-2786		4,397 ± 29				Kirkbride *et al.* (2006)
	Mean	Flood in Jökulsá á Fjöllum	4,413 ± 21	4,876–5,210	5.0	IntCal04	
27	SUERC-2789		4,085 ± 30				Kirkbride *et al.* (2006)
	SUERC-2791		4,044 ± 30				Kirkbride *et al.* (2006)
	Mean	Flood in Jökulsá á Fjöllum	4,064 ± 21	4,443–4,784	4.6	IntCal04	

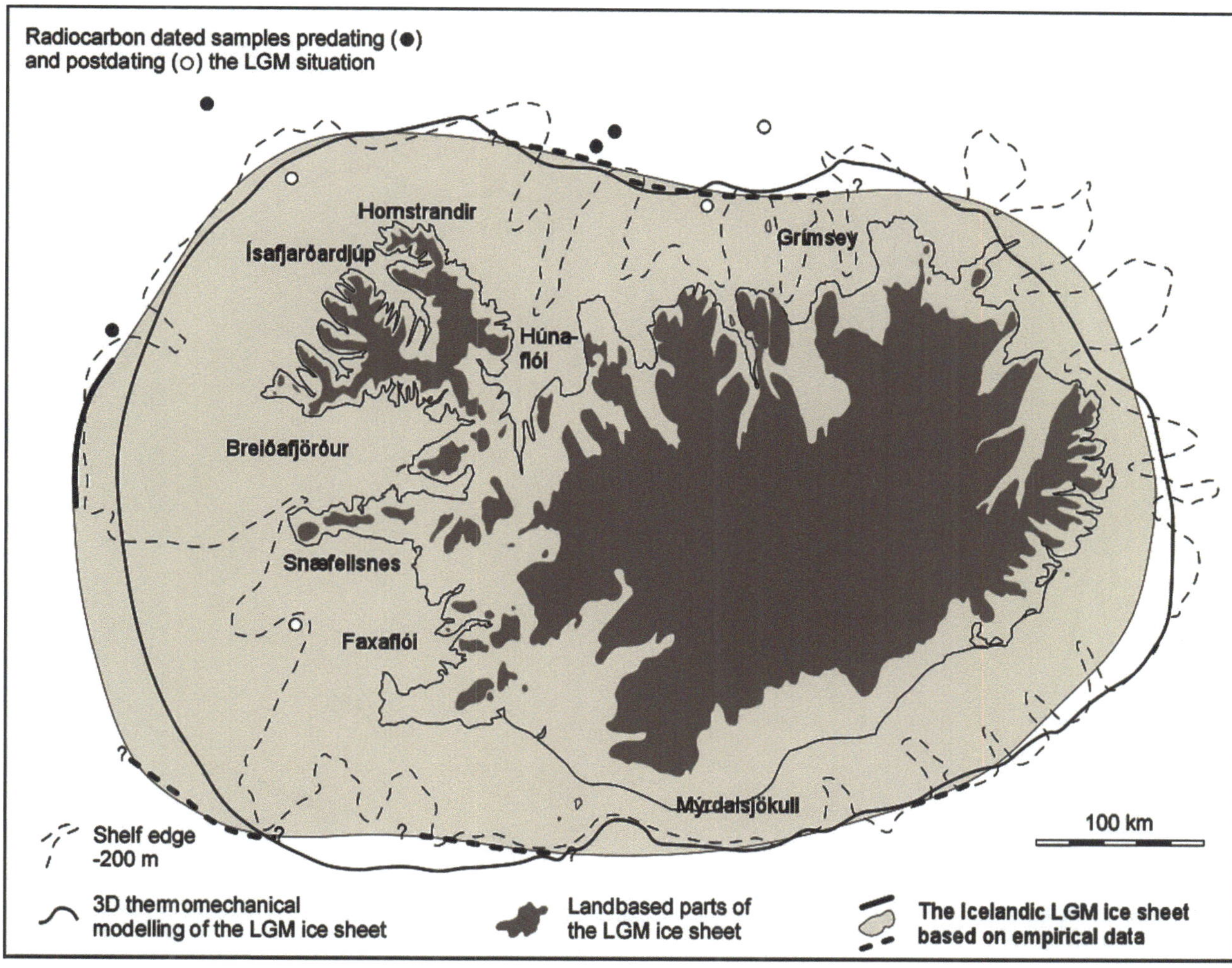

Fig. 4.1. The Icelandic ice sheet at LGM. Modified from Norðdahl et al. (2008).

1999; Eiríksson *et al.*, 2000; Andrews *et al.*, 2000; Geirsdóttir *et al.*, 2002). Ingólfsson and Norðdahl (2001) suggested that a rapid rise of global sea level, caused by melting of the large Laurentide, Scandinavian and Barents Sea ice sheets, floated marine-based parts of the Icelandic ice sheet and caused its collapse through intensive calving. The shelf off West Iceland was deglaciated before 14.9 cal. kyr BP (Syvitski *et al.*, 1999), and coastal areas in West Iceland were ice-free but submerged at 14.6 cal. kyr BP (Table 4.1;4; Ingólfsson and Norðdahl, 2001). Marine limit (ML) shorelines in West Iceland occur at 150 and 105 m a.s.l. at Stóri-Sandhóll and Stóra-Fellsöxl (Fig. 4.2), respectively, and dated to 14.6 cal. kyr BP. In Northeast Iceland, ML shorelines occur at altitudes between 50 and 60 m a.s.l. on Melrakkaslétta (Fig. 4.2) and have been dated to 14.7 cal. kyr BP (Table 4.1;5) (Pétursson, 1991). Marine sediments of Bølling age are widespread in coastal West and Southwest Iceland (Fig. 4.2). Subfossil mollusc fauna (*Macoma calcarea* community) contained within these sediments, as well as oxygen isotope ratios in shells, suggest sea surface temperatures similar to the present, indicating relatively warm Atlantic seawater around Iceland in Bølling times (Ingólfsson, 1988).

Raised shorelines occur up to 70 m a.s.l. in Southwest Iceland, up to 110 m a.s.l. in West Central Iceland, up to 90 m a.s.l. in Northwest and North Iceland, and as high as 60 m a.s.l. in Northeast Iceland. These undated shorelines have tentatively been assigned to the Bølling Chronozone, based on geomorphological and stratigraphical correlations. Additionally, extensive subaerial lava flows, originating from the Þeistareykir volcano and the Krafla area in North Iceland, show that interior parts of the Northern Volcanic Zone became ice-free in Bølling/Allerød times (Fig. 4.2). Norðdahl *et al.* (2008) concluded that the Icelandic ice sheet retreated to positions well inside the present coastline from Southwest to Northeast Iceland in early Bølling times, but coastal areas in South and Southeast Iceland were deglaciated somewhat later. The Icelandic ice sheet was reduced to about 25% of its LGM size at 13.9 cal. kyr BP (Fig. 4.2). Following this fast deglaciation and reduction of glacier load, there was a rapid fall in RSL.

A large outlet glacier in Borgarfjörður, West Iceland, advanced just after 14.0 cal. kyr BP (Table 4.1;6). The glacier overran and deformed marine sediments of Bølling age, now exposed in the lower Borgarfjörður area, and formed the spectacular Skorholtsmelar end moraine complex, where Borgarfjörður opens to the south (Ingólfsson, 1988). This is the only known glacier advance of this age in Iceland. Norðdahl *et al.* (2008) suggested that this advance might not have been climatically driven but that a Borgarfjörður ice stream had surged when the Icelandic ice sheet was being

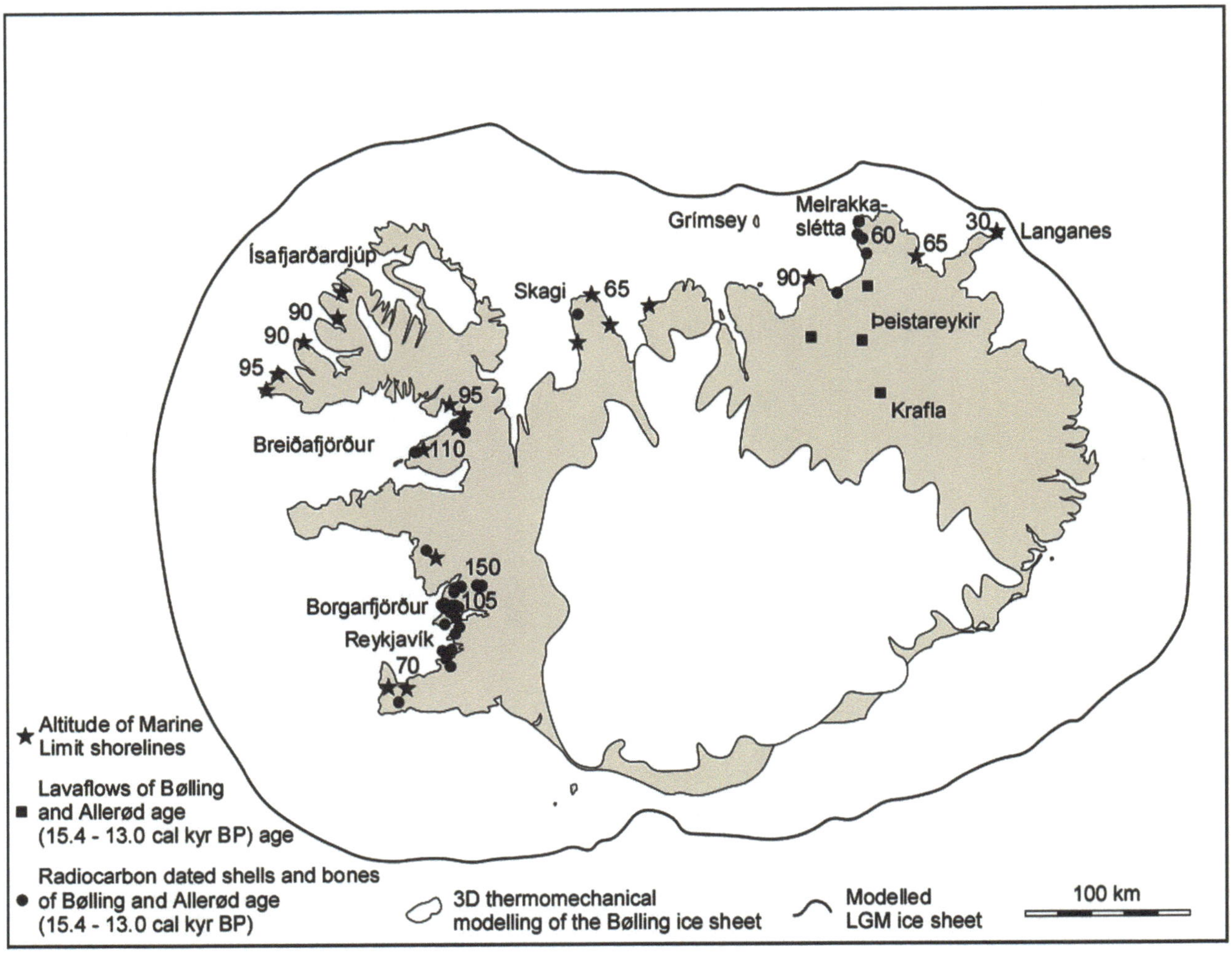

Fig. 4.2. The Icelandic ice sheet after the Bølling deglaciation. Modified from Norðdahl et al. (2008).

stabilized following an extensive and rapid early Bølling collapse.

4.4. Allerød (13.6–13.0 cal. kyr BP) Environmental Developments

Fossiliferous marine sediments of Allerød age occur in coastal Southwest, West and Northeast Iceland and have mainly been studied in Breiðafjörður, Borgarfjörður and Reykjavík (Fig. 4.2). Allerød sediments in the Breiðafjörður area contain arctic mollusc species such as *Portlandia arctica* and foraminifera faunal assemblage of *Elphidium excavatum–Casidulina reniforma*. Marine shells have been radiocarbon dated to 13.1 cal. kyr BP (Table 4.1;7) (Ásbjörnsdóttir and Norðdahl, 1995).

The Allerød marine sediments in the lower Borgarfjörður area show that RSL rose during the later part of the Allerød Chronozone. This transgression has been interpreted to signify both rise of eustatic sea level and subsidence of the Iceland crust caused by expansion of the Icelandic ice sheet (Norðdahl and Pétursson, 2005). The sedimentology and mollusc fauna indicate a near-shore marine environment with relatively rapid sedimentation and/or presence of brackish water, most likely due to inflow of glacial meltwater (Ingólfsson, 1988). The appearance of arctic mollusc species, *P. arctica* and *Buccinum groenlandicum*, towards the end of the Allerød Chronozone, radiocarbon dated to 12.9 cal. kyr BP (Table 4.1;8), shows that cold water characterized the marine environment in the outer parts of the Borgarfjörður region. The Fossvogur sedimentary sequence in Reykjavík, deposited close to the transition between the Allerød and the Younger Dryas Chronozones, reveals a history of glacier retreat and rise of RSL. A weighted mean of 30 radiocarbon dates (Sveinbjörnsdóttir et al., 1993) suggests deposition at around 13.0 cal. kyr BP (Table 4.1;9). The mollusc fauna of the Fossvogur sediments contains boreal species that occur today in the waters around Iceland. The upper part of the sequence demonstrates increasing proximity to a tidewater glacier and its subsequent retreat in early Younger Dryas times (Geirsdóttir and Eiríksson, 1994). Combined, the Borgarfjörður and Fossvogur data suggest rapid cooling of Southwest Iceland coastal waters close to the Allerød–Younger Dryas transition.

Pollen, diatom and carbonate analysis on sediments retrieved from Lake Torfadalsvatn on the Skagi peninsula in North Iceland show expansion of grassland/grass tundra in mid-Allerød, and that shrub and dwarf-shrub tundra vegetation was established by the end of the Allerød (Rundgren, 1995, 1999). The Lake Torfadalsvatn data show that late Allerød warmest-month temperature may have been as high as 10°C.

4.5. The Younger Dryas (13.0–11.5 cal. kyr BP) Glacial and Climatic Event

The Icelandic Younger Dryas ice sheet expanded across coastal sites that had been ice-free since the initial Bølling deglaciation. Truncated raised shorelines in mouths of fjords and valleys around Iceland demonstrate the extent of both the marine environment and the glaciers that prevented shorelines to be formed farther inland. The Younger Dryas glacial event is also bracketed by terminal moraine systems and ice-proximal delta and sandur deposits. An overview of the Younger Dryas glacier expansion is shown in Fig. 4.3.

Northeast of Reykjavík, the termination of a Younger Dryas glacier advance is marked by ice-contact deltas and related shoreline features radiocarbon dated to 12.0 cal. kyr BP (Table 4.1;10) (Ingólfsson *et al.*, 1995), while truncated shorelines show that the glaciers terminated in the Hvalfjörður and tributary valleys in the upper Borgarfjörður area. Truncated shorelines also suggest that there was a Younger Dryas expansion of glaciers in West Central Iceland and along the southern (Breiðafjörður) perimeter of Northwest Iceland (Norðdahl and Pétursson, 2005). Sediment cores retrieved from the Ísafjarðardjúp basin in Northwest Iceland reveal expansion of glaciers into the trunk fjord (Geirsdóttir *et al.*, 2002). Promontories along the western and northeastern coast of Northwest Iceland remained mostly ice-free at the same time (Hjort *et al.*, 1985).

The Skagi peninsula in North Iceland probably hosted cirque glaciers during the Younger Dryas glacial event, whereas its lowlands in the northern part were ice-free (Ingólfsson *et al.*, 1997). Primary occurrences of the Skógar-Vedde tephra in ice-lake sediments in Fnjóskadalur show that in Younger Dryas times there were numerous ice-free enclaves in North Central Iceland (Norðdahl and Hafliðason, 1992). A large outlet glacier or ice stream in the Eyjafjörður probably reached the island of Hrísey (Fig. 4.3). Younger Dryas raised beaches in the Öxarfjörður area (11.7 cal. kyr BP) (Table 4.1;11) suggest that the ice margin was situated inside the present coastline there, and truncated shorelines suggest that tip of the Langanes peninsula remained ice-free in Younger Dryas (Pétursson, 1991; Norðdahl and Hjort, 1995). Truncated shorelines south of the Héraðsflói bay in East Iceland suggest that outlet glaciers were grounded in the eastern fjords, while many promontories and headlands remained ice-free (Norðdahl and Einarsson, 2001). The ice sheet probably extended beyond the present coast in Southeast Iceland.

The Búði moraine marks the Younger Dryas glacier extent inside the present coast of Central South Iceland

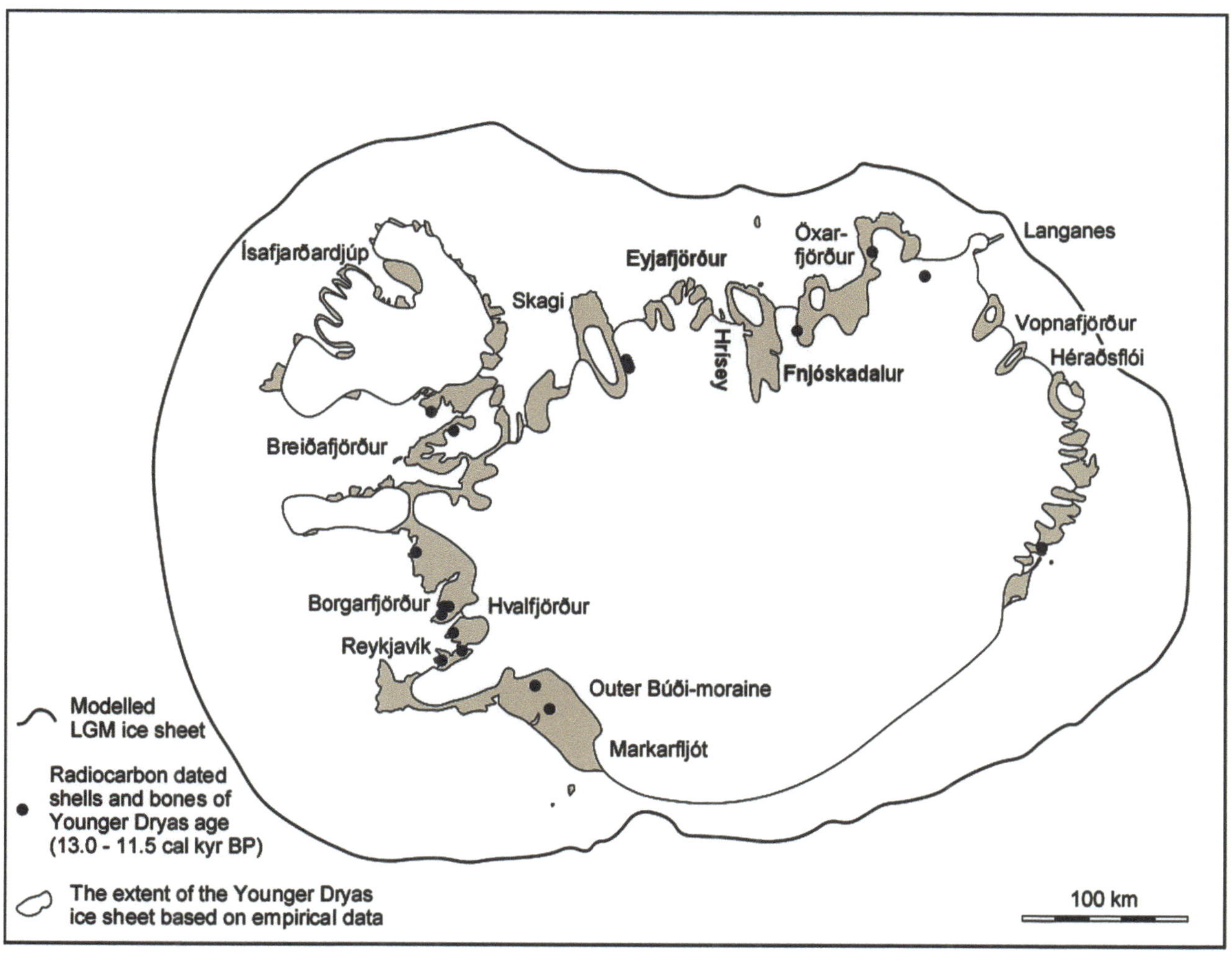

Fig. 4.3. Younger Dryas ice extent on Iceland. Modified from Norðdahl et al. (2008).

(Norðdahl and Pétursson, 2005). Recent investigations show that the Búði moraine is complex and composed of two different sets of moraines of different age. The outer set of moraines was formed during a glacier advance preceding the formation of the inner set, which has been dated to 11.2 cal. kyr BP (Table 4.1;12), i.e. most likely in Younger Dryas time (Norðdahl *et al.*, 2008). A moraine buried some 20 m below the surface of the Markarfljót sandur in South Iceland has been correlated with the outer set of the Búði moraines (Haraldsson, 1981).

Biostratigraphical data highlighting the Allerød—Younger Dryas–Holocene climate transitions come from sediment cores from Lake Torfadalsvatn in North Iceland (Rundgren, 1995, 1999). They show a rapid climatic cooling at the onset of the Younger Dryas, expressed in sudden fall in the concentration of pollen and reduction in lake productivity. The dominant vegetation at northern Skagi changed from dwarf-shrub tundra to grass tundra, reflecting a shift from mild to cold climatic conditions with low mean summer temperature and indicating perennial sea-ice coverage off North Iceland. The grass tundra on the Skagi peninsula persisted through the Younger Dryas and developed into dwarf-shrub tundra and subsequently into dwarf-shrub and shrub tundra after the Younger Dryas–Holocene transition. Sediment core data from the shelf off West and North Iceland have also been interpreted to signify a Younger Dryas cooling (Eiríksson *et al.*, 2000; Jennings *et al.*, 2000).

4.6. Early Preboreal (11.5–10.1 cal. kyr BP) Glaciation

A step-wise recession of the ice sheet after the Younger Dryas is reflected in series of terminal moraines in the highlands. The outermost post-Younger Dryas moraines and the inner Búði moraines (Fig. 4.4) in South Central Iceland were formed during a short lived re-advance of the Icelandic ice sheet at about 11.2 cal. kyr BP. Glaciers in East Iceland reached the sea in the central fjords at 11.2 cal. kyr BP (Table 4.1;13) (Norðdahl and Einarsson, 2001), whereas a major outlet glacier occupying the lake Lögurinn basin in Fljótsdalur terminated some 20 km inside the present coast and shells from marine sediments deposited north of the ice margin, when RSL was at about 30 m a.s.l, have been radiocarbon dated to 10.7 cal. kyr BP (Table 4.1;14). Glaciers in Vopnafjörður terminated in the inner parts of the fjord and raised beaches outside the ice-marginal zone have been dated to 11.1 cal. kyr BP (Table 4.1;15) (Sæmundsson, 1995). Moraines and other ice-marginal features 3–5 km inland of raised beaches at 30–35 m a.s.l. in the Langanes area delimit the Preboreal

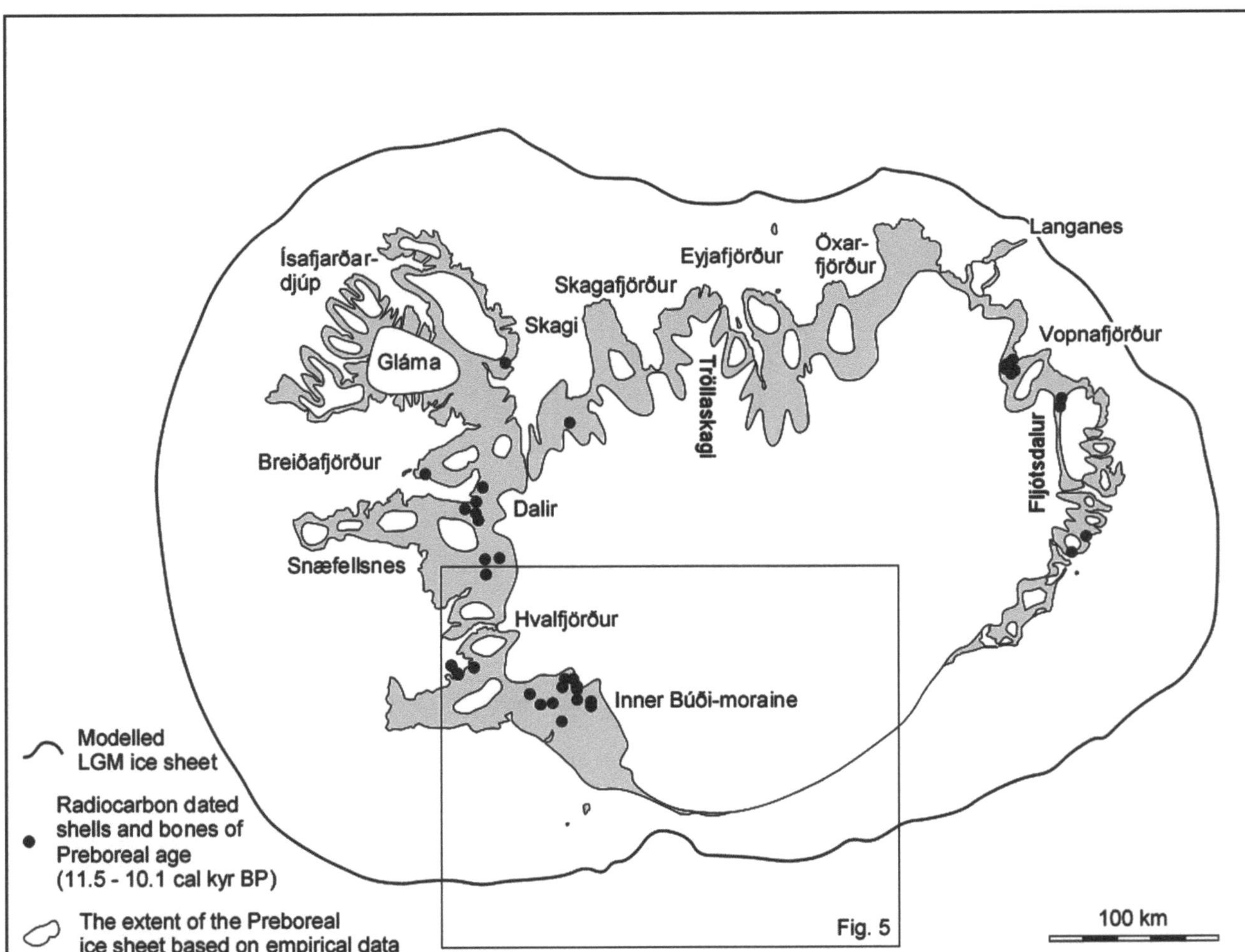

Fig. 4.4. Early Preboreal ice extent on Iceland. Modified from Norðdahl et al. (2008).

extent of the ice sheet (Fig. 4.4), and the ice margin in the Öxarfjörður area was also situated inland of the Preboreal shorelines (Norðdahl and Hjort, 1995). At 11.1 cal. kyr BP (Table 4.1;16), the outlet glacier in Eyjafjörður terminated about 50 km inside its Younger Dryas position, while glaciers in tributary valleys reached the sea close to the present coastline (Figs. 4.3 and 4.4). A major outlet glacier in Skagafjörður, Central North Iceland, reached the head of the fjord when the Skagi peninsula was mostly ice-free (Ingólfsson *et al.*, 1997). There is a paucity of data on the Preboreal extent of the ice sheet in West Iceland. Glaciers probably reached the sea in the Dalir area when RSL was at 30–35 m a.s.l, and an outlet glacier terminated in the innermost part of Hvalfjörður when RSL was situated at about 65 m a.s.l. Outlet glaciers extending from an ice cap situated above the Gláma area in Northwest Iceland reached the sea along the northern shore of Breiðafjörður. Tröllaskagi and Flateyjarskagi peninsulas in North Iceland, the Snæfellsnes peninsula in West Iceland and mountainous areas in East and Southwest Iceland probably carried numerous local corrie and valley glaciers in early Preboreal times (Norðdahl and Pétursson, 2005).

The spatial difference between the Younger Dryas and Preboreal ice sheets in Iceland was small except in North Iceland, where major outlet glaciers had retreated 30–50 km relative to their Younger Dryas positions (Figs. 4.3 and 4.4). The early Preboreal ice sheet was about 20% smaller than 800 years earlier (Hubbard *et al.*, 2006; Norðdahl *et al.*, 2008). After 11.2 cal. kyr BP the Icelandic ice sheet retreated very rapidly, and at 8.7 cal. kyr BP (Table 4.1;17) the ice sheet had disintegrated (Kaldal and Víkingsson, 1991).

4.7. Deglaciation of Central South Iceland

Here, we focus on late-glacial deglaciation of the area south, west and north of the Mýrdalsjökull area, while data are still lacking for the area southeast, east and northheast of the Mýrdalsjökull massif. The model of Hubbard *et al.* (2006) suggests a limited extent of the ice sheet in Central South Iceland and the possibility that during the LGM situation the Mýrdalsjökull massif acted as a barrier to effective southward drainage of the interior part of the ice sheet (Fig. 4.5). Both the empirical and the model data indicate that the shelf edge probably acted as an effective controller for the extent of the LGM ice sheet off Central South Iceland. During the Bølling Chronozone, the Mýrdalsjökull massif and the Torfajökull central volcano north thereof seem to have retained considerable ice masses. As noted above, two successive post-Bølling re-advances of the Icelandic ice sheet formed the South Icelandic Búði moraine (Norðdahl *et al.*, 2008) comprising spectacular and complex sets of end moraines that have been traced for about 25 km across the South Icelandic Lowland (Kjartansson, 1958; Fig. 4.5 and 4.6). The formation of the two sets of end moraines, the so-called *inner* (younger) and *outer* (older) moraines that are about 3 km apart in the Holt area (Fig. 4.6), has been attributed to glacier re-advances during the Younger Dryas and Preboreal Chronozones, respectively.

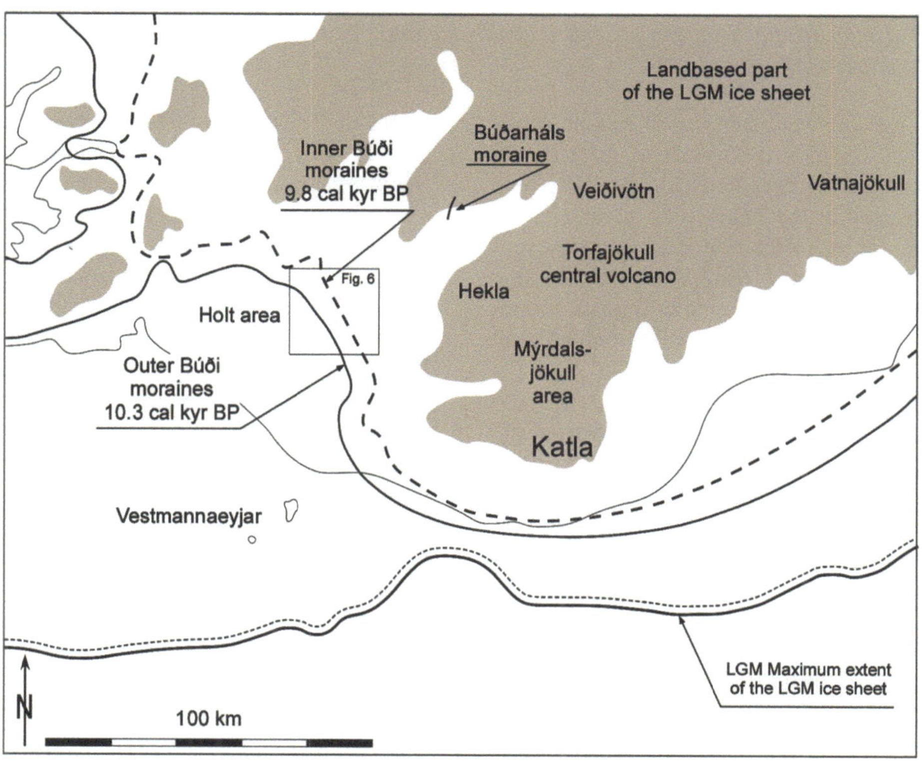

Fig. 4.5. Central South Iceland showing land-based part (grey shade) and extent of the Icelandic LGM ice sheet, and the approximate position of the ice sheet margin during the Younger Dryas (10.3 cal. kyr BP) and Preboreal (9.8 cal. kyr BP) chronozones with important place names mentioned in the text.

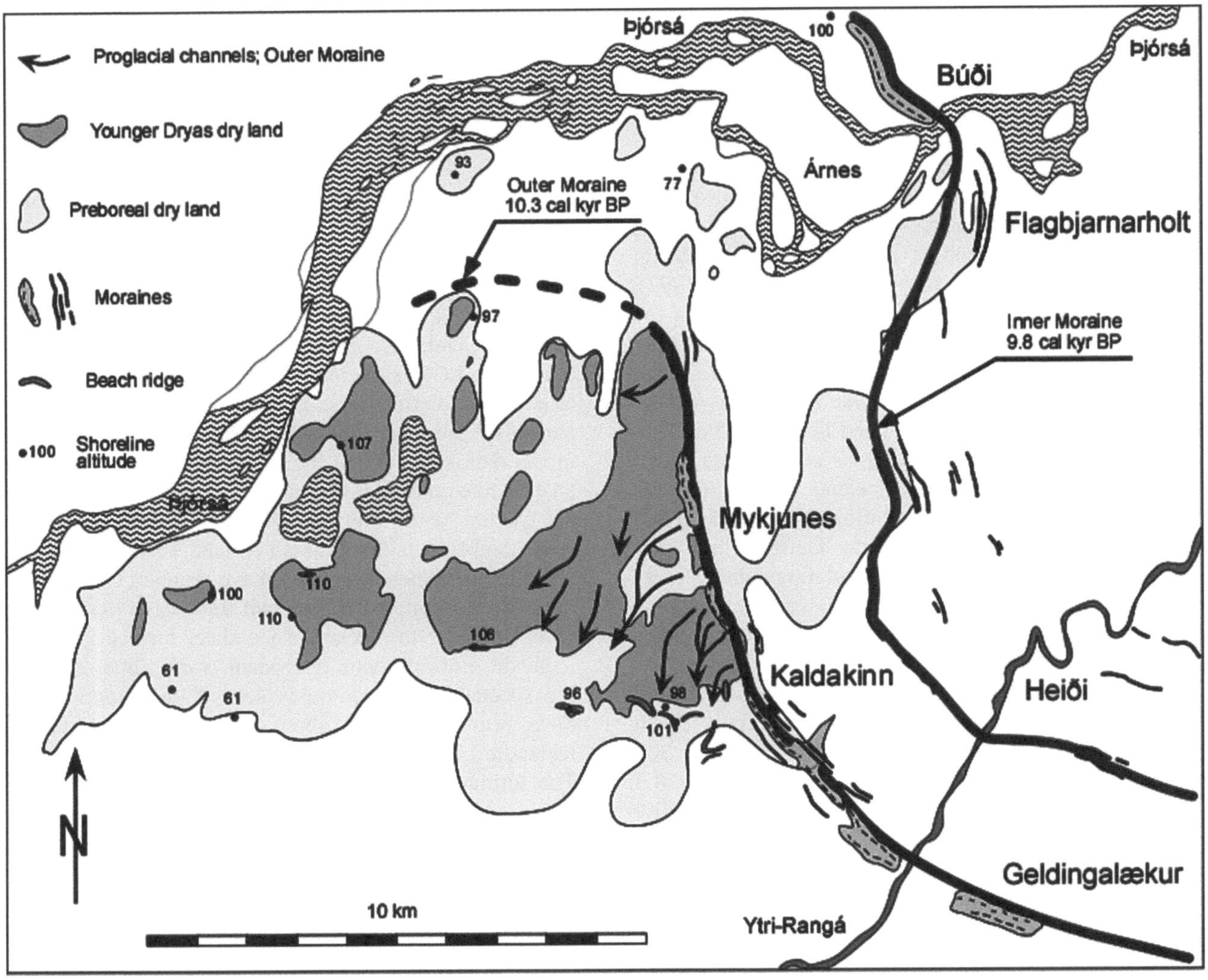

Fig. 4.6. A detailed map of the Holt area, Central South Iceland showing the position of the Outer and Inner Búði moraines and raised shoreline features.

A weighted mean age of seven samples of marine shells collected from the Búði moraine proper at river Þjórsá indicates an early Preboreal age (11.2 cal. kyr BP; Table 4.1;12) to the formation of the inner Búði moraines. The outer Búði moraines were formed during a glacier event preceding the formation of the early Preboreal moraine and have been attributed to the Younger Dryas Chronozone and given an age of 12.0 cal. kyr BP (Norðdahl and Pétursson, 2005).

The Younger Dryas glacial re-advance of the Icelandic ice sheet also saw a vigorous eruption of the Katla volcano and formation of a basaltic andesite and rhyolitic Skógar-Vedde tephra (Norðdahl and Hafliðason, 1992). This tephra is a very important component of Ash zone I found in deep-sea sediments throughout the North Atlantic Ocean and in lacustrine and sublittoral sediments in Scandinavia and the British Isles (Mangerud *et al.*, 1984; Kvamme *et al.*, 1989; Turney, 1998). The Skógar-Vedde tephra has been radiocarbon dated on land in Norway to 12.1 cal. kyr BP (Table 4.1;18 (Bard *et al.*, 1994) and in sediment cores retrieved from the bottom of the North Atlantic Ocean to 12.1 cal. kyr BP, based on NGRIP Ice core chronology (Hald *et al.*, 2007). Outside Fnjóskadalur in North Iceland primary occurrence of the Skógar-Vedde tephra has been verified, e.g. in cores retrieved from Lake Torfadalsvatn in North Iceland (Björck *et al.*, 1992) and Lake Hestvatn in South Iceland (Harðardóttir *et al.*, 2001).

The outer Búði moraines, which were formed in shallow water (about 40 m deep) and in front of grounded ice margin, are most prominent between Geldingalækur and Kaldakinn (Fig. 4.6), where they are predominantly made of stratified coarse grained fluvioglacial sediments. Further to the north, around Mykjunes, the moraines are made of stratified sand, mainly consisting of black volcanic glass shards originating from phreatomagmatic eruptions. Connected to the outer moraines are raised shorelines and beach ridges at elevations between 96 and 110 m a.s.l. outlining what was dry land at that time (Fig. 4.6). The inner Búði moraines were formed in front of a grounded ice margin in about 20–40 m deep water at the Búði waterfall, where the spectacular moraine north of River Þjórsá is predominantly made of stratified ice-contact diamicts (Hjartarson and Ingólfsson, 1989; Geirsdóttir *et al.*, 1997, 2000). South of the Búði waterfall the inner moraines are made of low relief, though distinct, ridges (Fig. 4.7) that can be traced between Flagbjarnarholt and Heiði where the moraines disappear underneath mid-Holocene Hekla lava flows (Fig. 4.6). Raised shorelines occur at about 100 m a.s.l. close to the inner Búði moraine.

Fig. 4.7. Part of the Inner Búði moraine near Heiði, east of Ytri-Rangá is correlated with the Preboreal advance of the Icelandic ice sheet there.

Because of decreased magnitude of glacio-isostatic uplift towards southwest, the shorelines occur at successively lower altitudes and roughly outline what was dry land in the Holt area at that time (Fig. 4.6).

During further deglaciation of Central South Iceland the glacier retreat progressed across dry land and an area that today is to a great extent covered with Holocene lava flows originating from the Veiðivötn area and Mt. Hekla. The earliest post-Búði indications for a temporary position of the retreating glacier margin are about 20 km long moraine ridges in Fitjaskógar, and a very prominent boulder rich moraine that can be traced for about 10 km along the Búðarháls ridge (Kaldal and Víkingsson, 1991), some 50–60 km northeast of the Búði moraine in the Holt area (Fig. 4.5). At Búðarháls, marginal sediment terraces containing black coloured pumice and glass shards have been chemically correlated with the Saksunarvatn tephra, thus dating the moraine and the position of the ice margin there to 10.1 cal. kyr BP (Table 4.1;19). Glacial striae in the area between Mýrdalsjökull and Vatnajökull show that the direction of ice movement successively changed from a westerly direction towards a north-westerly direction as the glacier retreated into this area (Kaldal and Víkingsson, 1991). The crater row of the great Þjórsárhraun lava flow is situated about 35 km north of Mýrdalsjökull and the craters were certainly ice-free at 8.7 cal. kyr BP (Table 4.1;17) when the lava flow was emitted. Hence, the interior highland area north of Mýrdalsjökull was deglaciated and the last remnants of the Icelandic ice sheet had disintegrated between 10.1 and 8.7 cal. kyr BP (Kaldal, 1993).

4.8. Relative Sea Level Changes

The Icelandic LGM ice sheet had a substantial grounded component (Fig. 4.1). The initial retreat of the ice sheet was probably controlled by rapidly rising global sea levels due to melting of the large North American, Scandinavian and Barents Sea ice sheet (Fairbanks, 1989; Tushingham and Peltier, 1991), causing the grounding line of the Icelandic ice sheet to move landwards. Because of the sensitivity of the Icelandic crust to variations in the glacial load, RSL history closely mirrors post-LGM glacial oscillations (Sigmundsson, 1991; Ingólfsson *et al.*, 1995; Ingólfsson and Norðdahl, 2001; Norðdahl and Einarsson, 2001). The following compilation of RSL changes in Iceland is mainly based on reviews by Norðdahl (1991), Norðdahl and Pétursson (2005), and Norðdahl *et al.* (2008).

In West Iceland the ML shoreline at about 150 m a.s.l. was formed at about 14.6 cal. kyr BP (Fig. 4.8). At 14.0 cal. kyr BP (Table 4.1;20), RSL was situated close to 50 m a.s.l., and between 14.0 and 13.4 cal. kyr BP (Table 4.1;20–21) it had been lowered down to or slightly below the present sea level in Northeast and West Iceland (Pétursson, 1991; Ingólfsson and Norðdahl, 2001). This signifies an extremely rapid isostatic rebound with a rebound half-life between 200 and 300 years, which is confirmed by RSL studies from West and North Iceland (Ingólfsson *et al.*, 1995; Rundgren *et al.*, 1997). Some 13.9 cal. kyr BP, extensive calving along with climatic amelioration, had reduced the Icelandic ice sheet to about 20–25% of its LGM size (Norðdahl *et al.*, 2008).

Deterioration of the climate in Allerød and Younger Dryas times caused renewed growth of the ice sheet and a glacial expansion with considerable isostatic depression of the crust and consequently a formation of shorelines now occurring as high as 60 m a.s.l. in Southwest Iceland. The Younger Dryas RSL has been dated to 12.0 cal. kyr BP in West Iceland (Ingólfsson, 1988). The subsequent retreat of the ice margin again caused isostatic uplift and a fall of RSL by up to 40 m in West Iceland. Early Preboreal glacier expansion and increased glacio-isostatic loading, which culminated at 11.2 cal. kyr BP, induced isostatically and eustatically controlled rise of RSL by as much as 25 m and the formation of distinct and extensive shorelines at 40 m a.s.l. in Southwest Iceland (Ingólfsson *et al.*, 1995; Norðdahl and Pétursson, 2005). The glaciers retreated towards and into the highlands and, in response to reduced glacier load and isostatic uplift, RSL fell below present sea level. At 10.7 cal. kyr BP (Table 4.1;22) RSL passed below the −2.5 m level in Reykjavík, while minimum position of RSL was reached at 10.1 cal. kyr BP, when it was at −40 m and peat accumulated at Hraunin (Fig. 4.8) on the floor of the Faxaflói Bay in Southwest Iceland (Ingólfsson *et al.*, 1995). Since then,

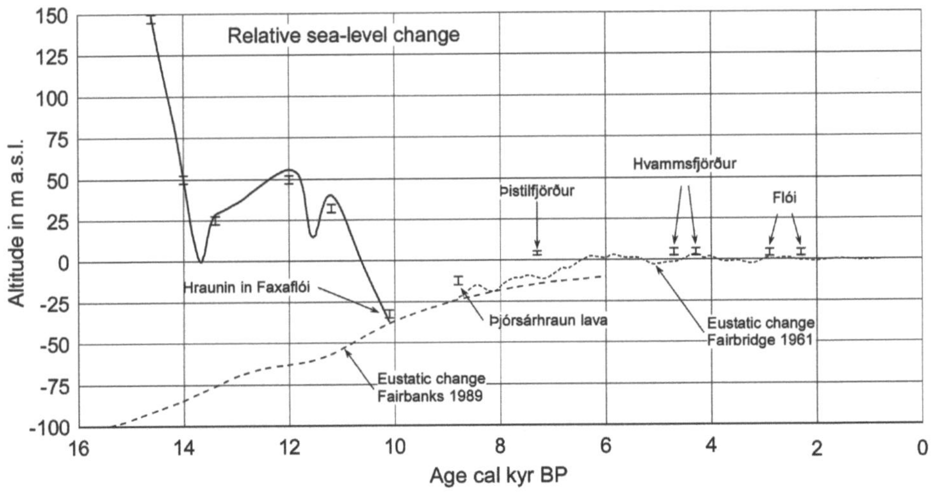

Fig. 4.8. Conceptual RSL curve for deglaciation and Holocene in Iceland, based mainly on field data from West and Southwest Iceland. Modified from Norðdahl and Pétursson (2005) and Norðdahl et al. (2008).

sea level changes have been controlled primarily by eustatic changes, causing a continuous rise in sea level from about −40 m to present sea level (Fairbridge, 1961; Fairbanks, 1989; Tushingham and Peltier, 1991) (Fig. 4.8). In South Iceland the Þjórsárhraun lava flow entered the sea at 8.7 cal. kyr BP (Hjartarson, 1988) when RSL was situated at −15 m (Einarsson, 1978). In Þistilfjörður in Northeast Iceland, a raised shoreline at about 4 m a.s.l. shows a middle-Holocene transgression at 7.3 cal. kyr BP (Table 4.1;23) (Richardson, 1997) (Fig. 4.8). Two sets of beach ridges situated between 3 and 5 m a.s.l. in Hvammsfjörður in West Iceland represent transgressions occurring both before and after 4.5 cal. kyr BP and at Stokkseyri, South Iceland, there are indications of sea level on two occasions reaching between 2 and 6 m a.s.l at 2.9 and 2.3 cal. kyr BP (Table 4.1;24–25) (Símonarson and Leifsdóttir, 2002) (Fig. 4.8).

4.9. Holocene Glacier History

Glaciers in Iceland are generally considered to have retreated towards the interior highlands in the early Holocene and to have attained a minimum extent during the Holocene Thermal Maximum (HTM), which ended 5.0 cal. kyr BP (Gudmundsson, 1997; NGRIP members, 2004; Flowers *et al.*, 2008; Ran *et al.*, 2008). The Neoglaciation of Iceland began after the HTM, 5.0 cal. kyr BP, as marked by glacier advances and ice cap growth of e.g. Langjökull (Flowers *et al.*, 2008), Regnbúðajökull (Kirkbride and Dugmore, 2006), cirque glaciers on the Tröllaskagi peninsula in North Iceland (Stötter *et al.*, 1999), and possibly the Drangajökull ice cap in Vestfirðir (Principato *et al.*, 2006). Jökulhlaup deposits north of Vatnajökull have been dated to 5.0 and 4.6 cal. kyr BP respectively (Table 4.1; 26–27), indicating that the Vatnajökull ice cap existed and had a significant volume at that time (Kirkbride *et al.*, 2006). In addition, there is geomorphological evidence of mid-Holocene glacial activity of southeastern outlet glaciers of Vatnajökull shortly after the HTM (Rose *et al.*, 1997). Well-preserved lateral moraines at Sólheimajökull, a southwestern outlet glacier from Mýrdalsjökull, have been tephrochronologically dated to the mid-Holocene in agreement with the pattern of glacier fluctuations from other parts of the country (Figs. 4.9 and 4.10; Dugmore, 1989; Mackintosh *et al.*, 2002; Rose *et al.*, 1997).

The terrestrial geological record for the late Holocene is more complete, and the pattern of glacier fluctuations is better constrained both in time and space. Neoglacial advances appear to cluster in two intervals: the Subatlantic period (2.6–2.0 cal. kyr BP; Björnsson, 1979; Gudmundsson, 1997; Principato, 2008; Björnsson and Pálsson, 2008) and more pronounced during the Little Ice Age (LIA). Most authors suggest that the LIA in Iceland began in the late sixteenth century AD and lasted until c. AD 1900 (Grove, 2004; Björnsson and Pálsson, 2008), although Massé *et al.* (2008) suggest that the LIA had already begun c. AD 1300 based on increased abundance in marine sediment cores of a biomarker produced by sea-ice algae. However, the general pattern is overprinted with major regional differences because of the large climatic gradients in Iceland with humid oceanic conditions in the south and dry, cold conditions in the interior and the north, which explains different responses and dynamics of the glaciers. For instance, outlet glaciers from the Eyjafjallajökull ice cap in South Iceland advanced during the Medieval Warm Period (Kirkbride and Dugmore, 2008), a period characterized by mild climate and limited sea ice cover (Massé *et al.*, 2008).

The most extensive and widespread Holocene glacier advances occurred during the LIA, when most glaciers reached their Neoglacial maximum (Björnsson, 1979; Gudmundsson, 1997; Grove, 2004; Björnsson and Pálsson, 2008). This agrees well with sea surface temperature reconstructions off North Iceland, demonstrating that the LIA was the coldest period since the HTM (Sicre *et al.*, 2008). Small valley glaciers and minor outlet glaciers reached their LIA maximum around AD 1750, whereas the large lobate outlet glaciers from the ice caps peaked around AD 1890 (Thorarinsson, 1943; Evans *et al.*, 1999; Kirkbride and Dugmore, 2001; Grove, 2004; McKinzey *et al.*, 2005; Björnsson and Pálsson, 2008). The timing and magnitude of LIA advances are often well

Fig. 4.9. Oblique view towards the east of Sólheimajökull, an outlet glacier of the Mýrdalsjökull ice cap. A spectacular system of Holocene lateral moraines, deposited by an expanded Sólheimajökull, is located above the glacier tongue (arrow). Dugmore (1989) suggested moraine ages of 7.0–4.5 cal. kyr BP for the outermost ridges. The image is an aerial orthophotograph from 2001 draped on a digital elevation model with 1.5 × vertical exaggeration. The glacier is c. 700 m wide at the margin.

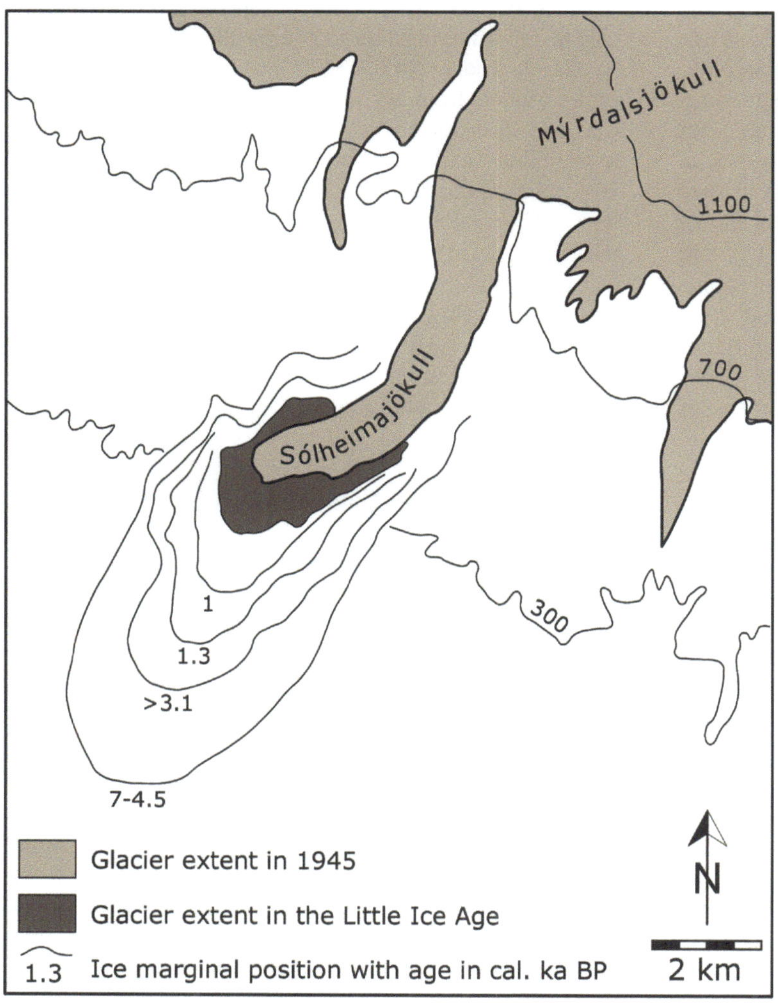

Fig. 4.10. The extent of Holocene advances of Sólheimajökull as reconstructed from lateral moraines. Ages are based on tephrochronology and ^{14}C datings of sedimentary sections in the area. Elevations are given in metres a.s.l. Modified after Dugmore and Sugden (1991).

known from written accounts because in many cases farmland or even farms were overridden or had to be abandoned due to the climate deterioration. For instance Breiðamerkurjökull, a southern outlet glacier from Vatnajökull, advanced at least 7.5 km during the LIA destroyed the farm Fell in AD 1869 by flooding the farmland before it eventually was overridden by the advancing glacier (McKinzey *et al.*, 2005). Similarly, Kaldalónsjökull, an outlet from the Drangajökull ice cap in Northwest Iceland, destroyed farms during its LIA advances in the eighteenth century (Eythorsson, 1935).

4.10. Pre-LIA Glacial History of the Mýrdalsjökull Area

The active Katla volcano underlies the Mýrdalsjökull ice cap and it has had at least 300 eruptions since 9.3 cal. kyr BP, or more than three eruptions per century (Óladóttir *et al.*, 2005, 2007). The morphological character of tephras from Katla reveals whether the eruptions were phreatomagmatic (subglacial) or magmatic (ice-free). Thus, the tephra stratigraphy gives a unique possibility to reveal the history of the Mýrdalsjökull ice cap. Óladóttir *et al.* (2007) concluded that the phreatomagmatic activity was consistent throughout the last 8,400 years, indicating that during all that time the Katla caldera must have been covered with ice. However, they also found indications of a limited HTM ice cover. For the sake of arguments, it must be emphasized that phreatomagmatic tephras could also have formed if the Katla caldera hosted a crater lake, and that the presence of phreatomagmatic tephras only indicates an ice-covered caldera but not necessarily that the outer caldera slopes were covered with ice too. Reconstructions of the extent of the ice cap are, therefore, based on other indicators such as end moraines and successions of glacial sediments. Jackson *et al.* (2005) measured grain size distributions in a loess-tephra succession at Hólmsá east of Mýrdalsjökull, covering the time since 8.4 cal. kyr BP. They associated a higher frequency of coarse grains around 5.5 cal. kyr BP with more windy and cool environmental conditions, which might reflect the build-up of the Mýrdalsjökull ice cap after the HTM. Their findings agree well with the volcanological study of Óladóttir *et al.* (2007).

The spectacular moraines at Sólheimajökull (Fig. 4.9) have been intensively studied by Dugmore (1989), Dugmore and Sugden (1991), Mackintosh *et al.* (2002), and Casely and Dugmore (2004). By using tephrochronology and radiocarbon dating, Dugmore (1989) suggested that the lateral moraines become successively younger towards the present glacier and that they were formed by glacier advances at four occasions; between 7.0 and 4.5 cal. kyr BP, before 3.1 cal. kyr BP, between 1.4 and 1.2 cal. kyr BP, and at 1.0 cal. kyr BP (Fig. 4.10). Intriguingly, there is no evidence of synchronous advances of other major outlet glaciers from Mýrdalsjökull (Krüger, 1994; Schomacker *et al.*, 2003; Casely and Dugmore, 2004). Therefore, the advances of Sólheimajökull have been interpreted as 'anomalous', occurring out of phase with other outlet glaciers and during periods with relatively mild climate. Dugmore and Sugden (1991) suggested that the 'anomalous' advances could be explained by topography and ice-divide migration, as Sólheimajökull drains a larger proportion of the ice cap, when the ice-cap volume is reduced. In periods with great ice-cap volumes, more ice is drained towards the north and northeast. This conceptual model is still debated and, to our knowledge, similar 'ice-divide migrations' have not yet been described from other glaciers in Iceland. However, the current warming and ascent of the equilibrium-line altitude might reveal if Sólheimajökull really responds 'anomalously' to climate change.

4.11. Little Ice Age Glacial History of the Mýrdalsjökull Area

Most of the outlet glaciers of Mýrdalsjökull appear to have reached their Holocene maximum during the LIA. Evans *et al.* (1999) used Schmidt hammer rebound values and lichenometry to estimate the ages of end moraines at Öldufellsjökull and Sandfellsjökull on the east side of Mýrdalsjökull. They concluded that the retreat from the most distal end moraine at Sandfellsjökull began in AD 1895 and at Öldufellsjökull as late as AD 1920. At Sléttjökull, the north margin of the ice cap, a photograph shows that a retreat from the outermost end moraine had just begun in AD 1906, suggesting that the glacier reached its LIA maximum around AD 1900 (Sapper, 1909; Krüger, 1994). Krüger (1987) suggested that this advance overrode an older end moraine from an extensive LIA advance c. AD 1750, when the glacier margin reached almost the same position as during the LIA maximum.

Casely and Dugmore (2004) investigated the western outlet glaciers from Mýrdalsjökull. Based on geomorphology and tephrochronology, they concluded that Tungnakvíslajökull and Krossárjökull had their LIA (and Holocene) maxima during the early–mid-nineteenth century. Sólheimajökull appears to have had three maxima during the LIA: one at around AD 1705 and the others around AD 1794 and AD 1820 (Thorarinsson, 1943). These advances were, however, not the largest Holocene advances (Fig. 4.10). A brief review of the fluctuations of Mýrdalsjökull during historical times is also given by Sigurðsson (this volume).

4.12. Summary and Conclusions

Sediment cores retrieved from the shelf around Iceland and sections in coastal areas hold evidence showing that the Icelandic ice sheet increased in size and advanced across the present coastline to reach the edge of the shelf during its LGM maximum extent. Due to extensive and extremely rapid environmental changes during late Weichselian and early Holocene the Icelandic ice sheet retreated and extensive areas became ice-free, which were subsequently overridden by the ice sheet during repeated late-glacial glacier advances. These changes in the extent and volume of the ice sheet were accompanied by repeated changes in the RSL between 14.6 and 10.7 cal. kyr BP. The Holocene thermal and vegetational optimum occurred before 6.0 cal. kyr BP with the Neoglaciation of

Iceland beginning later than 5.0 cal. kyr BP. The Mýrdalsjökull massif has most likely carried glaciers throughout the Holocene with a temporary maximum extent during the LIA around AD 1900. Below is a short summary of the main conclusions concerning the post-LGM environmental history of Iceland:

- About 25.3 cal. kyr BP the Icelandic ice sheet advanced and reached the break of the shelf around Iceland (Fig. 4.1). The c. 1,500–2,000 m thick ice sheet totally covered the island except a few high coastal mountains with ice-free enclaves.
- Between 14.9 and 13.9 cal. kyr BP the marine-based part of the ice sheet collapsed and rapidly retreated from the shelf and onto present-day dry land (Fig. 4.2). The rapid deglaciation allowed formation of greatly elevated ML shorelines in early Bølling times at about 14.6 cal. kyr BP (Fig. 4.8). Decreased volume and isostatic load of the ice sheet caused accelerated isostatic uplift, and RSL fell and reached a temporary minimum at about 13.9 cal. kyr BP (Fig. 4.8).
- Between 13.9 and 11.5 cal. kyr BP a climatic deterioration induced positive mass-balance change with increasing ice volume and isostatic lowering during the Allerød, a development that culminated in Younger Dryas time (at about 12.0 cal. kyr BP) when the margin of the ice sheet reached a position close to the coastline of that time (Fig. 4.3). RSL rose and reached a temporary maximum position at the same time (Fig. 4.8).
- Between 11.5 and 10.1 cal. kyr BP the ice sheet retreated and subsequently advanced to a distinct Preboreal position (at 11.2 cal. kyr BP) just inside the Younger Dryas position of the ice margin (Fig. 4.3 and 4.4). About 10.1 cal. kyr BP the ice margin had retreated to a position some 50–60 km northeast of the Búði moraine in South Central Iceland (Fig. 4.5). During this period RSL first fell and then rose again in phase with variations in the glacio-isostatic load and extent of the ice sheet to finally fall towards and eventually below the altitude of present sea level about 10.7 cal. kyr BP. RSL reached its minimum position of c. −40 m in the Faxaflói area at 10.1 cal. kyr BP (Fig. 4.8).
- Between 10.1 and 8.7 cal. kyr BP the interior highland area north of Mýrdalsjökull was deglaciated and the last remnants of the Icelandic ice sheet melted (Fig. 4.5). After 10.1 cal. kyr BP sea level changes were controlled by eustasy alone, a development displaying a more or less continuous rise of sea level from about −40 m up to present sea level. Since 7.3 cal. kyr BP there are signs of sea level being above present sea level on at least five occasions (Fig. 4.8).
- The Neoglaciation of Iceland began after 5.0 cal. kyr BP and glacier advances cluster in two main intervals: the Subatlantic period, between 2.6 and 2.0 cal. kyr BP and the LIA, between AD 1300 and AD 1920. At most localities, the LIA advances were the most extensive during the Holocene.
- At Sólheimajökull, lateral moraines originate from glacier advances dating back to the mid-Holocene. Tephrochronology has proven to be a valuable tool in dating these landforms. There is, however, still a need for more accurate datings.
- The tephra stratigraphy around Mýrdalsjökull extends back to at least 8.4 cal. kyr BP. The morphology of these Holocene phreatomagmatic tephras suggests that the Katla caldera was covered with ice throughout the last 8.4 cal. kyr BP. The ice cover may, however, have been restricted to the caldera rim during some of this interval and the morphology do not reveal information on the spatial extent of the ice cover.
- The LIA advances of the outlet glaciers from Mýrdalsjökull peaked c. AD 1900, although there were also extensive LIA advances culminating around AD 1750.

References

Andrews, J.T., Harðardóttir, J., Helgadóttir, G., Jennings, A.E., Geirsdóttir, Á., Sveinbjörnsdóttir, Á.E., Schoolfield, S., Kristjánsdóttir, G.B., Smith, L.M., Thors, K., & Syvitski, J., 2000. The N and W Iceland Shelf: insights into Last Glacial Maximum ice extent and deglaciation based on acoustic stratigraphy and basal radiocarbon AMS dates. Quaternary Science Reviews 19, 619–631.

Ásbjörnsdóttir, L. & Norðdahl, H., 1995. Götungar í sjávarsetlögum við Mela á Skarðsströnd. In: Hróarsson, B., Jónsson, D., Jónsson, S.S. (Eds), Eyjar í Eldhafi. Reykjavík, Gott mál, 179–188.

Bard, E., Arnold, M., Mangerud, J., Paterne, M., Labeyrie, L., Duprat, J., Mélières, M.-A., Sönstegaard, E., & Duplessy, J.-C., 1994. The North Atlantic atmosphere-sea surface ^{14}C gradient during the Younger Dryas climatic event. Earth and Planetary Science Letters 126, 275–287.

Björck, S., Ingólfsson, Ó., Hafliðason, H., Hallsdóttir, M., & Anderson, N.J., 1992. Lake Torfadalsvatn: a high resolution record of the North Atlantic ash zone I and the last glacial–interglacial environmental changes in Iceland. Boreas 21, 15–22.

Björnsson, H., 1979. Glaciers in Iceland. Jökull 29, 74–80.

Björnsson, H. & Pálsson, F., 2008. Icelandic glaciers. Jökull 58, 365–386.

Casely, A.F. & Dugmore, A.J., 2004. Climate change and 'anomalous' glacier fluctuations: the southwest outlets of Mýrdalsjökull, Iceland. Boreas 33, 108–122.

Dugmore, A., 1989. Tephrochronological studies of Holocene glacier fluctuations in south Iceland. In: Oerlemans, J. (Ed), Glacier fluctuations and climate change. Dordrecht, Kluwer Academic Publishers, 37–55.

Dugmore, A.J. & Sugden, D.E., 1991. Do the anomalous fluctuations of Sólheimajökull reflect ice-divide migration? Boreas 20, 105–113.

Egloff, J. & Johnson, G.L., 1979. Erosional and depositional structures of the Southwest Iceland insular margin: Thirteen geophysical profiles. In: Watkins, J.S., Montadert, L., Dickerson, P.W. (Eds), Geological and geophysical investigations of Continental Margins. Tulsa, OK, AAPG, 43–63.

Einarsson, Th., 1967. Zu der Ausdehnung der Weichselzeitlichen Vereisung Nordislands. Sonderveröffentlichungen des Geologischen Institutes der Universität Köln 13, 167–173.

Einarsson, Th., 1978. Jarðfræði. Reykjavík, Mál og menning, 240 pp.

Eiríksson, J., Knudsen, K.L., Hafliðason, H., & Henriksen, P., 2000. Late-glacial and Holocene palaeoceanography of the North Iceland Shelf. Journal of Quaternary Science 15, 23–42.

Evans, D.J.A., Archer, S., & Wilson, D.J.H., 1999. A comparison of the lichenometric and Schmidt hammer dating techniques based on data from the proglacial areas of some Icelandic glaciers. Quaternary Science Reviews 18, 13–41.

Eythorsson, J., 1935. On the variations of glaciers in Iceland, some studies made in 1931. Geografiska Annaler 17, 121–137.

Fairbanks, R.G., 1989. A 17,000-year glacio-sea level record: influence of glacial melting rates on the Younger Dryas event and deep-ocean circulation. Nature 342, 637–642.

Fairbridge, R.W., 1961. Eustatic changes in sea level. Physics and Chemistry of the Earth 4, 99–185.

Flowers, G.E., Björnsson, H., Geirsdóttir, Á., Miller, G.H., Black, J.L., & Clarke, G.K.C., 2008. Holocene climate conditions and glacier variation in central Iceland from physical modeling and empirical evidence. Quaternary Science Reviews 27, 797–813.

Geirsdóttir, Á. & Eiríksson, J., 1994. Sedimentary facies and environmental history of the late-glacial glaciomarine Fossvogur sediments in Reykjavík, Iceland. Boreas 23, 164–176.

Geirsdóttir, Á., Harðardóttir, J., & Eiríksson, J., 1997. The depositional history of the Younger Dryas – Preboreal Búdi Moraines in South-Central Iceland. Arctic and Alpine Research 29, 13–23.

Geirsdóttir, Á., Harðardóttir, J., & Sveinbjörnsdóttir, Á.E., 2000. Glacial extent and catastrophic meltwater events during the deglaciation of Southern Iceland. Quaternary Science Reviews 21, 291–298.

Geirsdóttir, Á., Andrews, J.T., Ólafsdóttir, S., Helgadóttir, G., & Harðardóttir, J., 2002. A 36 Ky record of iceberg rafting and sedimentation from north-west Iceland. Polar Research 21, 291–298.

Grove, J.M., 2004. Little Ice Ages: Ancient and Modern 1, 13–60.

Gudmundsson, H., 1997. A review of the Holocene environmental history of Iceland. Quaternary Science Reviews 16, 81–92.

Hald, M., Andersson, C., Ebbesen, H., Jansen, E., Klitgaard-Kristensen, D., Risebrobakken, L., Salomonsen, G.R., Sarnthein, M., Sejrup, H.P., & Telford, R.J., 2007. Variations in temperature and extent of Atlantic Water in the northern North Atlantic during the Holocene. Quaternary Science Reviews 26, 3423–3440.

Haraldsson, H., 1981. The Markarfljót sandur area, Southern Iceland: sedimentological, petrographical and stratigraphical studies. Striae 15, 1–58.

Harðardóttir, J., Geirsdóttir, Á., & Sveinbjörnsdóttir, Á.E., 2001. Seismostratigraphy and sediment studies of Lake Hestvatn, southern Iceland: implications for the deglacial history of the region. Journal of Quaternary Science 16, 167–179.

Håkansson, S., 1983. A reservoir age for the coastal waters of Iceland. Geologiska Föreningens i Stockholm Förhandlingar (GFF) 105, 65–68.

Hjartarson, Á., 1988. Þjórsárhraunið mikla – stærsta nútímahraun jarðar [The Thjórsár lava – the largest Holocene lavaflow on Earth]. Náttúrufræðingurinn 58, 1–16.

Hjartarson, Á. & Ingólfsson, Ó., 1989. Preboreal glaciation of Southern Iceland. Jökull 38, 1–16.

Hjort, C., Ingólfsson, Ó., & Norðdahl, H., 1985. Late quaternary geology and glacial history of Hornstrandir, Northwest Iceland: a reconnaissance study. Jökull 35, 9–29.

Hubbard, A., Sugden, J., Dugmore, A., Norðdahl, H., & Pétursson, H.G., 2006. A modelling insight into the Icelandic Late Glacial Maximum ice sheet. Quaternary Science Reviews 25, 2283–2296.

Hughen, K.A., Baillie, M.G.L., Bard, E., Bayliss, A., Beck, J.W., Bertrand, C.J.H., Blackwell, P.G., Buck, C.E., Burr, G.S., Cutler, K.B., Damon, P.E., Edwards, R.L., Fairbanks, R.G., Friedrich, M., Guilderson, T.P., Kromer, B., McCormac, F.G., Manning, S.W., Bronk Ramsey, C., Reimer, P.J., Reimer, R.W., Remmele, S., Southon, J.R., Stuiver, M., Talamo, S., Taylor, F.W., van der Plicht, J., & Weyhenmeyer, C.E., 2004. Marine04 marine radiocarbon age calibration, 0-26 cal kyr BP. Radiocarbon 46, 1059–1086.

Ingólfsson, Ó., 1985. Late Weichselian glacial geology of the lower Borgarfjörður region, western Iceland: a preliminary report. Arctic 38, 210–213.

Ingólfsson, Ó., 1988. Glacial history of the lower Borgarfjördur area, western Iceland. Geologiska Föreningens i Stockholm Förhandlingar (GFF) 110, 293–309.

Ingólfsson, Ó., 1991. A review of the Late Weichselian and early Holocene glacial and environmental history of Iceland. In: Maizels, J., Caseldine, C. (Eds), Environmental Change in Iceland: Past and Present. Dordrecht, Kluwer Academic Publishers, 13–29.

Ingólfsson, Ó. & Norðdahl, H., 2001. High relative sea level during the Bølling interstadial in Western Iceland: a reflection of ice-sheet collapse and extremely rapid glacial unloading. Arctic, Antarctic, and Alpine Research 33, 231–243.

Ingólfsson, Ó., Norðdahl, H., & Haflíðason, H., 1995. A rapid isostatic rebound in south-western Iceland at the end of the last glaciation. Boreas 24, 245–259.

Ingólfsson, Ó., Björck, S., Haflíðason, H., & Rundgren, M., 1997. Glacial and climatic events in Iceland reflecting regional North Atlantic climatic shifts during the Pleistocene–Holocene transition. Quaternary Science Reviews 16, 1135–1144.

Jackson, M.G., Oskarsson, N., Trønnes, R., McManus, J.F., Oppo, D.W., Grönvold, K., Hart, S.R., & Sachs, J.P., 2005. Holocene loess deposition in Iceland: evidence for millennial-scale atmosphere–ocean coupling in the North Atlantic. Geology 33, 509–512.

Jennings, A., Syvitski, J., Gerson, L., Grönvold, K., Geirsdóttir, Á., Harðardóttir, J., Andrews, J.T., & Hagen, S., 2000. Chronoogy and paleoenvironments during the late Weichselian deglaciation of the south-west Iceland shelf. Boreas 29, 167–183.

Kaldal, I., 1993. Fróðleiksmolar um gamla gjósku í Búðarhálsi. Geoscience Society of Iceland, Spring Meeting 1994, p. 19.

Kaldal, I. & Víkingsson, S., 1991. Early Holocene deglaciation in Central Iceland. Jökull 40, 51–66.

Kirkbride, M.P. & Dugmore, A.J., 2001. Can lichenometry be used to date the "Little Ice Age" glacial maximum in Iceland? In: Ogilvie, A.E.J., Jónsson, T. (Eds), The iceberg in the mist: Northern research in pursuit of a "Little Ice Age". Dordrecht, Kluwer Academic Publishers, 151–167.

Kirkbride, M.P. & Dugmore, A.J., 2006. Responses of mountain ice caps in central Iceland to Holocene climate change. Quaternary Science Reviews 25, 1692–1707.

Kirkbride, M.P. & Dugmore, A.J., 2008. Two millennia of glacier advances from southern Iceland dated by tephrochronology. Quaternary Research 70, 398–411.

Kirkbride, M.P., Dugmore, A.J., & Brazier, V., 2006. Radiocarbon dating of mid-Holocene megaflood deposits in the Jökulsá á Fjöllum, North Iceland. The Holocene 16, 605–609.

Kjartansson, G., 1958. Jarðmyndanir í Holtum og nágrenni [The Geology of Holt in Rangárvallasýsla, SW Iceland], 22 pp. Department of Agriculture, Reports 11. University Research Institute, Reykjavík.

Kjartansson, G., 1966. Nokkrar nýjar C14-aldursákvarðanir [Some new C14 datings in Iceland]. Náttúrufræðingurinn 36, 126–141.

Krüger, J., 1987. Træk af et glaciallandskabs udvikling ved nordranden af Mýrdalsjökull, Island. Dansk Geologisk Forening, Årsskrift for 1986, 49–66.

Krüger, J., 1994. Glacial processes, sediments, landforms, and stratigraphy in the terminus region of Mýrdalsjökull, Iceland. Folia Geographica Danica 21, 1–233.

Kvamme, T., Mangerud, J., Furnes, H., & Ruddiman, W.F., 1989. Geochemistry of Pleistocene ash zones in cores from the North Atlantic. Norsk Geologisk Tidsskrift 69, 251–272.

Mackintosh, A.N., Dugmore, A.J., & Hubbard, A.L., 2002. Holocene climatic changes in Iceland: evidence from modeling glacier length fluctuations at Sólheimajökull. Quaternary International 91, 39–52.

Mangerud, J., Andersen, S.T., Berglund, B.E., & Donner, J.J., 1974. Quaternary stratigraphy of Norden, a proposal for terminology and classification. Boreas 3, 109–126.

Mangerud, J., Lie, S.E., Furnes, H., Kristiansen, I.L., & Lömo, L., 1984. A Younger Dryas ash bed in western Norway, and its possible correlations with tephra in cores from the Norwegian Sea and the North Atlantic. Quaternary Research 21, 85–104.

Massé, G., Rowland, S.J., Sicre, M.-A., Jacob, J., Jansen, E., & Belt, S.T., 2008. Abrupt climate changes for Iceland during the last millennium: evidence from high resolution sea ice reconstructions. Earth and Planetary Science Letters 269, 565–569.

McKinzey, K.M., Ólafsdóttir, R., & Dugmore, A.J., 2005. Perception, history, and science: coherence or disparity in the timing of the Little Ice Age maximum in southeast Iceland? Polar Record 41, 319–334.

Norðdahl, H., 1991. A review of the glaciation maximum concept and the deglaciation of Eyjafjörður, North Iceland. In: Maizels, J.K., Caseldine, C. (Eds), Environmental changes in Iceland: past and present. Dordrecht, Kluwer Academic Publishers, 31–47.

Norðdahl, H. & Einarsson, Th., 2001. Concurrent changes of relative sea-level and glacier extent at the Weichselian – Holocene boundary in Berufjörður, Eastern Iceland. Quaternary Science Reviews 20, 1607–1622.

Norðdahl, H. & Hafliðason, H., 1992. The Skógar tephra, a Younger Dryas marker in North Iceland. Boreas 21, 23–41.

Norðdahl, H. & Hjort, C., 1987. Aldur jökulhörfunar í Vopnafirði, Abstract volume, Reykjavík, Jarðfræðafélag Íslands, 18–19.

Norðdahl, H. & Hjort, C., 1995. Lateglacial raised beaches and glacier recession in the Thistilfjördur – Bakkaflói area, North-eastern Iceland. Jökull 43, 32–44.

Norðdahl, H. & Pétursson, H.G., 2005. Relative sea level changes in Iceland. New aspect of the Weichselian deglaciation of Iceland. In: Caseldine, C., Russell, A., Harðardottir, J., Ó. Knudsen (Eds), Iceland – modern processes and past environments. Developments in Quaternary Science 5, 25–78.

Norðdahl, H., Ingólfsson, Ó., Pétursson, H.G., & Hallsdóttir, M., 2008. Late Weichselian and Holocene environmental history of Iceland. Jökull 58, 343–364.

NGRIP (North Greenland Ice Core Project) members, 2004. High-resolution record of Northern Hemisphere climate extending into the last interglacial period. Nature 431, 147–151.

Óladóttir, B., Larsen, G., Thórdason, T., & Sigmarsson, O., 2005. The Katla volcano, S-Iceland: Holocene tephra stratigraphy and eruption frequency. Jökull 55, 53–74.

Óladóttir, B., Thordarson, T., Larsen, G., & Sigmarsson, O., 2007. Survival of the Mýrdalsjökull ice cap through the Holocene Thermal Maximum: evidence from sulphur contents in Katla tephra layers (Iceland) from the last ~8400 years. Annals of Glaciology 45, 183–188.

Ólafsdóttir, Th., 1975. Jökulgarður á sjávarbotni út af Breiðafirði [A moraine ridge on the Iceland shelf, west of Breiðafjörður]. Náttúrufræðingurinn 45, 31–36.

Pétursson, H.G., 1986. Kvartærgeologiske undersøkelser på Vest-Melrakkaslétta, Nordøst-Island. Unpublished cand. real. thesis. University of Tromsø, Norway, 157 pp.

Pétursson, H.G., 1991. The Weichselian glacial history of West Melrakkaslétta, North-eastern Iceland. In: Maizels, J.M., Caseldine, C. (Eds), Environmental changes in Iceland: past and present. Dordrecht, Kluwer Academic Publishers, 49–65.

Principato, S., 2008. Geomorphic evidence for Holocene glacial advances and sea level fluctuations on eastern Vestfirðir, northwest Iceland. Boreas 37, 132–145.

Principato, S.M., Geirsdóttir, Á., Jóhannsdóttir, G.E., & Andrews, J.T., 2006. Late Quaternary glacial and deglacial history of eastern Vestfirðir, Iceland, using cosmogenic isotope (^{36}Cl) exposure ages and marine cores. Journal of Quaternary Science 21, 271–285.

Ran, L., Jiang, H., Knudsen, K.L., & Eiríksson, J., 2008. A high-resolution Holocene diatom record from the North Icelandic shelf. Boreas 37, 399–413.

Reimer, P.J., Baillie, M.G.L., Bard, E., Bayliss, A., Beck, J.W., Bertrand, C.J.H., Blackwell, P.G., Buck, C.E., Burr, G.S., Cutler, K.B., Damon, P.E., Edwards, R.L., Fairbanks, R.G., Friedrich, M., Guilderson, T.P., Hogg, A.G., Hughen, K.A., Kromer, B., McCormac, F.G., Manning, S.W., Ramsey, C.B., Reimer, R.W., Remmele, S., Southon, J.R., Stuiver, M., Talamo, S., Taylor, F.W., van der Plicht, J., & Weyhenmeyer, C.E., 2004. IntCal04 terrestrial radiocarbon age calibration, 0-26 cal kyr BP. Radiocarbon 46, 1029–1058.

Richardson, S., 1997. Deglaciation and shoreline displacement adjacent to a spreading ridge, NE Iceland. Unpublished Ph.D. Thesis, Department of Geography, Royal Holloway, University of London, 408 pp.

Rose, J., Whiteman, C.A., Lee, J., Branch, N.P., Harkness, D.D., & Walden, J., 1997. Mid- and late-Holocene vegetation, surface weathering and glaciations, Fjallsjökull, southeast Iceland. The Holocene 7, 457–471.

Rundgren, M., 1995. Biostratigraphic evidence of the Allerød–Younger Dryas–Preboreal oscillation in Northern Iceland. Quaternary Research 44, 405–416.

Rundgren, M., 1999. A summary of the environmental history of the Skagi peninsula, northern Iceland, 11300–7800 BP. Jökull 47, 1–19.

Rundgren, M. & Ingólfsson, Ó., 1999. Plant survival in Iceland during periods of glaciation? Journal of Biogeography 26, 387–396.

Rundgren, M., Ingólfsson, Ó., Björck, S., Jiang, H., & Haflidason, H., 1997. Dynamic sea-level change during the last deglaciation of northern Iceland. Boreas 26, 201–215.

Sapper, K., 1909. Bemerkungen über einige südisländische Gletscher. Zeitschrift für Gletscherkunde 3, 297–305.

Schomacker, A., Krüger, J., & Larsen, G., 2003. An extensive late Holocene glacier advance of Kötlujökull, central south Iceland. Quaternary Science Reviews 22, 1427–1434.

Sicre, M.-A., Yiou, P., Eiríksson, J., Ezat, U., Guimbaut, E., Dahhaoui, I., Knudsen, K.-L., Jansen, E., & Turon, J.-L., 2008. A 4500-year reconstruction of sea surface temperature variability at decadal time-scales off North Iceland. Quaternary Science Reviews 27, 2041–2047.

Sigmundsson, F., 1991. Post-glacial rebound and asthenosphere viscosity in Iceland. Geophysical Research letters 18, 1131–1134.

Símonarson, L.A. & Leifsdóttir, Ó.E., 2002. Late-Holocene sea-level changes in South and Southwest Iceland reconstructed from littoral molluscan stratigraphy. The Holocene 12, 149–158.

Steindórsson, S., 1963. Ice age refugia in Iceland as indicated by the present distribution of plant species. In: Löve, Á., Löve, D. (Eds), North Atlantic biota and their history. Oxford, Pergamon Press, 302–320.

Stötter, J., Wastl, M., Caseldine, C., & Häberle, T., 1999. Holocene palaeoclimatic reconstruction in northern Iceland: approaches and results. Quaternary Science Reviews 18, 457–474.

Stuiver, M. & Reimer, P.J., 1993. Extended C-14 database and revised Calib 3.0 C-14 age calibration program. Radiocarbon 35, 215–230.

Sveinbjörnsdóttir, Á.E., Eiríksson, J., Geirsdóttir, Á., Heinemeier, J., & Rud, N., 1993. The Fossvogur marine sediments in SW Iceland – confined to the Allerød/Younger Dryas transition by AMS ^{14}C datings. Boreas 22, 147–157.

Syvitski, J., Jennings, A.E., & Andrews, J.T., 1999. High-resolution seismic evidence for multiple glaciation across the Southwest Iceland Shelf. Arctic, Antarctic, and Alpine Research 31, 50–57.

Sæmundsson, Th., 1995. Deglaciation and shoreline displacement in Vopnafjörður, North-eastern Iceland. Lundqua Thesis 33, 1–106.

Thorarinsson, S., 1943. Vatnajökull. Scientific results of the Swedish-Icelandic investigations 1936-37-38. Geografiska Annaler 1–2, 1–54.

Turney, C.S.M., 1998. Extraction of rhyolitic component of Vedde microtephra from minerogenic lake sediments. Journal of Paleolimnology 19, 199–206.

Tushingham, A.M. & Peltier, W.R., 1991. Ice-3G: a new global model of Late Pleistocene deglaciation based upon geophysical prediction of post-glacial relative sea level change. Journal of Geophysical Research 96, 4497–4523.

5

Variations of Mýrdalsjökull during Postglacial and Historical Times

Oddur Sigurðsson

Icelandic Meteorological Office, Grensásvegur 9, IS-108 Reykjavík, Iceland
(e-mail: osig@os.is)

5.1. Introduction

Mýrdalsjökull, the fourth largest ice cap in Iceland ($597\,km^2$ in 1999), is located in the southern part of Iceland, where the mean annual temperature is the highest in the country. The elevation range is moderate (100–1500 m a.s.l.), indicating that the glacier owes its existence to very high precipitation, which ranges annually from approximately 2000 mm at sea level to 5000 mm or more at the summit of the glacier. Mýrdalsjökull is assumed to have more or less disappeared during the early Holocene, in common with most other Icelandic glaciers (Björnsson, 1979; Black *et al.*, 2006). During the settlement period about AD 900 the main ice caps were at the same locations as presently, but they were considerably smaller. During most of the past millennium glaciers in Iceland have increased in volume more or less continuously, reaching their maximum postglacial extent at about AD 1890. The maximum of glacier extent during the middle of the 18th century described by Thorarinsson (1943) is of relatively limited importance because it was primarily in surge-type glaciers. Clear indications of advancing glaciers during the latter half of the 18th century are described in Pálsson (2004). During the 20th century glaciers reversed this trend, retreating by roughly as much as they had advanced during the 3 previous centuries (Sigurðsson, 2005).

Mýrdalsjökull and its surroundings constitute one of the most dramatic areas in the world in terms of natural hazards. Two of the world's largest lava fields produced in single eruptions in historical time, namely the Eldgjá eruption of c. 934 AD and Laki eruption of AD 1783, are in the immediate vicinity (Gunnlaugsson *et al.*, 1984). There is reason to believe that no human being has ever witnessed a greater river discharge than what the inhabitants of the districts surrounding Mýrdalsjökull saw during jökulhlaups associated with the eruptions of Katla. These floods and eruptions transformed a prosperous farming district into a wasteland, now called Mýrdalssandur, on the east side of Mýrdalsjökull. Even during the present time, this is still the most serious natural hazard area of Iceland (Guðmundsson and Gylfason, 2005; Elíasson *et al.*, 2006).

During most of the history of Icelanders, the glaciers have been increasing in size, and in several well-documented cases they advanced over inhabited areas. Early scientific understanding of glacier fluctuations was incomplete because climate was only one of the factors controlling the variations of glacier termini. The local inhabitants knew that the glaciers were growing over time, while the main authorities on glaciers during the 18th and 19th centuries, for example Pálsson (2004) and Thoroddsen (1911), were of the opinion that glaciers were more or less of the same extent as during the settlement period c. 1000 years earlier.

Mýrdalsjökull is well suited for scientific investigations as it is relatively accessible. It has a high annual mass turnover, making changes in the ice cap and its outlet glaciers large and obvious. However, substantial non-climatic forcing of the mass balance complicates the task of glaciological researchers because outlet glaciers have retreated after eruptions during a period of general glacier advance. An example of this is the limited extent of Sólheimajökull in 1783 (Thorarinsson, 1943) probably because of the mass loss in the two great eruptions of Katla in 1721 and 1755. Each of these two eruptions may have melted more than 5% of the total mass of the ice cap.

5.2. Glacier Place Names

Through the ages different names, such as Eyjafjallajökull, Sólheimajökull, Höfðárjökull and Kötlugjárjökull, have been applied to Mýrdalsjökull as a whole (Sigurðsson and Williams, 2008). Different parts of the ice cap have been named separately, particularly during the last century. Most of the place names are derived from the river that issues from the glacier, the land adjacent to the glacier or its appearance or nature. Names have changed over time but the ones that appear in bold font in Fig. 5.1 are the ones presently used and preferred.

5.3. Causes of Changes in Glacier Mass

In general, glaciers are of two types with regard to the reaction of the terminus to variations in mass balance:

- Non-surge-type
- Surge-type

The terminus of a surge-type glacier ordinarily retreats regardless of climate and mass balance and advances only occasionally at a regular or quasi-regular interval of time

DEVELOPMENTS IN QUATERNARY SCIENCES
VOLUME 13 ISSN 1571-0866

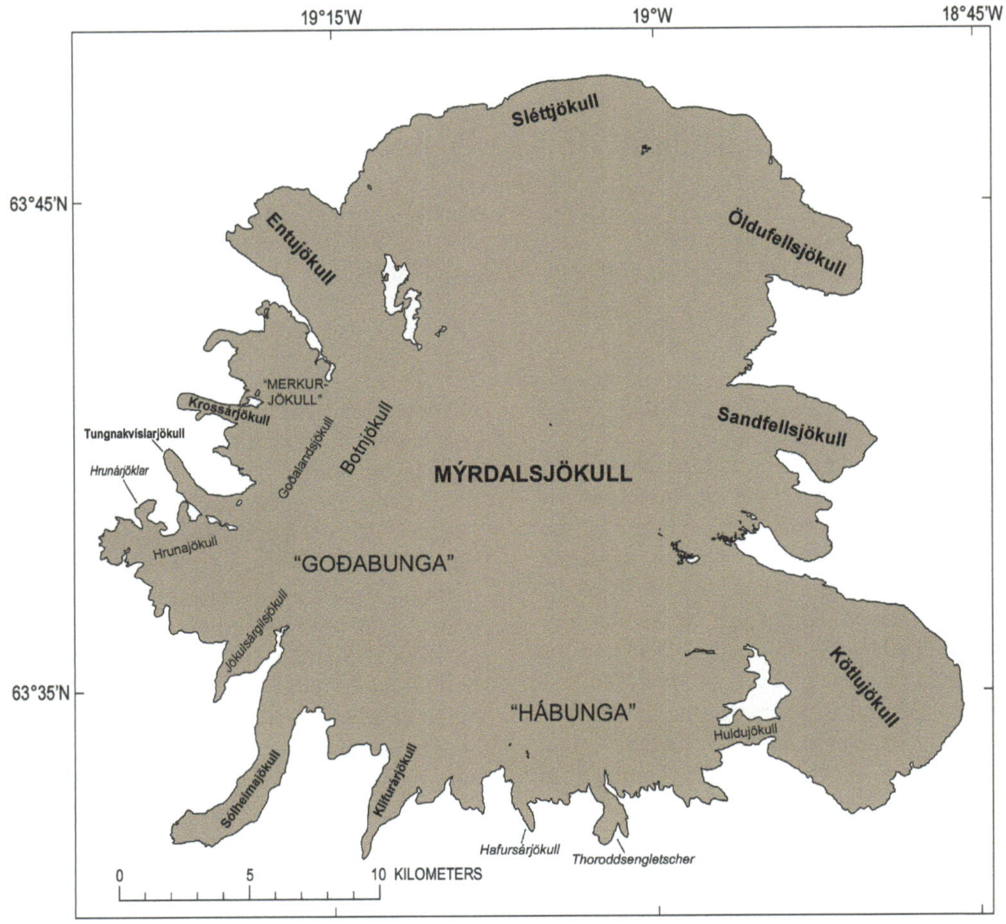

Fig. 5.1. Map of Mýrdalsjökull with the place names currently used for outlet glaciers and areas on the ice cap.

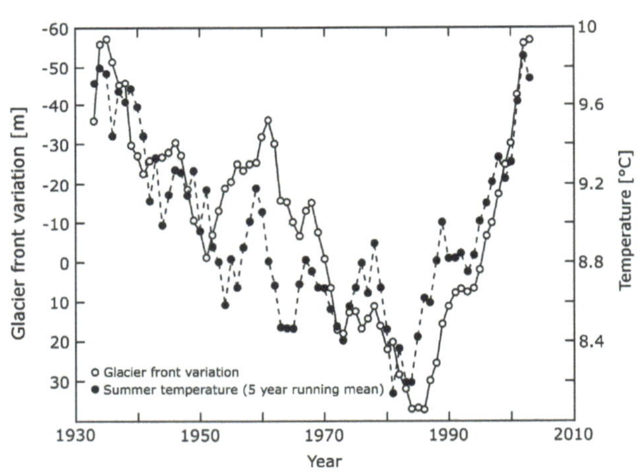

Fig. 5.2. Annual variations of the terminus of Sólheimajökull outlet glacier compared with mean summer temperature in Stykkishólmur, western Iceland.

(usually on the order of decades). The terminus of a non-surge-type glacier will react more or less directly to changes in mass which in turn depends primarily on mean summer temperature (Fig. 5.2). Two surge-type outlet glaciers have been identified in Mýrdalsjökull: Öldufellsjökull at the northeast margin and the westernmost part of Sléttjökull at the northwest margin (Fig. 5.1; Björnsson *et al.*, 2003). Krüger (1995) shows evidence of annual moraines at the eastern part of Sléttjökull, which indicates that it is not a surge-type glacier. The other outlet glaciers are considered to be non-surge-type until evidence appears to the contrary.

The climate-induced oscillations in mass balance of glaciers in Iceland are primarily based on two factors:

- air temperature during the summer
- precipitation during the winter

In Iceland, decadal variations of temperature are much greater than those of precipitation (Jóhannesson and Sigurðsson, 1998). Therefore, temperature is the most important climatic factor controlling the variations of glaciers.

In Iceland several non-climatic factors also affect the mass balance of glaciers. The most important of those are:

- volcanic activity
- geothermal activity
- debris on the glacier surface
- iceberg calving into water bodies.

All of these processes are present at Mýrdalsjökull, although iceberg calving occurs only occasionally. Volcanism, however, is extremely significant. The Katla volcano, located under Mýrdalsjökull, has erupted on average two times per century during historical time (Larsen, 2000). The eruptions have varied greatly in size and with various impacts on the glacier. Tómasson (1996)

Fig. 5.3. Water filled depression within the Katla caldera. Photo: Oddur Sigurðsson, August 29th, 1986.

estimated the total volume of meltwater discharged during the eruption of 1918 to be 8 km^3. The eruptions of 1660, 1721, and 1755 lasted much longer than the 1918 eruption (Larsen, 2000). According to descriptions (Steingrímsson, 1907–1915), at least some of the jökulhlaups during the 17th and 18th centuries were even larger than the one in 1918. This indicates that single eruptions in the Katla volcano may have melted between 5 and 10% of the total mass of the Mýrdalsjökull ice cap.

More or less permanent cauldrons in the surface of Mýrdalsjökull have been known for decades (Þórarinsson, 1957; Björnsson *et al.*, 2000; Scharrer *et al.*, 2008). These have occasionally deepened, corresponding with small- or medium-size jökulhlaups (Sigurðsson *et al.*, 2000; Sigurðsson and Einarsson, 2005). Usually the water in the jökulhlaups is contaminated with soluble compounds that have a geothermal origin, indicating that the most likely cause of melting is geothermal activity at the glacier bed. Cauldrons constitute approximately 1–2% of the glacier area and are at relatively high elevations (Figs. 5.3 and 5.4; Björnsson *et al.*, 2000). Geothermal energy melts the equivalent of the net mass increase within the catchment of the cauldrons. The annual, climatically induced net mass balance at this elevation may be expected to be on the order of 3.5–4.5 m water equivalent at 1350 m a.s.l. (Brandt *et al.*, 2005). This mass combined, which is annually melted by geothermal energy, is on the order of 20×10^6 m^3 w.eq. which requires energy of approximately 160 MW.

During Katla's explosive volcanic eruptions, a large amount of tephra is deposited on the glacier surface. In the accumulation area, the tephra is mostly covered by snow during the subsequent winter. Gradually, the tephra reappears in the ablation zone and, if in limited amounts, it is washed off into crevasses. In greater amounts, it stays on the surface until it is transported by the ice flow to the glacier terminus and incorporated into a morainic deposit or transported onto or across the outwash plain. Where the tephra layer on the surface reaches appreciable thickness it impedes melting, hampering the retreat of the terminus. This is particularly typical for the Kötlujökull outlet glacier (Fig. 5.5). Tephra would not be expected to have a hampering effect during the advance phases of glacier fronts.

Where outlet glaciers run between steep hillsides there will frequently be landslides or rockslides that are

Fig. 5.4. Fresh crevasses around cauldrons in the Katla caldera. Photo: Oddur Sigurðsson, August 4th, 1999.

deposited onto the glacier surface (Figs. 5.6 and 5.7; Sigurðsson and Williams, 1991). Such debris will impede melting and slow down retreat to some extent.

When Katla erupts, melting a substantial volume of glacier ice, it is the Kötlujökull outlet glacier that is most seriously impacted. According to maps (AMS, 1951), Kötlujökull had not recovered in 1946 from the eruption in 1918. The Kötlujökull outlet glacier started to advance at approximately 1970 at the same time as Sólheimajökull (Sigurðsson, 1996). This indicates that the glacier recovered from the substantial mass loss in 1918 in approximately 50 years or less.

5.4. Postglacial Oscillations of Mýrdalsjökull

5.4.1. Prehistoric Time

The Holocene had its warmest period in the northern hemisphere during the early millennia (IPCC, 2007). Temperatures during the earlier half of the Holocene were considerably higher than even the AD 20th century. During the culmination of the warm interval, glaciers in Iceland were much smaller than at the present time and at least one of the big ice caps, Langjökull, disappeared completely in the period approximately 7800–5600 years BP (Black *et al.*, 2006). The maximum postglacial glacier extent in Iceland occurred around 1890 (Sigurðsson, 2005). Mýrdalsjökull also reached its maximum postglacial size around this time (Einarsson, 1966; Krüger and Humlum, 1981). Therefore glaciers have, by their advances, obliterated a substantial portion of the evidence

Fig. 5.5. Katla caldera was identified in 1815 as a huge volcanic crater by Pálsson (2004). Kötlujökull is the main eastern outlet glacier of Mýrdalsjökull. Photo: Oddur Sigurðsson, November 10th, 1976.

Fig. 5.6. Rock slide on 'Jökulsárgilsjökull' on the south side of Mýrdalsjökull. Photo: Oddur Sigurðsson, November 10th, 1976.

Fig. 5.7. Rock slide on an outlet glacier by Mosakambur on the south side of Mýrdalsjökull. Tephra band visible at the glacier margin to the right but not on the outlet glacier. Photo: Oddur Sigurðsson, September 14th, 2007.

of fluctuations before 1890, and we have to rely on scarce stratigraphical and other geological information.

Dugmore and Sugden (1991) argued that the Sólheimajökull outlet glacier had a much greater extent during prehistoric and even early historic times than during the Little Ice Age. This is supported by the historic information that Jökulsá á Sólheimasandi changed course c. 1690 AD probably because the front of Sólheimajökull had a more advanced position before that (Magnússon, 1955). Remnants of subglacial till, identified by fabrics, striations, etc., 3.5–5.5 km in front of Kötlujökull have been interpreted as evidence of a large advance of the glacier 1610 years ago (Schomacker *et al.*, 2003). This might have been contemporaneous with an advance around 1200–1400 years BP discussed by Dugmore and Sugden (1991). An advance of this very large magnitude caused by changes in climate calls for a cooling of a few degree Celsius and/or a huge increase in precipitation of which there is no other evidence in Iceland, not even during the Little Ice Age. Another possibility of obtaining an advance of 3.5–5.5 km would be to increase greatly the catchment area at high elevations. The advanced position of Sólheimajökull at this time already occupies a substantial part of this catchment area. This implied advance of Kötlujökull does not seem to have a counterpart on the northern side of Mýrdalsjökull, however, which rules out climatic reasons. Mixtures of till-like material and soil, such as in thez Höfðabrekkujökull jökulhlaup deposit by the airstrip east of Vík, can be deposited in big slabs with undisturbed stratigraphy far away from the glacier by large jökulhlaups (Jónsson, 1995). Such features, which are common in the Mýrdalssandur area, seem to be a more plausible explanation for the deposits described in Schomacker *et al.* (2003).

5.4.2. Historical Time

In Iceland, historical time spans the past 11 centuries. The oldest manuscripts do mention glaciers and glacial rivers (which also appear in place-names like *Jökulsá* which means glacier river). This suggests that glaciers were more or less at the same locations as they are at present time (Thoroddsen, 1911, Thorarinsson, 1943; Pálsson, 2004). The historic record can offer surprising clues to changes in glaciers and glacial rivers. For example, the peculiar location of the boundary between Rangárvallasýsla and Vestur-Skaftafellssýsla counties indicates a much more advanced Sólheimajökull than at present. This particular boundary was actually determined at the corrections of the quadrants (landsfjórðungaskipting) about the AD 965. The border was originally placed at the main river, Jökulsá á Sólheimasandi (formerly called Fúlilækur), which also determined the boundary between the farms Skógar and Sólheimar according to the book of settlement, Landnámabók (1986). Great floods in the AD 1179 (Sturlunga saga, 1988) indicate that a major jökulhlaup occurred and therefore the existence of a substantial Mýrdalsjökull. The oldest mention of Mýrdalsjökull is in the *Njáls saga* (there named Eyjafjallajökull), *Árna saga biskups* and *Konungsannáll* (there named Sólheimajökull) all of which were written in the late 13th century (Sigurðsson and Williams, 2008). Little is known about the variations of the glacier during the first centuries after the settlement.

A sketch map of Sólheimajökull was drawn – probably in the year 1705 – by Árni Magnússon (1955) or more likely by his secretary Magnús Einarsson (Thorarinsson, 1939) accompanied by a description indicating that the river moved its course to its present position around the year 1690 (Fig. 5.8). This implies a retreat of the outlet glacier which formerly forced the river to run in the gully where the county boundary remained into the 20th century. In the same source (Magnússon, 1955) there is a statement that, according to the local people, Sólheimajökull was advancing at the turn of the 17th century as was the case with the other outlet glaciers of southern Mýrdalsjökull. At the same time farmers in the neighboring county complained

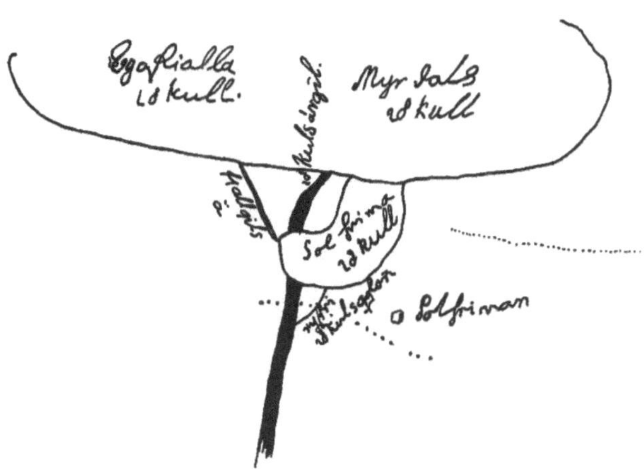

Fig. 5.8. Sketch map of the southern margin of Mýrdalsjökull drawn in 1705 by Einarsson (Magnússon, 1955).

that Eyjafjallajökull was advancing over grazing areas (Magnússon and Vídalín, 1980).

The more advanced position of Sólheimajökull in early historic time than at present is unusual among Icelandic glaciers. Dugmore and Sugden (1991) not only emphasize a climatically induced migration of the ice divide but also mention that this might be caused by volcanic or tectonic changes at the glacier bed. The present author finds the latter more plausible because of the eruptive history of the Katla volcano (Larsen, 2000). Additionally, most postglacial terminal moraines in Iceland that extend farther out than the Little Ice Age maximum are at active volcanoes as seen at Kvíárjökull, Svínafellsjökull, Gígjökull, and Sólheimajökull. No appreciable variations in glacier extent in recent times have been shown to be caused by climate-forced migration of an ice divide.

From the 18th century and onwards, records of the variations of Mýrdalsjökull variations are more complete. The sketch map of Sólheimajökull shows the terminus at the gully Fjallgil (Fig. 5.8). In 1783, the snout of Sólheimajökull did not quite reach to the mouth of the Jökulsárgil gully which indicates a retreat of approximately 2 km. For clarification it is important to note that when Sólheimajökull advances beyond Jökulsárgil, access to the grazing area Hvítmaga is seriously hampered. Then the farmers of Sólheimar have to drive their sheep across the glacier to and from summer grazing. The limited extent of Sólheimajökull in 1783 could be explained by the huge loss of ice mass due to the Katla eruptions of 1721 and 1755. In 1893 the terminus had advanced beyond the Jökulsárgil again, but some sources suggest that it was even more advanced in 1860, completely covering the hill Jökulhöfuð (Jökulhaus) (Thoroddsen, 1911). This was refuted by Eyþórsson (1959) who said that a part of the hill must have protruded out of the glacier surface. The terminus of Sólheimajökull retreated in the years following 1860 (Thorarinsson, 1939) probably as a result of the eruption of 1860. Outlet glaciers on the northwest side of Mýrdalsjökull advanced during the period 1855–1885 all the way to Bláfjöll and Smáfjöll. The valley by Fremri-Emstruá, which formerly was a grazing area for cattle, was completely covered by glacier (Vigfússon, 1892). Thoroddsen (1911) visited Mýrdalsjökull in 1893. He observed that in most places the margin was retreating, except for the advancing Kötlujökull. The margin of Mýrdalsjökull and Eyjafjallajökull are not assumed to have merged on Fimmvörðuháls in postglacial time (Sigurðsson, 2004).

The southern margin of Mýrdalsjökull was surveyed by the Danish General Staff in 1904 (1905). This was just after the maximum extent of glaciers in Iceland (Sigurðsson, 2005). The entire glacier was mapped from vertical aerial photographs, taken in 1945 and 1946 by the Army Map Service (Fig. 5.9; AMS, 1951).

Jón Eyþórsson (1931, 1963) started monitoring the front variations of Sólheimajökull in 1930. Since then the measurements have been carried out every fall except in 1942, 1946, and 1957 (Sigurðsson, 1998). During this time the front of the glacier has reacted more or less annually to the mean summer temperature (Fig. 5.2; Sigurðsson, 2006; Sigurðsson *et al.*, 2007). This, as well as the observations of Krüger (1995), implies that the reaction of the termini of

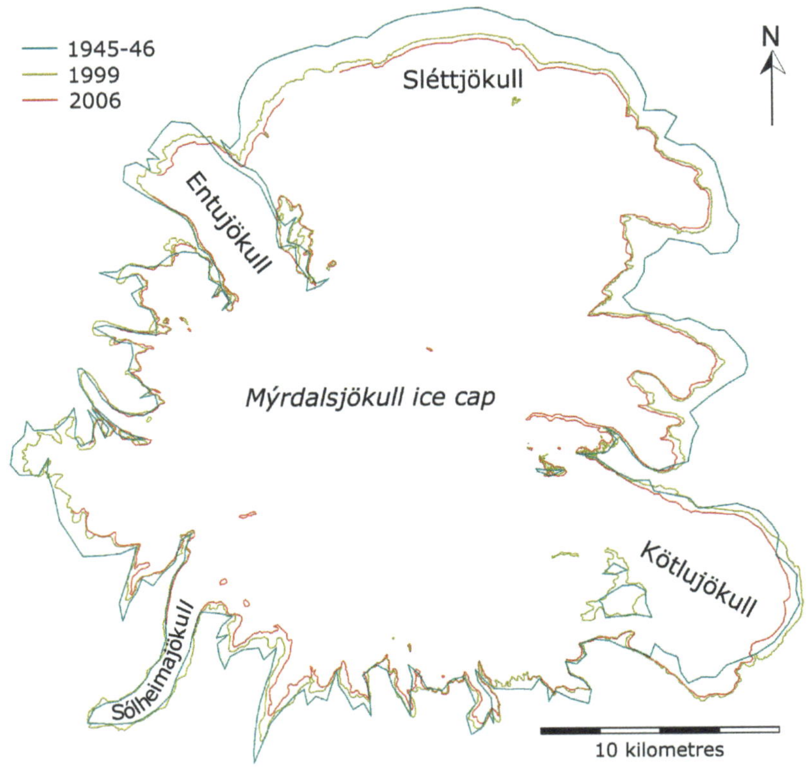

Fig. 5.9. The outlines of Mýrdalsjökull in 1945–1946 (blue) traced from AMS (1951), in 1999 (olive green) traced form vertical aerial orthophotographs, and in 2006 (red) traced from SPOT-5 satellite image. It is noteworthy that the 1945–1946 line is inside the 1999 line on Kötlujökull and Sólheimajökull indicating the effects of the 1918 Katla eruption.

the outlet glaciers from Mýrdalsjökull express climate changes promptly, except in the case of Öldufellsjökull and the western part of Sléttjökull which are, as previously noted, identified as surge-type glaciers. However, this does not apply to a period of a few decades following an eruption in Katla when the mass loss caused by volcanic activity overshadows the climatically forced mass balance.

During the first decades of the 20th century Mýrdalsjökull retreated slowly after the Little Ice Age maximum. During an abrupt period of climatic warming after 1925, the rate of retreat was quite high but slowed down during the 1950s and 1960s. The snout of Sólheimajökull retreated by 1 km between 1930 and 1969. During the period 1970–1995 glaciers advanced again. In the case of Sólheimajökull, the advance was 0.5 km. From 1995–2007, Sólheimajökull retreated by 0.5 km. This is the fastest retreat of this glacier since monitoring started in 1930 (Figs. 5.10 and 5.11). Krüger (1995) found that the eastern part of Sléttjökull, after a still stand during 1906–1928, retreated by 22 ma^{-1} during 1928–1937, by 34 ma^{-1} during 1937–1945, by 23 ma^{-1} during 1945–1960, by 19 ma^{-1} during 1960–1977, and by 8 ma^{-1} during 1977–1982. From 1982 to 1989 the front was stationary or slightly advancing. This is in general the same pattern as is seen at Sólheimajökull except that the advance during the period 1970–1989 was almost negligible at Sléttjökull.

5.5. Debris Known as 'Jökull'

The biggest jökulhlaups, caused by volcanic eruptions under Mýrdalsjökull, are enormous floods by any standard. The maximum discharge is on the order of several hundred–thousand $m^3 s^{-1}$ (Tómasson, 1996). This is probably more of a debris flood than water flood with a density greatly exceeding that of water. In some cases huge quantities of ice and other debris have been deposited on the Mýrdalssandur outwash plains. A contemporaneous source (Stefánsson, 1907–1915) describes heaps of rubble between the Háfell mountain and Hjörleifshöfði deposited by the Kötluhlaup of 1721 as being at least a couple of 100 m thick. More debris accumulated here during the next Kötluhlaup in 1755. In 1815, Sveinn Pálsson (2004) noted that only half of the ice deposited in 1755 at this site had melted. Everest and Bradwell (2003) found that still some blocks of this ice are surviving in the debris. Such jökulhlaup debris traditionally gets the name of 'jökull' ('glacier' in Icelandic). They may be deposited as far as 15 km away from the margin of the glacier. On Mýrdalssandur at least five such place-names are known: Höfðabrekkujökull, Austurjökull, Lambajökull, Lágjökull, and Jökull by Hafursey (Sigurðsson and Williams, 2008). Many similar deposits, historic, and prehistoric, are surely to be found in the vicinity of Mýrdalsjökull.

5.6. Tephra Horizons and Turnover Time of Ice Mass

Volcanic tephra layers are visible on the surface of Mýrdalsjökull as is the case on most other glaciers in Iceland (Larsen *et al.*, 1998). The most conspicuous tephra band can be traced on satellite images more or less all around the ice cap. Tephra layers are also to be found internally in the glacier. This has been used to estimate the specific mass-balance rate to 4.5 and 3.5 ma^{-1} at 1350 m a.s.l. on the southern and northern slopes, respectively, for the period 1918–1991 (Brandt *et al.*, 2005). On the south side the tephra layer is much closer to the margin than in the north. Above this line the glacier surface is more or less clean. Therefore, the tephra is

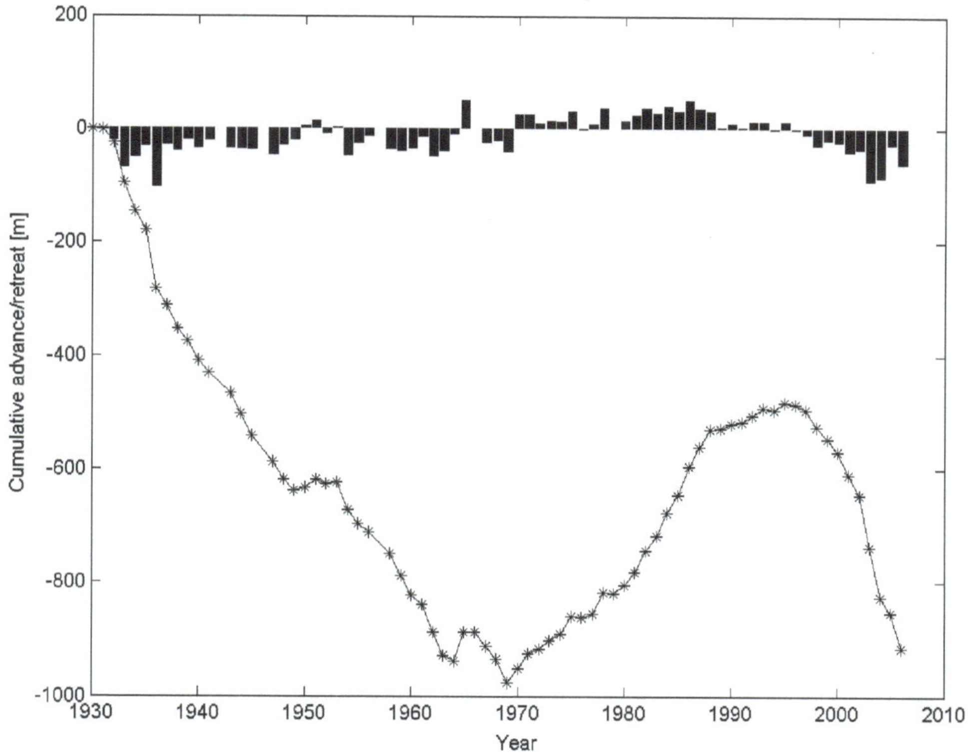

Fig. 5.10. Cumulative front variations of Sólheimajökull outlet glacier 1930–2006. From the database of the Iceland Glaciological Society.

Fig. 5.11. Vertical aerial photograph of Sólheimajökull. Two distinct tephra horizons obliquely cross the outlet glacier close to the terminus. Photo: USAF, August 29th, 1945.

Fig. 5.12. Oblique aerial photograph of the northern margin of Mýrdalsjökull, Sléttjökull. Note the three most obvious tephra horizons, the uppermost of 1918, the middle one presumably of 1860 and the lowermost one of 1823. Photo: Oddur Sigurðsson, August 19th, 2000.

Fig. 5.13. Oblique aerial photograph of Entujökull which now fills a valley earlier used as grazing area for cattle. Photo: Oddur Sigurðsson, August 4th, 1999.

interpreted as deposited during the Katla eruption of 1918. The tephra band is absent on some of the outlet glaciers on the south side of the ice cap (Fig. 5.7). Inside the northern margin of Mýrdalsjökull there are two more conspicuous, parallel bands of tephra that are most likely from the eruptions of 1823 and 1860, whereas the tephra of 1755 is not to be seen (Fig. 5.12). These bands are not easily identified elsewhere along the ice cap margin because they have merged with other surface debris or they have melted out of the glacier (Fig. 5.13). However, tephra bands presumably from 1823 to 1860, which have since been flushed out of the system, are identifiable on aerial photographs from 1945 close to the terminus of Sólheimajökull (Fig. 5.11), Entujökull and other places on the southwestern side of the ice cap. During the first decade of the 21st century, the tephra band of 1918 eruption is very close to the location of the presumed 1860 tephra band in 1945. All tephra bands run obliquely across Sólheimajökull indicating a higher velocity at the eastern margin of the outlet glacier. Outlet glaciers on the south side of Hábunga such as Hafursárjökull now have no such bands and, therefore, have a turnover time less than 90 years. For Sólheimajökull, the turnover time seems to be very close to 150 years. This indicates that the mass turnover time on the southern side of Mýrdalsjökull ranges from 80 to 150 years, but approximately 2 centuries at the northern side. This difference is probably because of greater precipitation on the south side. The mass turnover time is primarily a function of the size (length) of the glacier and velocity, which in turn is partly dependent on the amount of precipitation. The main outlet glaciers flow much faster than the ice between. Therefore, tephra bands are more well-preserved at the margin between outlet glaciers.

5.7. Conclusions

Most of the outlet glaciers of Mýrdalsjökull react promptly to changes in mass. It takes between three and five decades for the glacier to recover from volcanically induced mass loss. During the warmest periods of the Holocene, approximately 7800–5600 years BP, the ice cap may have more or less disappeared. During the settlement period (about AD 900) the ice cap had a considerably diminished extent compared to the present time, except for Sólheimajökull and possibly Kötlujökull. During the Little Ice Age, the glacier increased more or less continuously and reached its maximum volume around the year 1890. The 20th century was a time of general mass loss except for the period 1970–1995. During the first decade of the 21st century, Mýrdalsjökull experienced its fastest mass loss since the start of front-variation monitoring in 1930. The volume of the ice cap is replenished within approximately 80–200 years depending on location.

References

AMS, 1951. Höfdhabrekkujökull. US Army Map Service, Series C762, Scale 1:50 000, Sheet 5817 IV, based on vertical aerial photographs (stereoscopic overlap) from 1946.

Björnsson, H., 1979. Glaciers in Iceland. Jökull 29, 74–80.

Björnsson, H., Pálsson, F., & Guðmundsson, M.T., 2000. Surface and bedrock topography of the Mýrdalsjökull ice cap, Iceland: The Katla caldera, eruption sites and routes of jökulhlaups. Jökull 49, 29–46.

Björnsson, H., Pálsson, P., Sigurðsson, O., & Flowers, G.E., 2003. Surges of glaciers in Iceland. Annals of Glaciology 36, 82–90.

Black, J.L., Miller, G.H., & Geirsdóttir, Á., 2006. Diatoms as proxies for a fluctuating Holocene ice cap margin in Hvítárvatn, Iceland. 36th Annual International Arctic Workshop, March 16–18. Program and Abstracts. INSTAAR, University of Colorado at Boulder, p. 36.

Brandt, O., Björnsson, H., & Gjessing, Y., 2005. Mass-balance rates derived by internal tephra layers in Mýrdalsjökull and Vatnajökull ice caps, Iceland. Annals of Glaciology 42, 284–290.

Danish General Staff, 1905. Map No. 69 NV Scale 1:50 000. Surveyed in 1904.

Dugmore, A.J. & Sugden, D.E., 1991. Do the anomalous fluctuations of Sólheimajökull reflect ice-divide migration? Boreas 20, 105–113.

Einarsson, E.H., 1966. Suðurbrún Mýrdalsjökuls við Gvendarfell. Breytingar síðustu 100 ár o.fl. Jökull 16, 216–218.

Elíasson, J., Larsen, G., Guðmundsson, M.T., & Sigmundsson, F., 2006. Probabilistic model for eruptions and associated flood events in the Katla caldera, Iceland. Computational Geosciences 10, 175–200.

Everest, J. & Bradwell, T., 2003. Buried glacier ice in southern Iceland and its wider significance. Geomorphology 52, 347–358.

Eyþórsson, J., 1931. On the present position of the glaciers in Iceland. Some preliminary studies and investigations in the summer 1930. Vísindafélag Íslendinga X. Reykjavík. 35pp.

Eyþórsson, J., 1959. Skýringar og athugasemdir við nokkur atriði merkt * í bókinni. In: Ferðabók III by Þorvaldur Thoroddsen, Snæbjörn Jónsson & co. h.f., Reykjavík, 363–367.

Eyþórsson, J., 1963. Variation of Iceland glaciers 1931–1960. Jökull 13, 31–33.

Guðmundsson, M.T. & Gylfason, Á.G., 2005. Hættumat vegna eldgosa og hlaupa frá vestanverðum Mýrdalsjökli og Eyjafjallajökli. Ríkislögreglustjórinn, Reykjavík, Háskólaútgáfan, 210pp.

Gunnlaugsson, G.Á., Guðbergsson, G.M., Þórarinsson, S., & Einarsson, Þ., 1984. Skaftáreldar 1783–1784. Mál og menning, Reykjavík, Ritgerðir og heimildir, 442pp.

IPCC, 2007. Climate Change 2007. The physical science basis. In: Solomon, S., Qin, D., Manning, M., Marquis, M., Averyt, K., Tignor, M.M.B., Miller, H. LeRoy Jr., Chen, Z. (Eds), Cambridge, Cambridge University Press, 996pp.

Jóhannesson, T. & Sigurðsson, O., 1998. Interpretations of glacier variations in Iceland 1930–1995. Jökull 45, 27–33.

Jónsson, J., 1995. Höfðabrekkujökull. Náttúrufræðingurinn 65, 103–106.

Krüger, J. & Humlum, O., 1981. The proglacial area of Mýrdalsjökull with particular reference to Sléttjökull and Höfðabrekkujökull. General report on the Danish Geomorphological Expedition to Iceland, 1977. Folia Geographica Danica 15, 1. Copenhagen, 58pp.

Krüger, J., 1995. Origin, chronology and climatological significance of annual-moraine ridges at Mýrdalsjökull, Iceland. Holocene 5, 420–427.

Landnámabók., 1986. (originally written in the 12th century). Íslendingabók. Landnámabók. In: Benediktsson, J. (Ed), Hið íslenska fornritafélag, Reykjavík.

Larsen, G., 2000. Holocene eruptions within the Katla volcanic system, south Iceland: characteristics and environmental impact. Jökull 49, 1–27.

Larsen, G., Guðmundsson, M.T., & Björnsson, H., 1998. Eight centuries of periodic volcanism at the center of the Iceland hotspot revealed by glacier tephrostratigraphy. Geology 26, 943–946.

Magnússon, Á., 1955. (written in 1702–1712). Chorographica Islandica. In: Lárusson, Ó. (Ed), Safn til sögu Íslands og íslenskra bókmennta, annar flokkur I, 2. (Written in 1702–1714.). Reykjavík, Hið íslenska bókmenntafélag.

Magnússon, Á. & Vídalín, P., 1980. (written in 1702–1712). Jarðabók Árna Magnússonar og Páls Vídalín. (Farm account by Árni Magnússon and Páll Vídalín (information gathered in 1702–1714)), vols. 1–13, 2nd ed., Hið íslenska fræðafélag, Reykjavík.

Pálsson, S., 2004. Icelandic Ice Mountains. Draft of a physical, geographical and historical description of Icelandic ice mountains on the basis of a journey to the most prominent of them in 1792–1794 with four maps and eight perspective drawings. An annotated and illustrated English translation. In: Williams, R.S. Jr., Sigurðsson, O. (Eds), Hið íslenska bókmenntafélag, Reykjavík XXXVI, 183pp.

Þórarinsson, S., 1957. The jökulhlaup from the Katla area in 1955 compared with other jökulhlaups in Iceland. Jökull 7, 21–25.

Scharrer, K., Spieler, O., Mayer, Ch., & Münzer, U., 2008. Imprints of sub-glacial volcanic activity on a glacier surface-SAR study of Katla vocano, Iceland. Bulletin of Volcanology 70, 495–506.

Schomacker, A., Krüger, J., & Larsen, G., 2003. An extensive late Holocene glacier advance of Kötlujökull, central south Iceland. Quaternary Science Reviews 22, 1427–1434.

Sigurðsson, O., 1996. Nýir mælistaðir. Jöklarannsóknafélags Íslands. Fréttabréf 56, 7–8.

Sigurðsson, O., 1998. Glacier variations in Iceland 1930–1995- From the database of the Iceland Glaciological Society. Jökull 45, 3–25.

Sigurðsson, O., 2004. Gláma. Að vera eða vera ekki-jökull (Gláma. To be or not to be-a glacier). Náttúrufræðingurinn 72, 47–61.

Sigurðsson, O., 2005. Variations of termini of glaciers in Iceland in recent centuries and their connection with climate. In: Caseldine, C., Russell, A., Harðardóttir, J., Knudsen, Ó. (Eds), Iceland – Modern processes and past environments. Developments in Quaternary Science 5, Amsterdam, Elsevier, 241–255.

Sigurðsson, O., 2006. Jöklabreytingar 1930–1970, 1970–1995, 1995–2004 og 2004–2005. Jökull 56, 85–93.

Sigurðsson, O. & Williams, R.S., Jr., 1991. Rockslides on the terminus of ''Jökulsárgilsjökull'', Southern Iceland. Geografiska Annaler 73A, 129–140.

Sigurðsson, O., Zóphóníasson, S., & Ísleifsson, E., 2000. Jökulhlaup úr Sólheimajökli 18. júlí 1999 (The Jökulhlaup from Sólheimajökull on July 18, 1999). Jökull 49, 75–80.

Sigurðsson, O., & Einarsson, B., 2005. Jökulhlaupaannáll 1988–2004. Orkustofnun, OS-2005/031, Reykjavík, 46pp.

Sigurðsson, O., & Williams, R.S., Jr., 2008. Geographic Names of Iceland's Glaciers: Historic and Modern. United States Geological Survey Professional Paper 1746, 225pp.

Sigurðsson, O., Jónsson, T., & Jóhannesson, T., 2007. Relation between glacier-termini variations and summer temperature in Iceland since 1930. Annals of Glaciology 46, 170–176.

Stefánsson, S., 1907–1915. (written in the 1720s), IV. Skýrslur um Kötluhlaupið 1721. Viðbætir Sigurðar Stefánssonar sýslumanns um hlaupin 1660 og 1721. In: Thoroddsen, Þ. (Ed), Safn til sögu Íslands og íslenskra bókmennta að fornu og nýju, IV, I, 230–233. Kaupmannahöfn, Hið íslenska bókmenntafélag.

Steingrímsson, J., 1907–1915. (written in 1788), I. Um Kötlugjá. In: Thoroddsen, Þ. (Ed), Safn til sögu Íslands og íslenskra bókmennta að fornu og nýju, IV, I, 190–199. Kaupmannahöfn, Hið íslenska bókmenntafélag.

Sturlunga saga., 1988. (written in the 13th century) vols. I, II, III, Thórsson, Ö. (Ed), Svart á hvítu, Reykjavík, 937pp.

Thorarinsson, S., 1939. Vatnajökull. The scientific results of the Swedish-Icelandic investigations 1936-37-38. Chapter IX. The ice dammed lakes of Iceland with particular reference to their values as indicators of glacier oscillations. Geografiska Annaler 21, 216–242.

Thorarinsson, S., 1943. Vatnajökull. The scientific results of the Swedish-Icelandic investigations 1936-37-38. Chapter XI. Oscillations of the Iceland glaciers in the last 250 years. Geografiska Annaler 25, 1–56.

Thoroddsen, Þ., 1911. Lýsing Íslands. Jöklar. (Description of Iceland. Glaciers.) Vol. II, 1–68. Reykjavík, Sjóður Þorvaldar Thoroddsen.

Tómasson, H., 1996. The jökulhlaup from Katla in 1918. Annals of Glaciology 22, 249–254.

Vigfússon, S., 1892. Á Fjallabaksvegi. Árbók hins íslenzka fornleifafélags 1888–1892, 69–71. Reykjavík.

6

Ice-Marginal Environments: Geomorphic and Structural Genesis of Marginal Moraines at Mýrdalsjökull

Johannes Krüger[1,*], Anders Schomacker[2] and Ívar Örn Benediktsson[2]

[1]*Department of Geography and Geology, University of Copenhagen, Geocenter Copenhagen, Øster Voldgade 10, DK-1350 Copenhagen K., Denmark*
[2]*Institute of Earth Sciences, University of Iceland, Askja, Sturlugata 7, IS-101 Reykjavík, Iceland*
**Correspondence and requests for materials should be addressed to Johannes Krüger (e-mail: jk@geo.ku.dk)*

6.1. Introduction

Ice-marginal moraines at contemporary glacier margins have increasingly been investigated to serve as analogues to Pleistocene ice-marginal moraines (Kälin, 1971; Boulton, 1972, 1986; Heim, 1984; Humlum, 1985; Krüger, 1985, 1993, 1994, 1996; Croot, 1987, 1988; Eybergen, 1987; Evans, 1989; Shakesby, 1989; Lehmann, 1992; Hambrey and Huddart, 1995; Matthews *et al.*, 1995; Möller, 1995; Bennett *et al.*, 1996; Boulton *et al.*, 1996, 1999; Huddart and Hambrey, 1996; Lønne and Lauritsen, 1996; Hart and Watts, 1997; Winkler and Nesje, 1999; Lyså and Lønne, 2001; Krüger *et al.*, 2002; Bennett *et al.*, 2004; Benediktsson *et al.*, 2008; Roberts *et al.*, 2009). Four basic types of ice-marginal moraine ridge forming processes acting separately or in combination are recognized: (1) squeezing; (2) dumping; (3) pushing; and (4) glaciotectonic deformation. Squeezing out of material from beneath the ice margin requires only static loading of water-saturated material by the ice mass; dumping of supraglacial material requires a steep debris-charged glacier margin; pushing involves a forward movement by the glacier bulldozing the proglacial material from behind. Rapid glaciofluvial sedimentation against the ice margin and burial of the glacier snout so that the glacier pushes against a sediment face facilitates the formation of large push-moraines (Humlum, 1985; Boulton, 1986; Krüger, 1985, 1994). The mechanics of glaciotectonic deformation refer to replacement of proglacial or submarginal materials by excavating and elevating the material due to stresses imposed by static loading combined with glacier advance and involve ductile or brittle deformation or a combination of the two (Benn and Evans, 1998).

The landforms created by these processes can be described under four headings, although it should be noted that in nature some are not distinct, but composite. Squeeze moraines are minor ridges, rarely more than 1 m high, which are commonly subject to reworking by ice push (Price, 1970; Worsley, 1974; Benn, 1995; Benn and Evans, 1998). Dump moraines will form where the ice margin remains stationary during debris accumulation, and the moraine size is therefore related to the volume of supraglacial debris and the length of the stillstand period (Boulton and Eyles, 1979; Benn and Evans, 1998). Push moraines are small moraine ridges, usually less than 10 m in height, produced by minor glacier advances (Krüger, 1985; Bennett, 2001). Two basic types of glaciotectonic end-moraine ridges are recognized (Aber *et al.*, 1989; Aber and Ber, 2007): (1) hill–hole pairs and (2) composite ridges and thrust-block moraines. A hill–hole pair represents a basic combination of large-scale ice-scooped topographic depression and ice-shoved hill. The hill complex consists of a series of sub-parallel ridges. The topographic depression, which is of approximately the same area as the hill complex, is located on the upglacier side; it represents the source of material now in the hill (Aber and Ber, 2007). Composite ridges are ice-shoved ridges composed of great slices of up-thrust and commonly contorted sediments (Prest, 1983). Large composite ridges usually involve considerable disruption of pre-Quaternary bedrock, which may comprise a major volume of the ridges, whereas small composite ridges may or may not include consolidated bedrock; in fact, many are composed mostly of unconsolidated Quaternary strata (Aber and Ber, 2007). Composite ridges are usually associated with ice margins marking glacier stillstands or readvances (Aber and Ber, 2007).

Comprehensive reviews of the morphology, structural evolution and significance of ice-marginal moraine ridges are given by van der Meer (1987), Bennett (2001) and Aber and Ber (2007). Here, we review processes and implications of ice-marginal moraine formation along the margins of Kötlujökull and Sléttjökull, two major outlets from the Mýrdalsjökull ice cap (Fig. 6.1). Additionally, a short overview of neoglacial marginal moraines in front of Mýrdalsjökull is given.

6.2. Ice-marginal moraine formation at Kötlujökull

From the end of the 1970s – after a period of glacier retreat and dead-ice formation (Krüger *et al.*, this volume) – and throughout the 1980s the debris-charged NE front of the principal outlet glacier Kötlujökull on the east flank of the Mýrdalsjökull ice cap advanced across a

DEVELOPMENTS IN QUATERNARY SCIENCES
VOLUME 13 ISSN 1571-0866

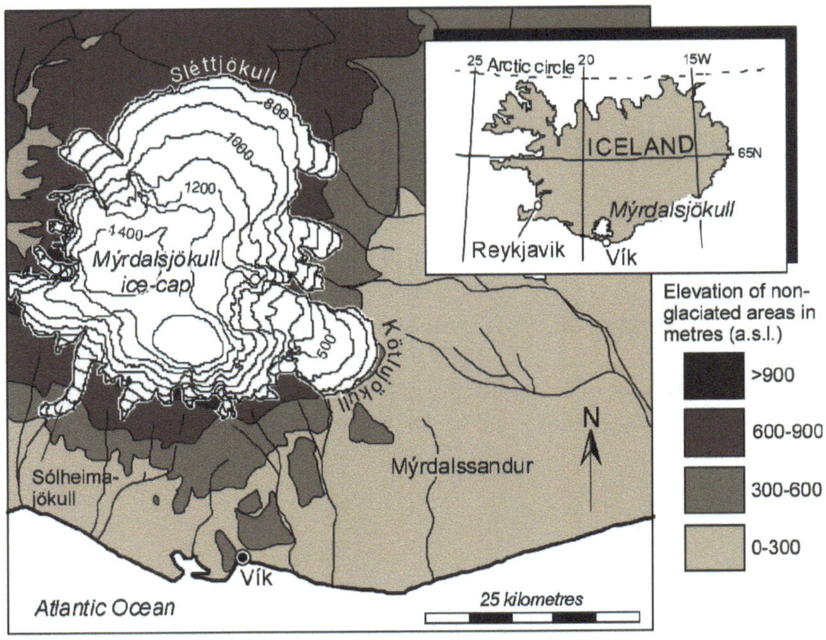

Fig. 6.1. Location map showing the Mýrdalsjökull ice cap with the prominent outlets Kötlujökull, Slettjökull and Sólheimajökull.

forefield of outwash fans, ice-cored moraine and ground moraine (Fig. 6.2).

The recorded frontal advance is shown in Fig. 6.3. From 1979 to 1982, the ice front advanced slowly by a mean of around 10 m/yr. During the advance, the debris-covered glacier surface appeared as a chaotic and karst-like ice-disintegration field. During the period between 1982 and 1984, the glacier advance accelerated and involved an average rate of annual frontal advance of 32 m. A mean annual ice-surface velocity of 40 m was recorded close to the ice front. During the following years (1984–1986), the rate of frontal advance decreased to 20 m/yr, but the mean ice-surface velocity was almost unchanged. Under such conditions, the ice creep could not adjust rapidly enough to the stresses within the marginal ice, so the ice fractured and moved along thrust-planes, and therefore, the glacier margin appeared as a frontal ice cliff, or even as an overhanging wall (Fig. 6.2B).

Between 1986 and 1989, the rate of frontal advance was only 4 m/yr. In 1989 the steepness of the frontal slope had reduced somewhat, suggesting that the glacier advance had stopped, probably around 1987. Within the study field the glacier had advanced to a position 80–120 m behind its 1955 position. Since 1989, the terminal 300–500 m of the debris-charged glacier has developed into a zone of dead-ice (Fig. 6.2C,D) (Krüger *et al.*, this volume). During the glacier advance, three basic types of processes were acting: (1) development of frontal ice-contact fans, (2) mass movement on the steep frontal ice slope and (3) ridge-shaped ice-marginal moraine development by pushing and thrusting. In the following these processes are reviewed in turn.

6.2.1. Frontal ice contact fans

As the glacier advanced, a series of minor ice-contact fans, termed *hochsander* fans (Gripp, 1975; Krüger, 1997), was produced in front of it (Fig. 6.4). In contrast to the large-scale fans fed by high-energy streams from underneath the glacier and therefore consisting of coarse-grained sediments of pebble to boulder size in their proximal part (Boothroyd and Ashley, 1975; Miall, 1977, 1983; Maizels, 1995), hochsander fans result from deposition of chiefly fine-grained material transported by low-energy streams from the supraglacial source area to the glacier forefield (Heim, 1983; Krüger, 1994, 1997; Kjær *et al.*, 2004).

Because of their low transporting capacity, the streams leave the coarser components of the debris mass on the glacier. Accordingly, the vertical alteration in texture of the fan sediments reflects changing influx of sediments, which in turn is coupled to processes acting in the sediment source area such as ablation, influx of rainwater, sudden drainage of pockets of water and re-sedimentation of debris by fall, slide and flow because of heavy rainfalls (Krüger, 1997; Krüger *et al.*, this volume). Between 1982 and 1989, the surface altitude of hochsander fans under formation had increased 2.5–3 m. During the same period, the proximal 100–125 m of the fans was pushed and successively overridden by the advancing glacier (Figs. 6.4 and 6.5). Hochsander fans are characteristic of advancing, or stationary, glaciers where the frontal slope is steep and covered by debris. Retreat phases curtail the production of hochsander fans, because the terminal ice stagnates and decreases in altitude, so that the supraglacial streams chiefly deposit their sediment load in broad depressions in the stagnant-ice field and not beyond the glacier (Krüger, 1994, 1997).

6.2.2. Mass movement on the frontal ice slope

Scree slopes that masked the glacier front to a considerable height were widely occurring along the advancing glacier margin throughout the 1980s. Material that flowed, rolled and fell down the ice front was released from series of dirt bands and from debris-laded thrust-planes cropping out on the free ice faces. The scree slopes were also fed by means of dumping of large quantities of

Fig. 6.2. (A) The debris-covered stagnant margin of Kötlujökull in 1977. In the foreground an abandoned boulder-paved meltwater gap cuts through a terminal moraine ridge from a glacier advance around 1955. (B) The same locality in 1986. In the background, the steep overhanging front of the advancing Kötlujökull is seen. The glacier has advanced about 160 m and a pushed end-moraine is under formation. In the foreground outwash fan sediments, 2 m thick, have been deposited across the boulder lag seen in Fig. 6.2(A). (C) The same locality in 2003. In the background, behind the new end-moraine, the debris-covered stagnant margin of Kötlujökull is seen. (D) The same locality in 2008.

materials continuously released from debris-laded thrust-planes along the bottom surface of overhanging ice plates, or it was delivered from the sediment cover above the free ice faces due to backwasting and undermining (Fig. 6.6). Channels incised into sediment and ice led sediment flows from source area to the base of the glacier front, where the material was deposited in single or multiple tongues.

6.2.3. Moraine-ridge formation

The most prominent ice-marginal ridges under formation were seen where the advancing glacier pushed against a rather thick body of hochsander fan deposits or a frontal ramp of mass-movement deposits. Where the glacier overrode ground moraine, the ice-marginal ridge was commonly small, probably because the glacier toe, when overriding ground moraine, was almost at the same level as the till surface beyond. The pushed frontal ridge commonly had a height of 4 m above the terrain beyond, and behind this, an ice-contact scree up to 15 m high had formed. At the very back, the glacier margin rose in the shape of a 2–15 m high cliff of ice (Fig. 6.7).

An idealized sequential model for frontal moraine development at Kötlujökull is shown in Fig. 6.8. (A) In 1982, when the glacier advanced relatively fast, the glacier toe pushed into a face of outwash sediments. The fan surface was ruptured along thrust-planes that inclined upglacier, and slabs of unfrozen or frozen sediments, depending on the time of the year, were raised in front of the glacier. From source areas on and above the frontal ice cliff material was delivered by dumping, and a frontal scree with avalanche tracks and sediment flow tongues developed. (B) As fast forward movement of the ice continued in 1984, new slabs appeared immediately in front of older slabs. The advancing ice front thrust and bulldozed the material into a ridge so rapidly that the ice-front scree under development behind the pushed ridge could not advance fast enough to completely transgress the ridge even though it was continuously fed by large quantities of material. (C) In 1986, when the glacier margin was relatively slow-advancing, the production of thrust-slabs of outwash deposits decreased. As a result of a relatively high ice–surface velocity, the ice fractured and moved along thrust-planes, so the glacier margin appeared as an overhanging wall. In that case, the ice-front scree was continuously fed by large quantities of material and advanced and buried the pushed ridge. During periods of both fast and slow ice advance, the glacier simply overrode the lower part of its own push moraine along with continued proglacial thrusting. This combined mechanism of proglacial thrusting and overriding during glacier advance explains why the pushed ridge did not increase in size during the period 1982–1986, even though the margin advanced more than 100 m during this period. (D) In 1989, the glacier advance and the production of thrust-slabs had stopped and the glacier margin was in a progress of lowering. The scree slope had advanced and completely transgressed the pushed ridge. (E) The present-day end product is not a push moraine, but it appears as a single-crested dump-moraine ridge, 4–7 m high (Fig. 6.9).

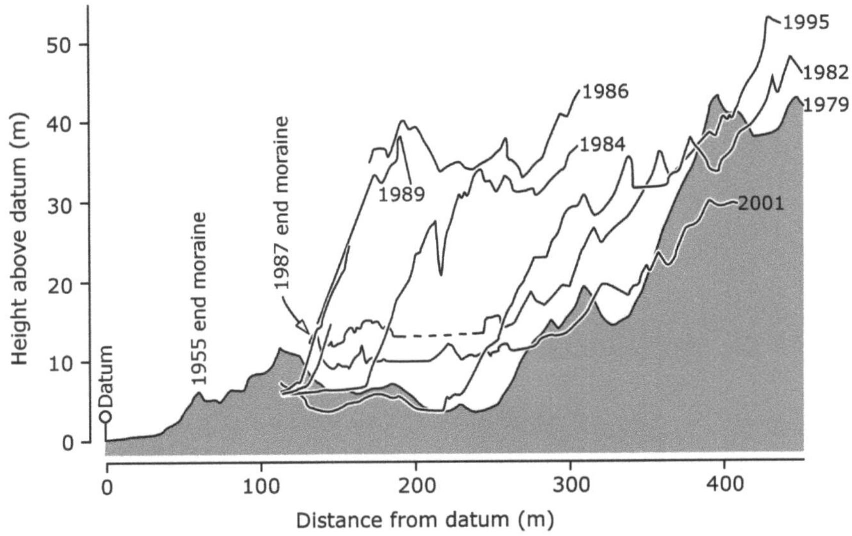

Fig. 6.3. Selected profiles illustrating the advance of Kötlujökull in the 1980s and the following stagnation and dead-ice moraine development in the 1990s (after Krüger et al., 2002).

Fig. 6.4. Supraglacially fed hochsander *fan in front of Kötlujökull as seen in 1982. Slabs of newly deposited fan sediments are pushed up along the advancing ice front.*

The ridge is in an advanced state of fluvial erosion. Drainage routeways from the dead-ice field behind the ridge are confined to distinct flat-bottomed channels deeply incised into the ridge and the outwash beyond it. Exposures in the end-moraine ridge demonstrate that the ridge core and the foreslope represent the lower part of the scree body seen in 1989 and therefore consist exclusively of front cliff-fall sediments and generally without evidence of thrusting. The top half of the 1989-scree slope has collapsed because of down-melting of the supporting glacier margin; the sedimentary architecture of the backslope of the ridge indicates collapse and slumping, and also the many sinkholes that appear at the foot of the backslope reflect the space vacated by melting of the supporting ice margin throughout the 1990s.

Without having observed the active mechanism of ridge formation, this ridge could mistakenly have been interpreted as a frontal dump moraine produced in association with a pause in a general ice recession. However, the present dump-moraine ridge marks the maximum extent of a sustained glacier advance, but during this advance, the basic mechanisms involved in the frontal moraine formation were pushing and thrusting.

6.3. Ice-marginal moraine formation at Sléttjökull

To the north, the Mýrdalsjökull ice cap appears as a gently sloping lobe, Sléttjökull, which terminates at 550–600 m a.s.l. on unlithified sediments of mainly glacial and glaciofluvial origin (Fig. 6.10). The ice surface is rather clean; almost all glacial debris is subglacially derived and concentrated basally as a thin layer immediately above the glacier sole (Humlum, 1981).

Fig. 6.5. View from the ice front of Kötlujökull across an abandoned hochsander fan in 1989. The fan has invaded an area of partially ice-cored hummocky moraine immediately beyond the glacier. Some 50–70 m in front of the ice margin the fan extends through a gap in a frontal moraine ridge from around 1955 and spreads out across an area of ground moraine. From the point at which the stream had flowed from the ice margin to its distal extremity, the fan is 380 m long and the difference in altitude is 14.6 m. The surface slopes are 1:22 for the upper reach of the fan and 1:28 for its lower portion.

Fig. 6.6. Morphology and acting processes on the terminal slope of Kötlujökull advancing across a hochsander fan under formation in 1984. (1) Source area with sediment flow deposits. (2) Exposed ice. (3) Materials that fall and roll down-slope. (4) Sediment flow. (5) Channelized sediment flow. (6) Sediment flow tongue. (7) Former meltwater-tunnel gate. (8) Present gate. (9) Proximal part of hochsander fan.

The part of the forefield that has been modelled by the glacier is bounded by a girdle of ice-marginal moraine ridges developed during a glacier advance around AD 1890. It separates a zone of fluted and partly drumlinized ground moraine located between the glacier and the marginal moraine from an extensive outwash plain named Mælifellssandur. In 1906 the ice edge was found only 50–140 m behind the most distal moraine ridge (Sapper, 1909). At present the glacier front is situated 1.4–1.7 km behind the girdle of ice-marginal moraine. Within the study field, the orientation of flutes indicates a dominant ice-flow direction towards the north-northeast (NNE).

The ground moraine consists of till that forms a continuous sheet with a thin pavement of gravel and stones typically 1–15 cm in size. It rests on thick beds of highly permeable outwash sediments or older till deposits. The diamict characteristics from selected stream-cut sections in the ground-moraine terrain are presented in a condensed stratigraphical log shown in Fig. 6.11. The lowest 80 cm of the log (Unit 1) is combined of

Fig. 6.7. The advancing, debris-mantled glacier margin as seen in 1982. A frontal push moraine is under development by proglacial thrusting of hochsander fan deposits.

alternating layers of sorted, horizontally laminated coarse-grained sand and fine gravel representing proglacial outwash deposits. This sediment association is overlain by two beds of diamict: a lower grey, massive, sandy-silty, matrix-supported, firm to extremely firm diamicts unit, about 20–30 cm thick, with a moderate content of clasts and draped by a poorly defined clast pavement (Unit 2) and an upper clast-paved, grey, massive, sandy-silty, matrix-supported, friable diamict, 50–60 cm thick, with a moderate content of clasts (Unit 3). Gravel comprises 15%, sand 42% and clay/silt 43% of the till (mean weight percentages of 13 samples). Clay-size particles constitute less than 8% of the matrix (Fig. 6.12).

Two clast fabrics sampled in Unit 2 and another two in Unit 3 show a rather spatially consistent orientation with the principal vectors in SSW to SW directions (Unit 2: 206° and 218°; Unit 3: 208° and 216°). The mean vector lies relatively close to ice-flow direction (as indicated by flute orientation) and with a shallow up-glacier plunge (5°). Both fabrics have a strong to very strong degree of clustering (S_1 values amount 0.71 and 0.83, respectively). For additional fabric data, see Krüger (1994), Kjær and Krüger (1998) and Kjær *et al.* (2003). Numerous bullet-nosed clasts, a few centimetres to one metre in size, with distinctly shaped stoss and lee sides occur deeply embedded in the ground-moraine surface as components of the upper diamicts, or randomly scattered within the diamicts. They show a consistent orientation with their upper stoss sides mainly facing up-glacier (206°).

The two till units resting on outwash deposits are interpreted as being deposited in a subglacial environment in connection with the two most prominent advances of Sléttjökull during the Little Ice Age (LIA), respectively, around 1750 and 1890 (Krüger, 1994; Schomacker *et al.*, this volume). Particularly diagnostic in substantiating this interpretation is the occurrence of flutes, which can be traced laterally beneath the glacier, the relatively uniform thickness and high degree of compactness of the diamicts, and the spatially consistent orientation of the sedimentary directional elements such as elongated clasts and stoss-and-lee morphologies, being parallel with ice-flow direction (Krüger, 1984, 1994; Dreimanis, 1989). There are good reasons for believing that the clast pavement on top of the exposed ground moraine represents not a residue from proglacial wash-out of fines but material that has escaped subglacial frictional deposition. First, an identical layer of clasts is seen melting out from beneath the glacier sole along the retreating ice edge in the 1970s and 1990s (Fig. 6.13), and a similar concentration of clasts has also been observed as part of the basal transport zone of the sliding glacier (Humlum, 1981). Second, a rather low fraction (43%) of clasts obtained from the pavement is striated, with most striations being on the lower surface, whereas the subsurface till yields a high content of striated clasts (68%), with striations being present on both the upper and lower surfaces and parallel to the ice-flow direction (Krüger, 1994).

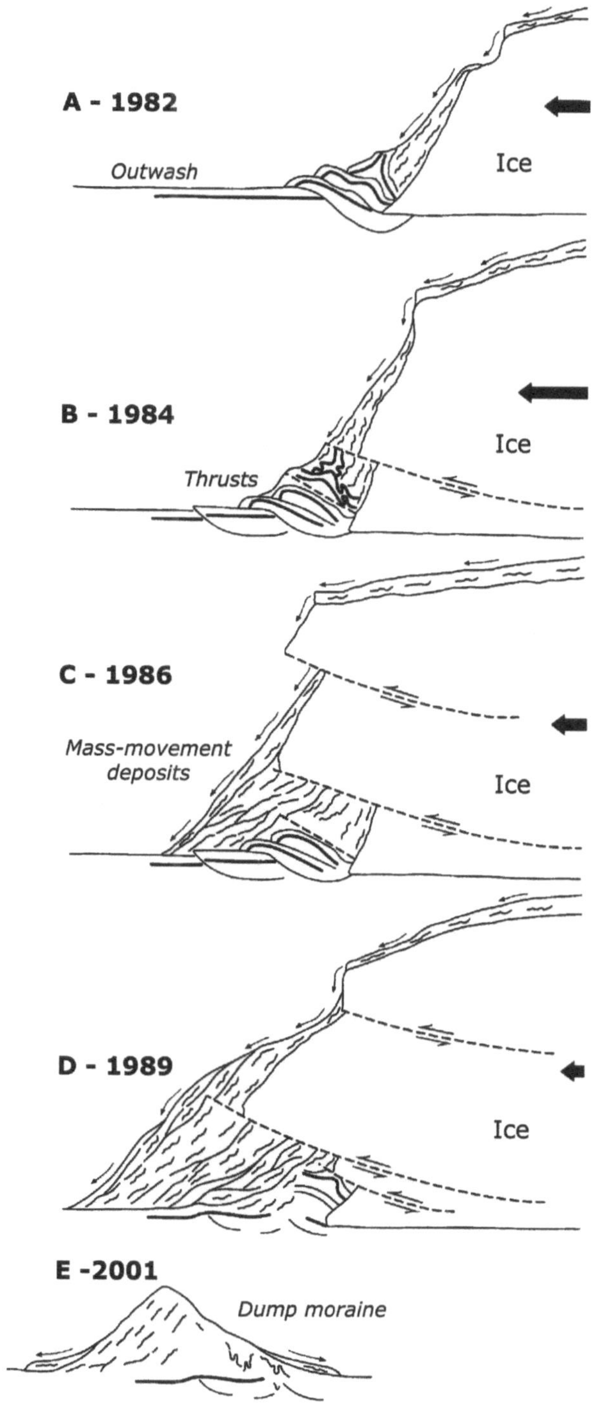

Fig. 6.8. Sequential model for push- and dump-moraine development along the advancing and finally stagnating margin of Kötlujökull. (A) Along the advancing glacier, a frontal push moraine is currently produced by mixing of thrust and folded outwash material with ice-front scree-derived materials. (B) During fast advance, the glacier simply overrides the lower part of its own push moraine along with recurrent thrusting of the outwash sediments. (C) During slow advance, the slightly pushed ice-front talus transgresses the ridge of pushed outwash sediments. (D) The glacier advance has stopped, and along the stagnating glacier margin, the ice-front scree has completely transgressed the pushed ridge. (E) The retreating glacier has left a dump-moraine ridge without evidence of thrusting in the majority of the ridge body (after Krüger et al., 2002).

Aerial photographs taken in 1937 (oblique), 1945 and 1960 illustrate continued glacier recession by frontal retreat with a well-defined ice edge up through the twentieth century. Observations in the field confirmed that the ice front was still retreating by the end of the 1970s. However, during the following period, 1982–1990, the glacier margin was roughly stationary, but since then the ice front has been retreating.

6.3.1. Annual moraines

Between the outer ice-marginal moraine system and the present glacier front, series of minor moraine ridges occur trending roughly parallel to the ice margin and almost perpendicular to the numerous flutes on the till surface. Most moraine ridges are continuous for more than 100–200 m, whereas others are more broken. A field survey distinguishing between relatively long and distinct moraine ridges and less conspicuous and more broken ridges clearly shows that the spacing of the distinct ridges is variable; close behind the outer ice-marginal moraine system, it is less than 10–20 m, increasing to 20–50 m in the area halfway to the 1960 glacier front. Near the 1960 glacier front, the spacing is somewhat variable and may decrease to 10–30 m. In the part of the forefield that was exposed in the 1970s, the spacing of the distinct ridges is 5–15 m, but in front of the present ice front, where the ridges have been produced around the turn of the century, it has increased to 10–20 m.

The ridges vary somewhat in size. Recent ridges produced since the mid-1990s are distinct and typically 0.3–0.8 m high. Further from the glacier, the ridges are less conspicuous, and many of them are only recognized as lines of coarser sediments most likely due to degradation. The ridges are generally asymmetric in cross-profile, their distal slopes being steeper and shorter than the proximal ones. A number of cross-sections excavated in the most recently produced ridges in 1982 (the 1979 ridge) and 2001 (the 1998 ridge) show that the ridges are mainly constructed of friable to firm basal till cut by one or two thrust-planes dipping up-glacier at 20° to 50° (Fig. 6.14). No remnants of glacier ice were found within the ridges. The distal flank of the ridges was draped by mass-movement deposits probably derived from gradual disintegration of the ridge crests. In contrast to this, the proximal slopes were capped by a clast-pavement suggesting that each moraine ridge represents a slab of clast-paved basal till. This is supported by clast-fabric data. Two clast fabrics sampled in one of the slabs dipping 19° up-glacier show a rather spatially consistent orientation with the principal vectors in SW (211°) and S (186°) directions and with a shallow up-glacier plunge (respectively, 27° and 19°) relative to the dip of the slab. Both fabrics have a strong degree of clustering (S_1 values amount 0.73 and 0.69, respectively). These fabric data correspond to those for the basal till in the glacier forefield (Fig. 6.11). The slab was most probably moved in a frozen state, because bodies of frozen till were found at several occasions 0.5 m below the terrain surface, few metres beyond the retreating glacier margin.

Three other features are indirect evidence that the ridge till was moved in a frozen state. First, silt horizons

Fig. 6.9. Single-crested dump moraine in front of Kötlujökull in 1997.

Fig. 6.10. View across the eastern region of Sléttjökull and its forefield of smooth, slightly undulating ground moraine, which slopes gently away from the glacier. The ground moraine, exposed by glacier retreat up through the twentieth century, has become more or less changed by terminoglacial and proglacial processes. Series of annual moraine ridges drape the ground moraine and ponding of water in topographic lows currently results in deposition of fine material. Flowing meltwater dissects the high-lying parts of ground moraine, while laterally migrating streams in low-lying areas systematically strip the ground moraine and deposit outwash sediments.

found in the ridge till were similar to those commonly found *in situ* in the basal till. They probably originate from subglacial shearing, a process that is favoured by the decreasing ice load due to glacier retreat, or they may originate from submarginal shearing along a discrete plane representing a former position of the 0°C isotherm during a limited winter re-advance (see later). Second, where the moraine ridges intersect flutes, the flutes increase in both height and width, suggesting that each moraine ridge represents one coherent anticline parallel to the glacier front. Third, it is seen in Fig. 6.23 that flutes crossing the moraine ridges at slightly oblique angles may show minor displacements along the crest of the moraine ridge, suggesting that the flutes were frozen to the glacier sole during winter re-advances and then carried forward and directed in harmony with the local ice movement during formation of the moraine ridge.

In 1982, when the general glacier retreat had stopped, the terminal 20 m of the glacier was covered by a snow-patch, about 1 m thick, which persisted late into the summer season. However, a ridge in the process of formation at the seasonally oscillating glacier terminus indicated the possibility of obtaining some information about the mode of ridge formation (Krüger, 1995). The ridge, about 1 m high and trending parallel to the ice margin, was seen 10–20 m up-glacier from the distal snow limit. The ridge consisted of a raised slab, 0.8–1 m thick, of frozen till, apparently basal till, dipping 30° up-glacier (Fig. 6.15). The texture of the till sediment accords with that for the basal till in the glacier forefield: gravel

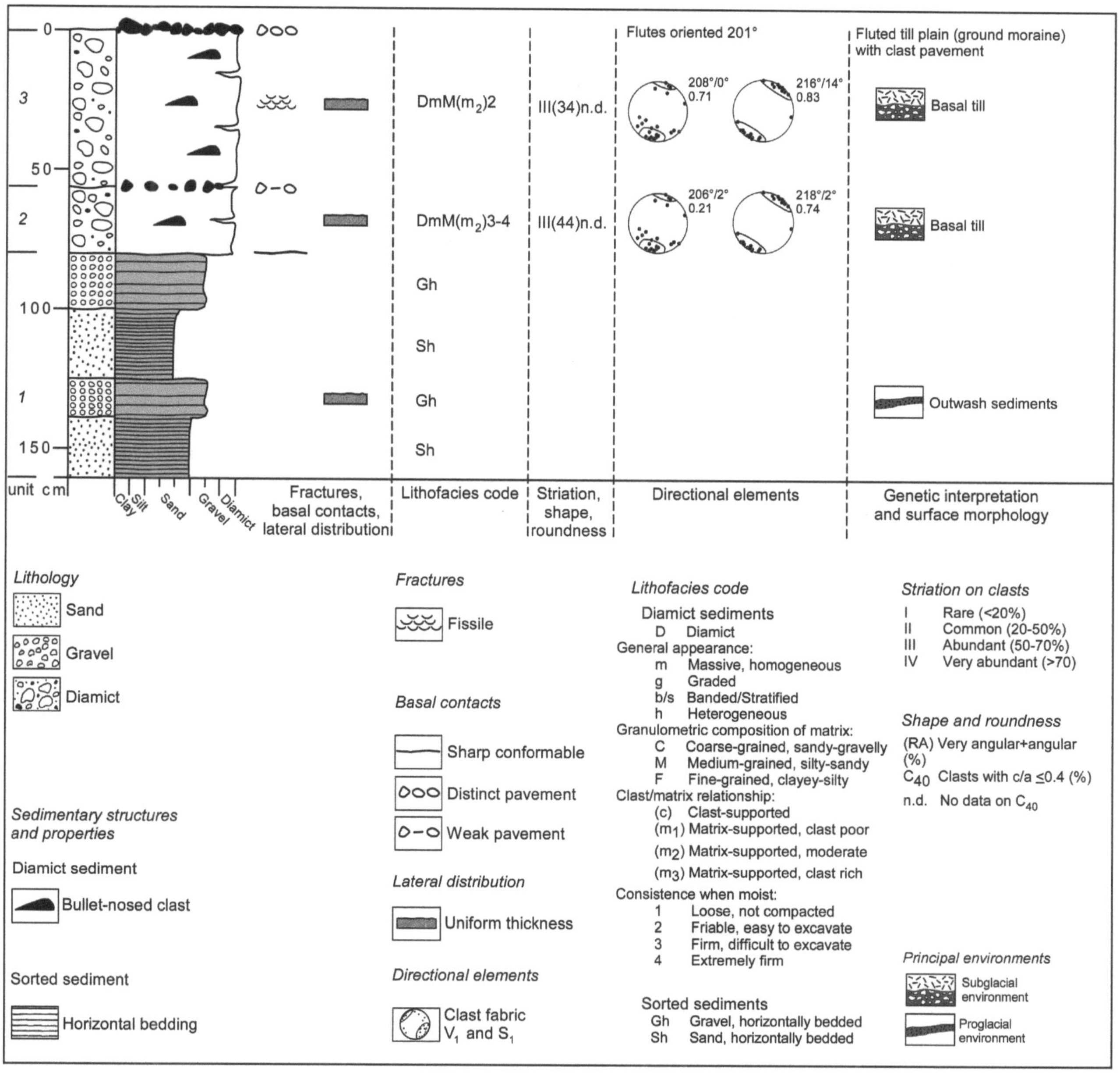

Fig. 6.11. Data chart comprising a description of the sedimentary succession recognized in the ground moraine landscape (based on data chart developed by Krüger and Kjær, 1999).

comprises 22%, sand 42% and clay/silt 36% (mean weight percentages of two samples). Clay-size particles constitute 8% of the matrix. Thin bands of silt occurring within the ridge till suggest that the till sediment was carried as a coherent mass. The proximal slope of the ridge was mantled by gravel and stones similar to those draping the ground moraine. This surface layer continued underneath the glacier ice suggesting that the till/ice contact represented the basal transport zone and the glacier sole. The unfrozen top surface of the ridge was somewhat disaggregated due to development of desiccation cracks and extension fractures.

Individual particles and clots of sediment released from the snout of the melting till slab rolled, or slid, down the foreslope of the ridge, producing tongues of mass-movement deposits with downslope size-sorting on the surface of the fronting snow-bank. A deformed depositional layering within the snow clearly showed that the snow-bank located against the frozen till slab had been up-arched somewhat due to the combined advance of the glacier toe and the adhered till slab. As the summer ablation progressed, the ridge sediment collapsed and flowed downslope in tongues, so that bodies of snow became incorporated into the ridge. The final ablation of snow left a main ridge consisting of the proximal part of the till slab and two secondary ridges composed of mass-movement deposits (Fig. 6.16).

The data obtained at Sléttjökull suggest the following model for moraine-ridge formation (Fig. 6.17): (A) Owing to the cold wave of the winter, a slab of sediment freezes on to the clast-loaded basal transport zone of the glacier. (B) During glacier advance in late winter, the sediment frozen to the glacier sole shears along a discrete plane that follows the 0°C isotherm. (C) As the ice-edge advance continues, the frozen-on sediment slab is moved as an integral part of the glacier foot and overrides the frontal ground. (D) During the following summer ablation, the glacier edge will leave a distinct ridge composed of a slightly deformed slab of sediment.

Thus, the mechanism of moraine ridge formation at Sléttjökull is primarily an interaction between basal freezing of overridden sediment below the glacier toe and transportation and deposition of the frozen sediment slab in harmony with the seasonal ice-edge fluctuations. This type of moraine ridges produced annually is likely to be characteristic of retreating temperate glaciers with thin glacier toes and where winters are cold.

Annual-moraine ridges are described from other glacier forefields in Iceland, where they are produced either by squeezing of semi-liquid till from beneath the glacier (Price, 1970), or by scraping, or pushing, of frontal sediments (Sharp, 1984; Boulton, 1986), or they may result from a mechanism that couples the ridge-forming process to submarginal deformation of basal till and bulldozing of the extruded sediments (Sharp, 1984). One possible explanation of this difference in mode of annual-moraine ridge production in Iceland could be ascribed to differences in ice surface slope and climate. At the north margin of Mýrdalsjökull, the terminal 10–15 m of the glacier is extremely thin in contrast to most other annual-moraine ridge localities described, for example, Fjallsjökull and Skálafellsjökull (Price, 1970; Sharp, 1984), and, therefore, permits limited basal freezing during winter. The surface of the terminal 100 m of the Sléttjökull glacier snout slopes at about 10° and ice thicknesses of 2–3 m have been recorded 12–15 from the ice edge; the maximum thickness of ice through which winter cooling may penetrate is roughly 15 m (Harris and Bothamley, 1984). A further argument is that Fjallsjökull and Skálafellsjökull terminate on the coastal lowlands, whereas the environments in the highland north of Mýrdalsjökull are controlled by a more continental climate with severe winters and colder and more prolonged springs.

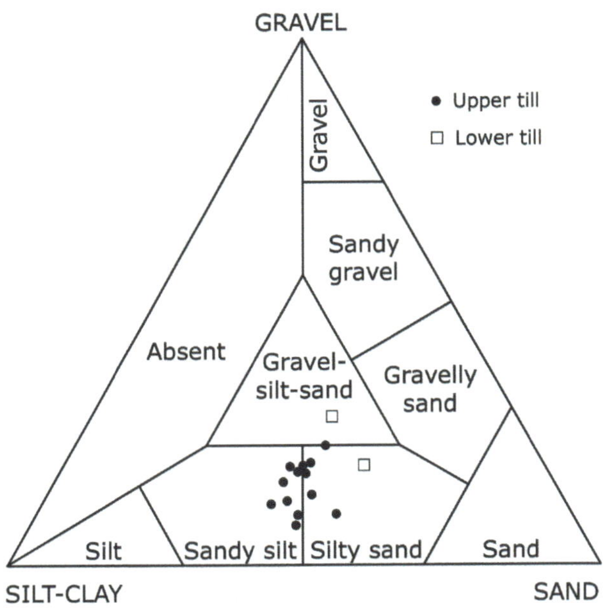

Fig. 6.12. Textural characteristics of Upper and Lower Till from the ground moraine terrain in front of Sléttjökull. Modified after Lawson (1979) (after Krüger, 1994).

6.3.2. A composite ridge formed at a stationary ice margin

During the period 1982–1989, the Sléttjökull glacier margin was roughly stationary. The glacier retreated slowly during summer and re-advanced during winter to much the same position as the previous winter (Fig. 6.18). The annual-moraine ridge formation continued, but instead of creating a series of discrete moraine ridges, the annual cycle of ice front activity produced a complex ridge, 3–5 m high, composed of imbricately stacked slabs of frozen, clast-paved basal till, or sorted deposits,

Fig. 6.13. Pavement of stones of medium size melting out from the retreating glacier toe.

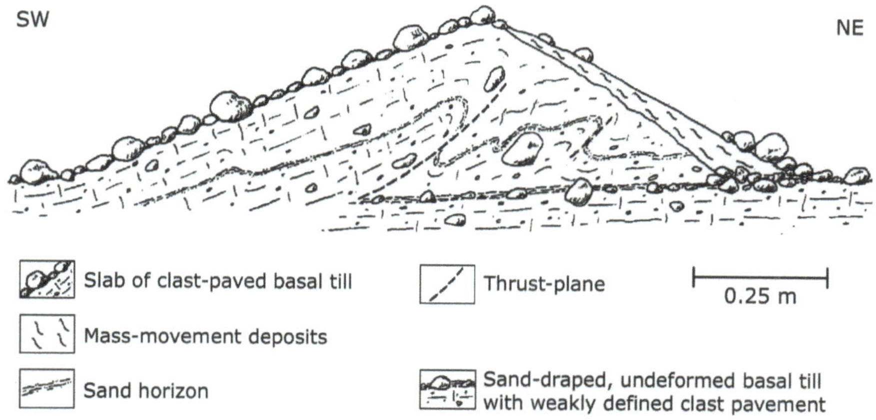

Fig. 6.14. Section in annual-moraine ridge formed during the winter 1979 (after Krüger, 1994).

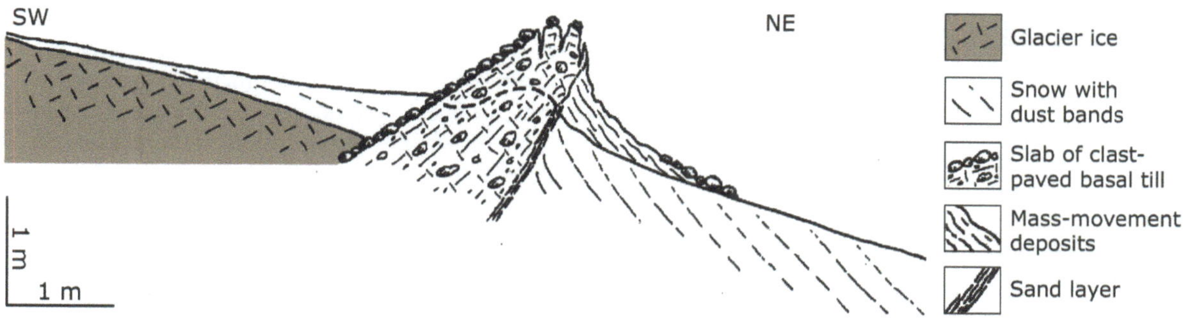

Fig. 6.15. Annual-moraine ridge under formation between glacier ice and a proglacial snowbank at the margin of Sléttjökull in 1982. Below the dashed line the till slab was frozen.

Fig. 6.16. Annual-moraine ridge with a core of frozen basal till and two secondary ridges in 1982. In the foreground, the glacier toe is seen and behind the person remnants of the rapidly melting snow-bank appears.

dipping up-glacier (Fig. 6.19). It is most likely that a slab of sediments had been added to the ridge each winter.

In 1986, a ridge, 1.5–2.5 m high, was under formation. It was combined of two to three closely spaced ridges, each of them steep-sided and asymmetric with a gentle proximal and steeper distal flank (Fig. 6.20A). The ridge crest nearest the glacier was the highest, and crest heights fell progressively outwards. From its proximal end to its distal extremity, the composite ridge measured 8–14 m. Its distal part was constructed of slabs, 0.5–0.9 m thick, of

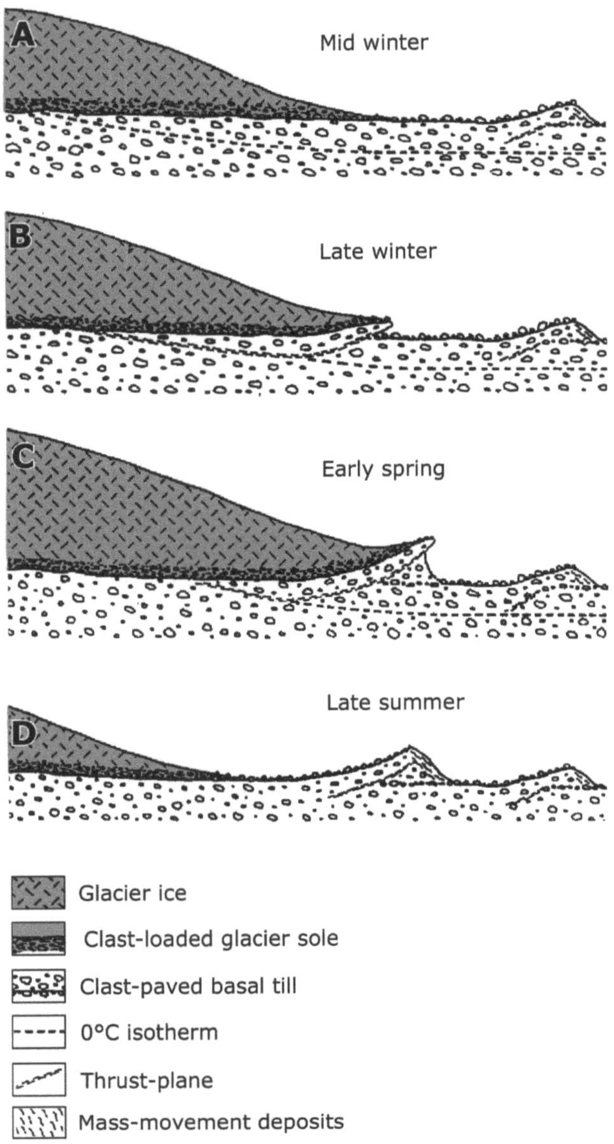

Fig. 6.17. Sequential model for annual-moraine ridge formation at Sléttjökull (after Krüger, 1995).

partly frozen till. A clast pavement on top of each till slab served as marker defining the individual slabs. Digging at different sites showed the folded sediments of laminated silt and sand – deposited in a shallow pond on top of the exposed ground moraine in the early 1980s, according to photographs taken in 1982 – may very well underlie the most distal slab. No marks of deep-seated proglacial thrusting were seen in front of the ridge; only a narrow zone, 4–5 m wide, was affected somewhat by thrusting, for example, some of the largest flutes were cut by numerous closely spaced thrust-planes dipping at 23–35° up-glacier. The proximal part of the composite ridge was snow-cored and consisted of a raised slab, 0.5–0.7 m thick, of basal ice and frozen basal till. As a result of the penetrating winter cold, a slab of till was frozen to the basal clast-loaded transport zone of the thin glacier toe and subsequently raised as a coherent mass that over-thrust the fronting snowbank and the proximal slope of the ridge.

In 1989, the composite ridge was significantly higher than in 1986, reaching 3.5–4 m above the ground moraine in front (Fig. 6.20B). From its proximal end to its distal extremity, the ridge measured 20–25 m. Since 1986, the front of the moraine ridge had moved 5 m down-glacier, probably as a result of internal thrusting during winter re-advances. Tongues and fans of sediment-flow deposits and downwash sediments originating from disintegration of the moraine-ridge crests extended from the foreslope of the ridge on to the fronting terrain. The frozen core only permitted the upper 1 m of the ridge to be investigated in detail. The interior of the ridge appeared as a pile of at least 5–7 clast-paved basal-till slabs, 0.3–0.4 m thick, dipping 24–35° upglacier. With the exception of the most distal slabs, which were hidden underneath scree deposits, the leading edges of individual slabs were expressed in the surface morphology as small ridges on top of the frontal moraine. The most proximal slab was frozen to the up-arched clast-loaded glacier sole and dipped beneath the glacier at about 30°. Narrow inter-ridge depressions on top of the composite ridge were partly filled with downwash sediments of stratified silt, sand or fine gravel and clusters of stones and boulders were common as a result of gravity sorting. In their up-glacier end, these deposits were disturbed and over-thrust by till slabs. Two types of sliding surfaces were distinguished: those separating till slabs and those representing thrust-planes, which cut through till slabs or inter-ridge deposits. Obviously, the first type was connected with the construction of the ridge by repeated stacking of till slabs, whereas the second type was produced by ice-push. No marks of thrusting were seen in front of the ridge.

An accurate assessment was obtained of the volume of sediments included in the 1989 ridge. Two lines aligned perpendicular to the ridge crests – a straight horizontal line drawn at the foot of the ridge and another line along the proximal till-slab profile – together with the upper surface of the moraine ridge, circumscribe an area proportional to the volume of a transversal ridge segment with unity width. The volume of the mass raised above the frontal terrain thus calculated was found to be 38 m^3. As the ice surface was generally clean, this volume of sediments must originate from the overridden ground moraine. Bodies of glacier ice were not identified within the ridge. With a mean till-slab thickness of 0.4–0.6 m, the volume of sediment within the ridge represented a horizontal shortening of the ground moraine of roughly 60–90 m. It is interesting to note that the distance between the 1984-ridge position and the front of the 1989 ridge was only 10 m. This discrepancy between the large volume of the composite ridge and the small lateral displacement of the ridge on top of the fronting ground moraine clearly shows that only a small portion of the material comprising the frontal moraine ridge can result from proglacial thrusting of the ground moraine. The internal structure of the ridge combined with its almost stationary position throughout the period of observation suggest that the growth of the ridge chiefly resulted from recurrent superposition of till slabs on the proximal side of the ridge.

In the summer of 1994, the terminal part of the glacier appeared 10–25 m behind the moraine-ridge complex. The composite ridge was 3–4.5 m high and occupied a 20–40 m wide zone. Its topmost parts consisted of several

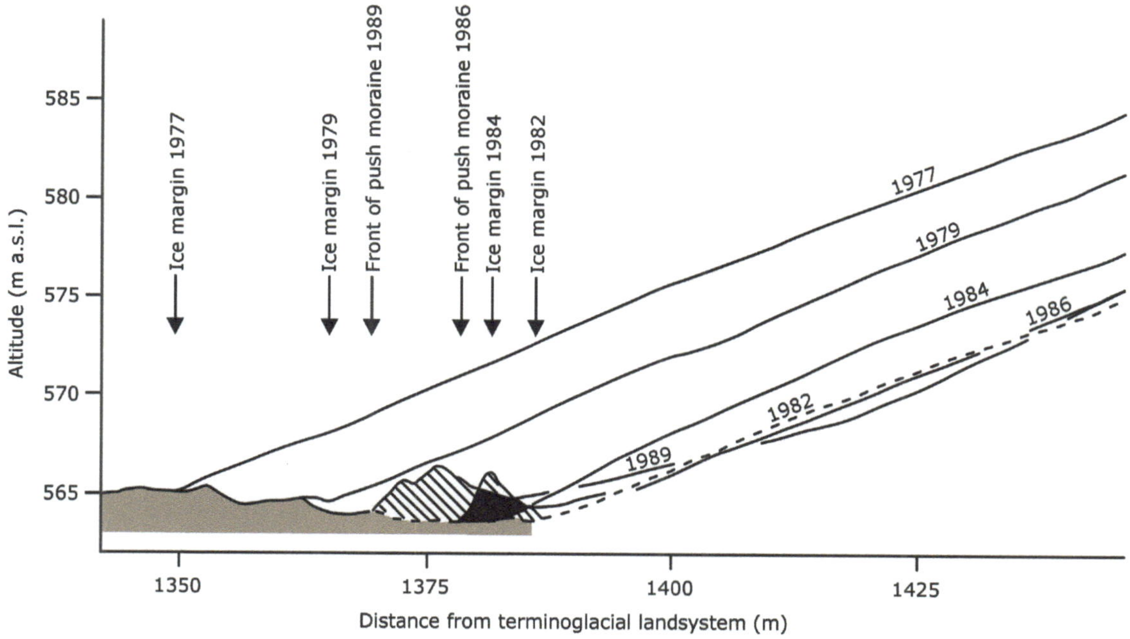

Fig. 6.18. Glacier longitudinal sections showing the retreat of Sléttjökull in the 1970s and the following stagnation and fluctuation of the glacier margin during the period 1982–1989 (after Krüger, 1993).

Fig. 6.19. Oblique air photograph from 2001 showing the ice margin of Sléttjökull and the forefield with the composite moraine ridge produced during the 1980s. In the foreground, annual-moraine ridges formed during the glacier retreat in the 1970s are seen. Between the composite ridge and the present ice edge, a number of annual-moraine ridges produced along with the glacier retreat in the 1990s occur.

small ridges chiefly composed of slabs of clast-paved till dipping up-glacier (Fig. 6.21). In distal parts of the ridge complex, snouts of till slabs were covered by scree deposits originating from disintegration of the ridge crests. Isolated ponds – the largest being 20 m in length, about 8 m wide and at least 2 m deep – located between discrete ridges suggest that the ridge core was still frozen. Thus, there was mainly no difference in the moraine-ridge composition in 1994 and 1989. However, sections cut into the ridge by meltwater streams show that some segments of the proximal ridge differed somewhat in composition. At some places, the most proximal part of the ridge contained a slab of glacier ice, 30 cm thick, dipping up-glacier at about 32°. The ice slab was concordantly draped by well-sorted sediments of chiefly planar-bedded, or trough cross-bedded sand, 10–25 cm thick. This surface layer was highly disturbed by slumping as a result of collapse of the ice-cored slope due to ice melting. A frozen layer of basal till, up to 30 cm thick, underlay the clast-loaded sole of the ice slab. The

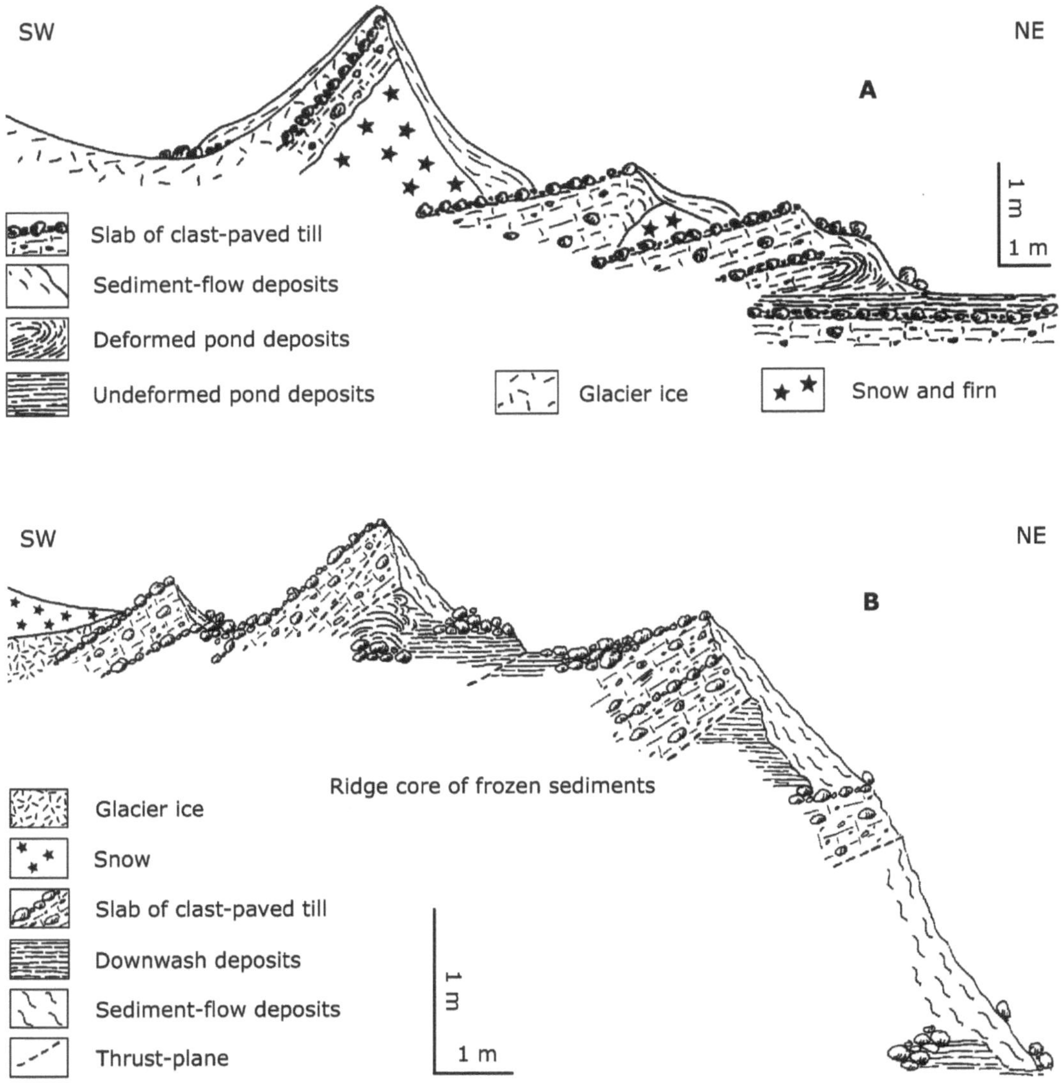

Fig. 6.20. Two sections in the composite moraine ridge being formed along the margin of Sléttjökull in respectively 1986 (A) and 1989 (B) (after Krüger, 1994).

Fig. 6.21. View along the crest of the composite moraine ridge in 1994 showing several small distinct ridges.

combined slab of glacier ice and associated sediments lay against snow buried underneath a veneer of scree deposits.

At localities where the proximal ridge was sand-draped, the depression between the composite moraine ridge and the glacier proper was occupied by shallow ponds, or braided streamlets, with sediments of sorted sand underlain by glacier ice. The buried ice, less than 1 m thick, was contiguous to both the glacier proper and the ice slab in the proximal ridge. Apparently, the ice slab in the proximal ridge was a result of up-arching of the thin sand-draped glacier foot in connection with a limited re-advance during the preceding winter. The sand drape on the proximal slope of the ridge was most probably deposited in a shallow meltwater basin located on the thin glacier foot during the 1993 ablation season. In places the steep-sloping sand drape on the proximal ridge was discordantly overlain by well-sorted sand in the shape of a diminutive kame terrace apparently produced early in the 1994 ablation season in connection with snow melting.

The question arises: why are glaciofluvial processes and associated deposits locally involved in the final stage of the composite moraine-ridge formation? The surface of the terminal part of Sléttjökull is almost clean. However, 500–700 m up-glacier from the terminus, volcanic ash of sand-size is seen melting out from two englacial bands, 2–5 cm thick. The outcropping bands appear as slightly undulating lines of sharp-pointed dirt cones extending in conformity with the ice front. From time to time, released materials collapse and flow down the steep foreslope of the cones, but melt- and rainwater flowing in sheets and rills on the glacier surface carry the material further down-glacier. In the 1980s, when the summers were relatively cold and the composite moraine ridge was under production along the stationary ice margin, fairly large segments of the glacier terminus were buried by snow, which persisted late into the ablation seasons and larger meltwater streams were absent on the glacier snout. But in the 1990s, the average summer temperature increased markedly. As a consequence, the glacier snout thinned, and the terminal part of the glacier was snow-free during the ablation periods, so that meltwater flowed in sheets and rills down the glacier snout. Locally, however, the new composite moraine ridge acted as a frontal dam, which gave rise to accumulation of re-deposited volcanic ash of sand size on the very thin glacier foot. In this fashion glaciofluvial processes and associated deposits were locally involved in the final stage of moraine-ridge production at Sléttjökull.

An idealized sequential model for the formation of the composite moraine ridge along the margin of Sléttjökull is shown in Fig. 6.22. (A) Around 1983, when the ice advanced slightly, an annual-moraine-like ridge was formed. (B) In the following years, the glacier edge became stationary, but an annual cycle of ice-front activity occurred; the glacier toe thinned and retreated slowly by a few metres during summer and re-advanced during winter to much the same position as the previous winter. As a result of winter coldness, a slab of basal till beneath the glacier toe freezes on to the clast-loaded basal transport zone of the sliding glacier, so that the ice toe and

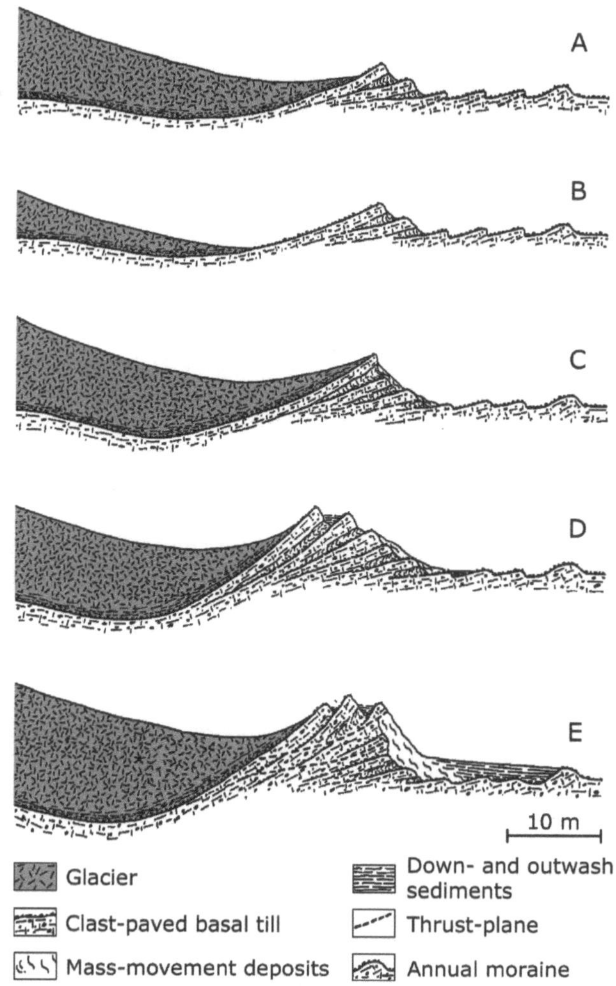

Fig. 6.22. Model for composite moraine ridge formation at Sléttjökull (after Krüger, 1994).

the till sediment frozen to the glacier sole are raised and superimposed on the proximal slope of the fronting ridge. (C) During the following years, a series of till slabs are stacked up in an imbricate structure as a result of seasonal ice-edge fluctuations. The horizons of clast-pavement separating the till slabs represent a residue resulting from melting of the basal transport zone of the up-arched glacier toe. (D) As superposition of till slabs build up the ridge to a certain point, internal deformations take place owing to the increased stress induced within the ridge by pushing as the ice advances in late winter. All things being equal, the seasonal freezing-on process will come to an end as soon as the terminal ice becomes sufficiently thick to prevent winter freezing from reaching the substratum; it takes place when the submarginal depression representing the source area of till slabs has been enlarged to a critical point.

Thus, the earlier mechanisms may result in a frontal moraine morphologically similar, but genetically distinct from conventional pushed moraines. Therefore, a genetic distinction should chiefly be made on the basis of differences in internal composition and stratigraphy. Push-moraine ridges are created by thrusting and stacking of wedge-shaped sediment slabs in front of advancing glaciers (Boulton, 1986; Aber *et al.*, 1989; Aber and Ber, 2007). Therefore, during the ridge evolution, new thrusts appear in front of older thrusts, and thus the most distal slab is the youngest one. Because the slabs are fragments of the overridden ground, the subsurface stratigraphy of the glacier forefield will be repeated in each individual sediment slab. Consequently, the depth of the basin behind the moraine ridge, which represents the source area for the slabs, is roughly equal to the thickness of the individual slabs.

During the evolution of marginal-moraine ridges, such as that described here from Sléttjökull, sediment slabs are superimposed on the proximal slope of the embryonic moraine ridge, and, as a consequence, the most proximal slope is the youngest one. Because the slabs originate from repeated removal of sediment from the substratum within a narrow marginal zone of the glacier, the combined series of slabs reflect a stratigraphic column where the oldest sediment strata are recognized in the youngest – and most proximal – slab, and the most distal slab represents the youngest sediment strata. This pattern of strata ages may or may not be interrupted by layers of glaciofluvial origin deposited during the ridge formation in accordance with the model of Matthews *et al.* (1995).

6.3.3. *Annual-moraine-ridge spacing as a usable climate signal*

The relationship of annual ice-front retreat to the ridge topography during two periods, 1977–1981 and 1995–2001, was examined to test the hypothesis that the ridges were formed annually. Figure 6.23 shows the crest-line positions of distinct and less conspicuous moraine ridges superimposed on a small drumlin emerging from the retreating ice front in 1982. The plan shape of the moraine ridges curves up-glacier on top of the drumlin and clearly follows the ice front, indicating that the ice margin closely responded to changes in the submarginal topography. The position of cairns established at the ice edge in 1977 and 1979 is shown, so that the ridges can be dated by comparing their location with ice margin positions. It appears that a moraine system combined of one distinct ridge and one to four less conspicuous ridges has been produced every year during the observation period. In the summer of 1982, the terminal 10–20 m of the surveyed part of the glacier was covered by a snowpatch, so the ridge produced during the winter 1982 was found some metres up-glacier from the distal snow limit.

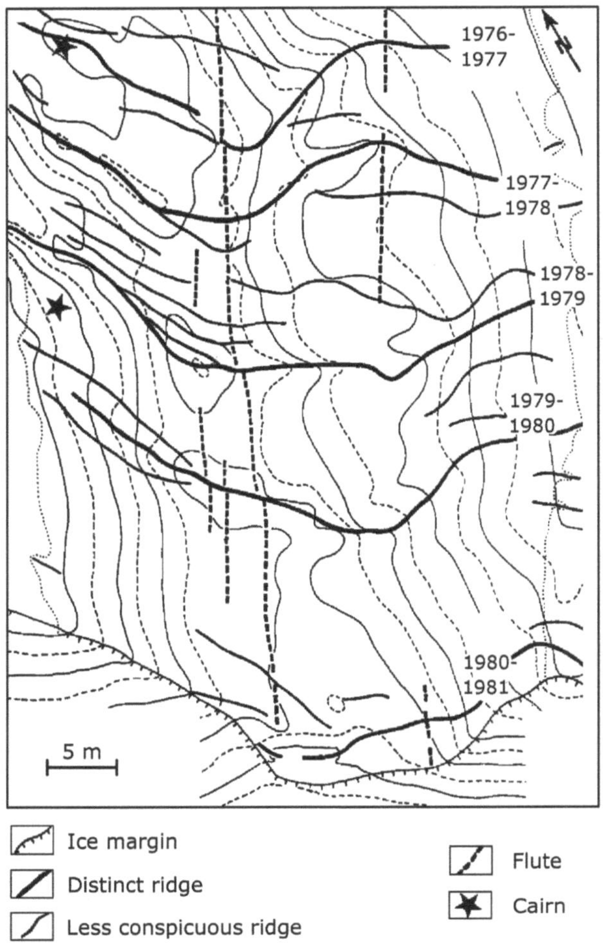

Fig. 6.23. Crest-line positions of annual-moraine ridges superimposed on a small drumlin emerging from the retreating glacier in 1982. Contours are at 0.25-m intervals (after Krüger, 1994).

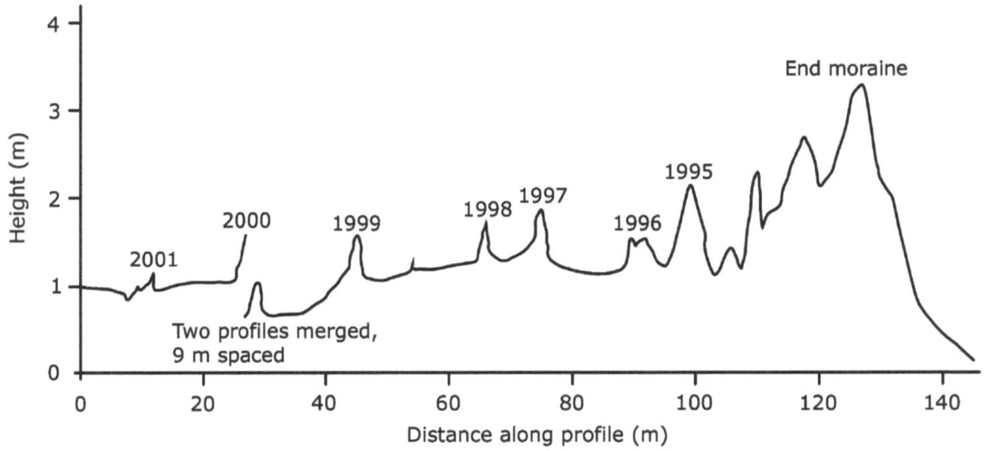

Fig. 6.24. Transect across the glacier forefield between the composite ridge produced during the glacier-edge stillstand in the 1980s and the 2001-ice front illustrating the year of production of the individual distinct moraine ridges (after Specht, 2003).

The same section of the glacier front was visited every summer during the period 1994–2001. Along with the glacier retreat and the production of moraine ridges, sticks numbered consecutively were placed on the individual ridge crests to indicate the year of production. Figure 6.24 shows a transect across the glacier forefield between the composite ridge produced in the 1980s and the 2001-ice front illustrating the year of production of the individual distinct moraine ridges. It is reasonable to conclude that during the two periods of glacier retreat, 1977–1981 and 1995–2001, the distinct moraine ridges formed annually.

In the area exposed before 1977, 20 distinct ridges appeared in the area de-glaciated between 1960 and 1977, 15 ridges were found between the 1945- and 1960-ice-margin positions and another eight ridges in the area exposed between 1937 and 1945 (Fig. 6.25). That is 43 ridges during 40 years, 1937–1977, or roughly one distinct ridge each year. If the ridges were indeed

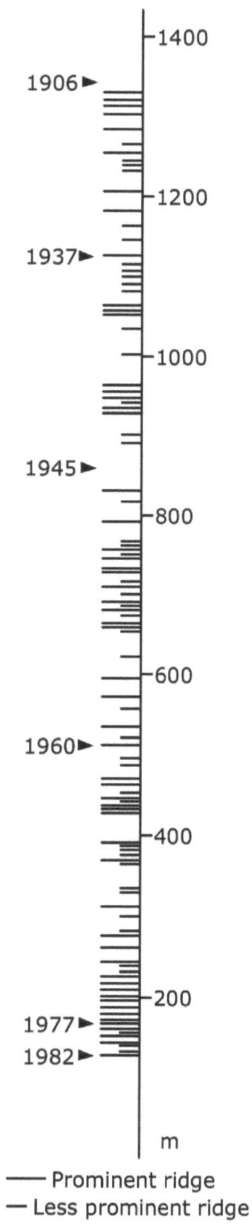

Fig. 6.25. Annual-moraine ridges recorded in the field along a traverse running from the composite marginal moraine ridge produced in the 1980s to the outermost marginal moraine system formed around 1890.

produced annually and without allowing for the possible effects of overriding and superimposition of ridges, they have obviously formed since the end of 1920s, because there are nine distinct ridges in the area de-glaciated between 1906 and 1937.

It is possible to calibrate the ice-front retreat of Sléttjökull against the trend of mean summer air temperature (June to September). A temperature curve covering the period 1906–2002 is modelled on the basis of temperature data partly from the forefield of Sléttjökull (August 2001 to August 2002) and Kötlujökull (August 1995 to April 1998 and August 2001 to August 2002) and partly from the Kirkjubæjarklaustur station (which goes back to 1932) situated 40 km eastwards (Specht, 2003).

First, temperatures from Sléttjökull were correlated ($R^2 = 0.96$) on an hourly basis for August 2001 to August 2002 with temperature data from Kötlujökull, and the monthly mean temperatures for August 1995 to August 2002 were calculated based on the equations from the regression analysis ($T_{SLÉTT(1)} = T_{KÖTLU} \times 1.0533 - 3.3826$, where $T_{SLÉTT(1)}$ is the air temperature at Sléttjökull and $T_{KÖTLU}$ is the air temperature at Kötlujökull). Second, the composite temperature series for August 1995 to August 2002 was correlated on a monthly basis with meteorological data from Kirkjubæjarklaustur ($R^2 = 0.96$). The equation from the last regression analysis ($T_{SLÉTT(2)} = T_{KIRK} \times 0.82 - 4.39$) allowed calculation of the mean monthly temperature at Sléttjökull for 1932–2002 and subsequently calculation of mean summer air temperature (June–September) (Fig. 6.26). Extrapolating monthly temperatures further back in time is possible by comparing with the meteorological data from the Stykkishólmur station in West Iceland (which goes back as far as 1845 and presents the longest continuous record of climatological data in Iceland), because of the strong correlation between main air temperatures from east and west Iceland (de Ruyter de Wildt *et al.*, 2003; Ravn, 2006).

If the Stykkishólmur and Kirkjubæjarklaustur records are taken as a general picture of temperature fluctuations in the forefield of Sléttjökull, it appears that from the record of colder temperatures, a sharp rise occurs in the late 1920s, as in other north Atlantic regions (Grove, 1988). Thus, it is reasonable to conclude that the annual-moraine-ridge production goes back to that time as documented by ridge chronology.

It is now possible to give an overall picture of the glacier retreat rates since 1906 (Fig. 6.27). Until the end of the 1920s, the glacier front was almost stationary, but in the subsequent period 1932–1937, the average frontal retreat was 37 m/yr which was the most rapid retreat in the twentieth century. During the period 1937–1945, the net retreat rate fell to 32 m/yr, followed by 23 m between 1945 and 1960 and 19 m between 1960 and 1977. A much lower rate of retreat of only 11 m/yr has been documented for the period 1977–1981. Between 1982 and 1990, the glacier front was almost stationary, which resulted in the production of the composite moraine ridge. In the subsequent period 1991–1995, the average frontal retreat was only 6 m/yr rising to 15 m/yr in the period 1995–2001.

It appears from Fig. 6.27 that the above glacier retreat rate pattern closely follows the mean summer (June to

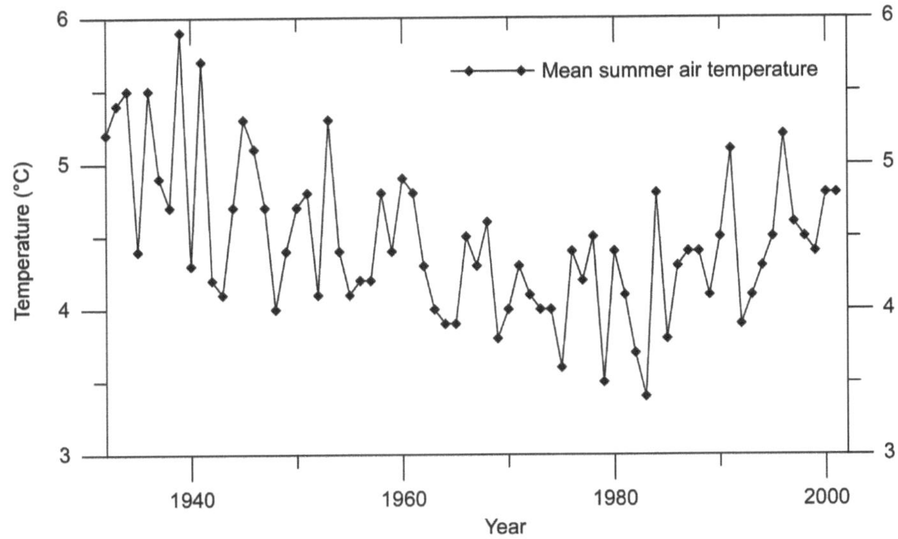

Fig. 6.26. Calculated mean summer air temperature (June–September) at Sléttjökull 1932–2002 based on temperature measured at Sléttjökull (August 2001 to August 2002), Kötlujökull (August 1995 to April 1998 and August 2001 to August 2002) and Kirkjubæjarklaustur (1932–2002; based on reports of the Icelandic Meteorological Office) (partly after Specht, 2003).

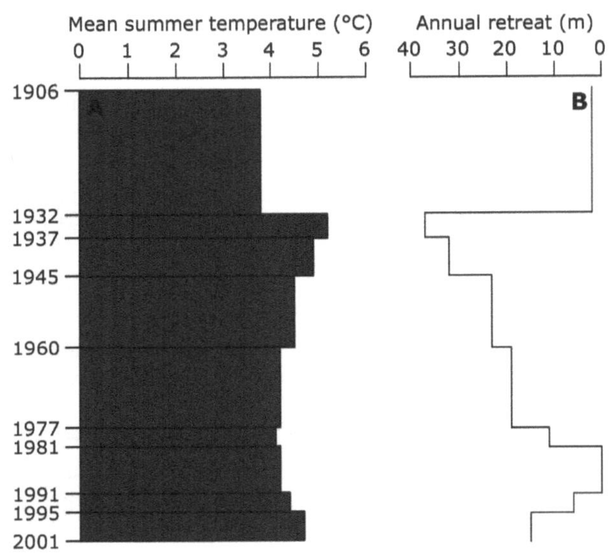

Fig. 6.27. Calibration of mean summer temperature (June to September) at Sléttjökull (A) against annual net retreat of Sléttjökull (B).

September) temperature. This suggests that even the short-term ice-front fluctuations of Sléttjökull are induced by climatic changes which mean that this part of the Mýrdalsjökull ice cap is not a surging glacier.

6.4. Neoglacial ice-marginal moraine ridges

Ice-marginal moraine ridges indicating the maximum extent of the Mýrdalsjökull ice cap in historical times are found bordering the glacier forefield in front of all the main outlets from the ice cap, such as the south flowing Sólheimajökull, the east flowing Kötlujökull, Sandfellsjökull and Öldufellsjökull and the north flowing Sléttjökull and Entujökull (Fig. 6.1) (Sigurðsson, this volume). In the following, we describe the neoglacial ice-marginal moraines in front of Sléttjökull and Sólheimajökull.

6.4.1. *Sléttjökull*

A girdle of ice-marginal moraine ridges extends 1.5–2 km beyond the present front of Sléttjökull and separates a landscape, mainly of streamlined ground moraine located between the glacier and the ice-marginal moraine from the Mælifellssandur outwash plain (Fig. 6.28). The ground moraine is heavily dissected by a network of active and abandoned meltwater channels and here and there laterally migrating streams have removed large areas of moraine terrain and segments of the ice-marginal ridges. A photograph taken by Sapper (1909) in 1906 documents that the ice front was standing in the immediate vicinity of the ice-marginal moraine ridges at that time suggesting that these ridges were produced around the turn of the century (Fig. 6.29).

The girdle of ice-marginal moraine ridges is 75–150 m wide and curves slightly, in the main, following the trend of the present glacier front. Southeast of Mælifell the marginal moraine consists of at least three well-defined sub-parallel rows of ridges with very uneven crests as high segments alternate with lower ones. The outer moraine ridge is 3–4 m high and has a sharp crest; the innermost ridges are generally lower and more gently sloping. Following the girdle of ice marginal moraine ridges westwards, the ridges climb two small basalt knobs, and 3–4 km west of Mælifell – passing at close range outside a 50 m high and almost 1 km long basalt hill – the series of ridges shows a well-marked concavity. Apparently, the broad glacier lobe was forced to split up into discrete sub-lobes as the glacier flowed across and between the basalt knobs.

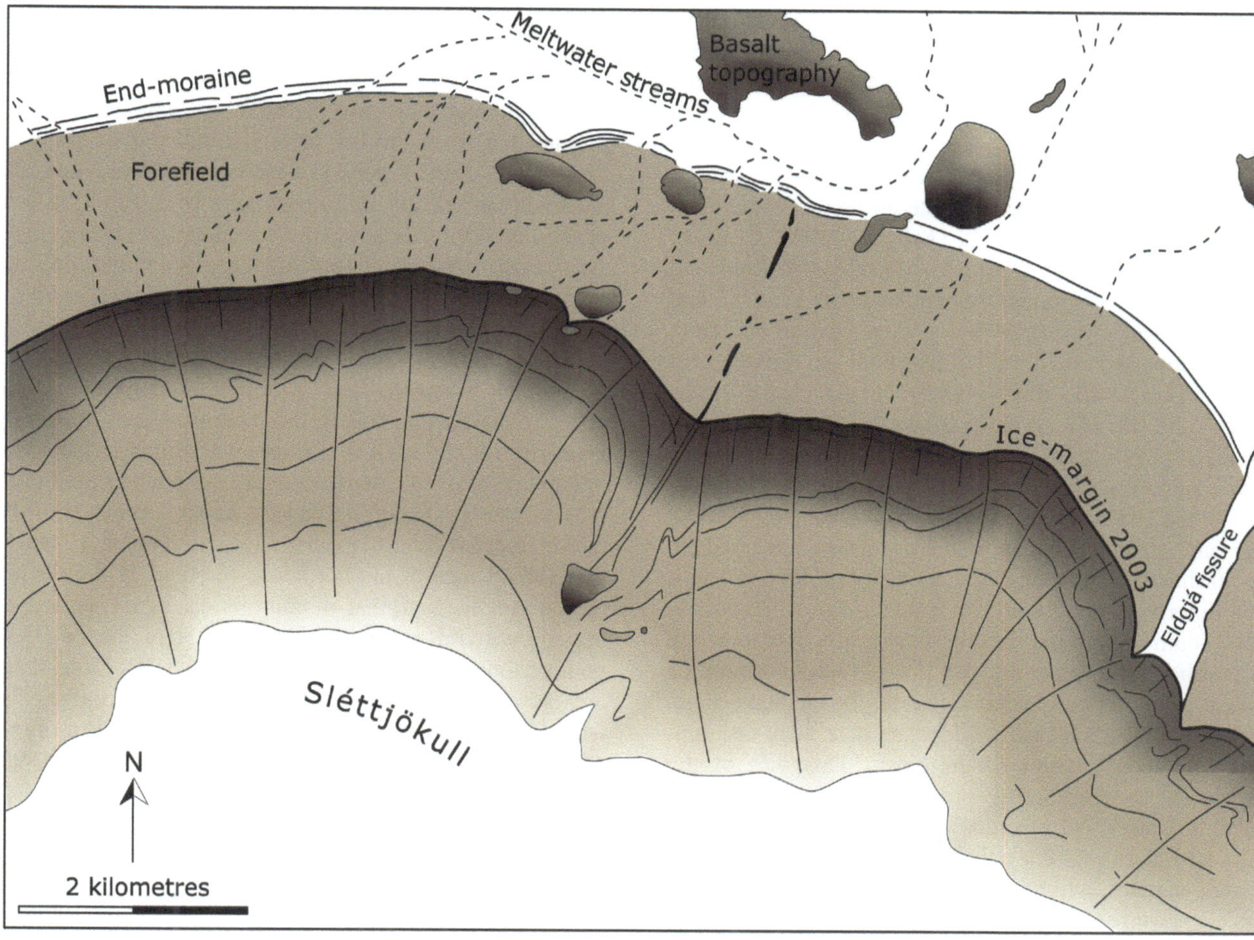

Fig. 6.28. Geomorphological outline map of the terminus region of Sléttjökull and its forefield. Mapped on the basis of field observations and satellite imagery from 2003.

Fig. 6.29. Photograph taken by K. Sapper in 1906 documents the ice-front position of Sléttjökull in the immediate vicinity of the outermost ice-marginal moraine system (after Sapper, 1909).

The moraine ridges are separated by elongated depressions, 80–100 m wide, of fluted ground moraine or flat-bottomed beds of outwash sediments joined to outwash fans beyond the ice-marginal ridge system. Numerous small basins are draped with downwash deposits, lacustrine sediments or aeolian sand. Tongues of sediment-flow deposits extend from the foreslope of the moraine ridges onto the fronting terrain. At many points, meltwater streams cut through the ridges; some of the gaps are narrow and cross single ridges, others appear as broad channels, 30–150 m wide, cut through the whole ridge system by proximal extensions of the Mælifellssandur outwash plain.

The internal architecture of one of the proximal ridges, a rather gently sloping ridge, in the girdle of ice-marginal ridges is shown in Fig. 6.30 (Krüger, 1994). Outwash sediments consisting of planar-bedded coarse sand and fine gravel with interbeds of tabular cross-bedded sand are discordantly overlain by a grey, extremely firm, matrix-supported basal till identified as Lower Till (Schomacker *et al.*, this volume). In the centre of the section, the till was strongly deformed and appeared as two clast-paved slabs suggesting that they represent a former ridge much similar to that produced along the ice front in the 1980s. The Lower Till is overlain by stratified sand and silt deposited in a shallow proglacial pond. In the northeastern part of the section, these deposits have been deformed along with the frontal ridge construction. In the southwestern part of the section, however, they are undisturbed and rest on the former ground-moraine surface and the up-glacier slope of the ancient ridge, indicating that the lacustrine deposits here were accumulated after construction of the ridge. This part of the succession is overlain by strongly deformed diamict, probably Lower Till mixed with lacustrine deposits. The dislocated diamict is cut by a prominent thrust-plane with sheared diamict, which can be traced laterally along the whole exposure. In the left-hand and central part of the section, the succession is capped by till strongly thrust and folded by ice push from a southwesterly direction. The up-glacier slope of the present ridge is capped by a layer of clast-paved, dark-grey, friable, matrix-supported basal till identified as Upper Till (Schomacker *et al.*, this volume), while the foreslope consist of sediment-flow deposits.

This exposure is interpreted as representing two different events of ice-marginal moraine ridge formation: One produced around 1890 and superimposed on remnants of a pre-existing frontal moraine (Krüger, 1988, 1994). First, the glacier advanced from a southwesterly direction across an outwash plain and deposited Lower Till. Then, the glacier retreated slightly and became almost stationary. Seasonal ice-edge fluctuations produced a frontal ridge chiefly combined of slabs of Lower Till. After a general deglaciation of the area, the ice re-advanced and deformed parts of Lower Till and pushed up the material against the already existing ridge. As the ice advance continued, the glacier overrode the ridge. The following retreat was interrupted by a small re-advance where a moraine ridge was superimposed on top of the overridden ridge.

Further westwards the system consist of a massive series of six parallel ridges; the innermost ones, however, appearing as very smoothed ridges with fluted surfaces, are interpreted as overridden ice-marginal moraine ridges (Fig. 6.31A); they probably indicate the maximum extent of the mid-eighteenth century glacier advance (Krüger, 1994). In the Bláfjöll area in the northwest forefield of Sléttjökull, the ancient ice-marginal moraine is easily recognized as mossgrown ridges, but extending 50 m beyond the younger ones from around 1890 (Fig. 6.31B). Thus, there are good reasons to presume that the ice-marginal landsystem indicates the two most prominent extensions of the glacier in historical times, around 1750 and 1890, and these ice-front positions were almost identical.

6.4.2. Sólheimajökull

Series of lateral and marginal moraine ridges mark the extent of neoglacial advances of Sólheimajökull, a southern outlet valley glacier of Mýrdalsjökull (Sigurðsson, this volume). During its LIA maximum, Sólheimajökull culminated approximately 1.6–2 km in front of its present (2007) margin. According to Thorarinsson (1943), Sólheimajökull had major advances around AD 1705, 1794 and 1820. It is most probable that Sólheimajökull reached its Neoglacial maximum around 1705. Since then, Sólheimajökull has generally been retreating, although punctuated by short intervals of

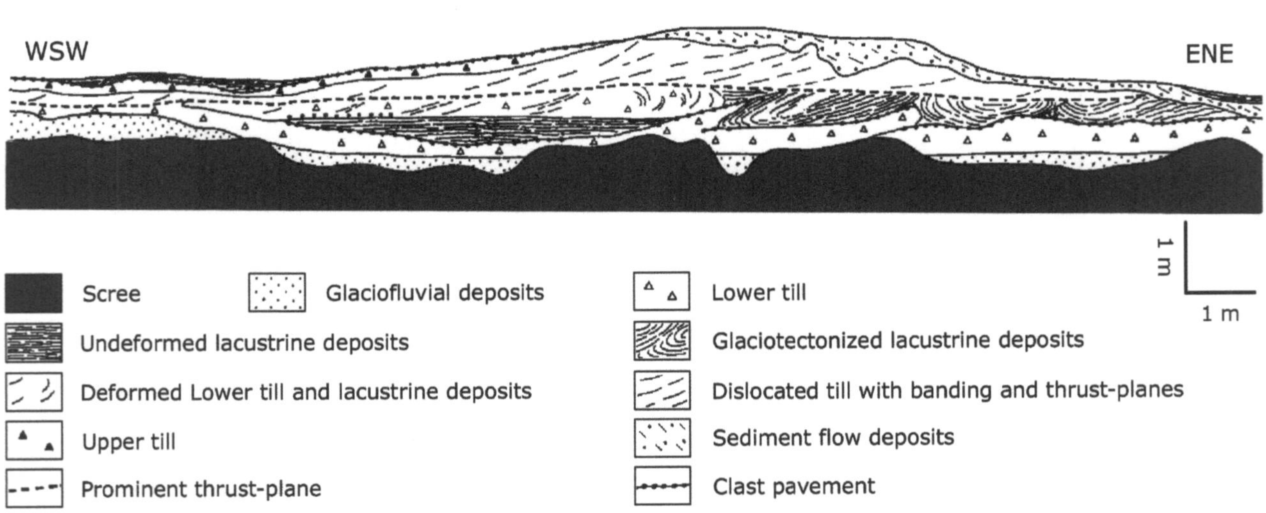

Fig. 6.30. Section in one of the innermost ridges included in the outer marginal moraine system from the Little Ice Age (after Krüger, 1994).

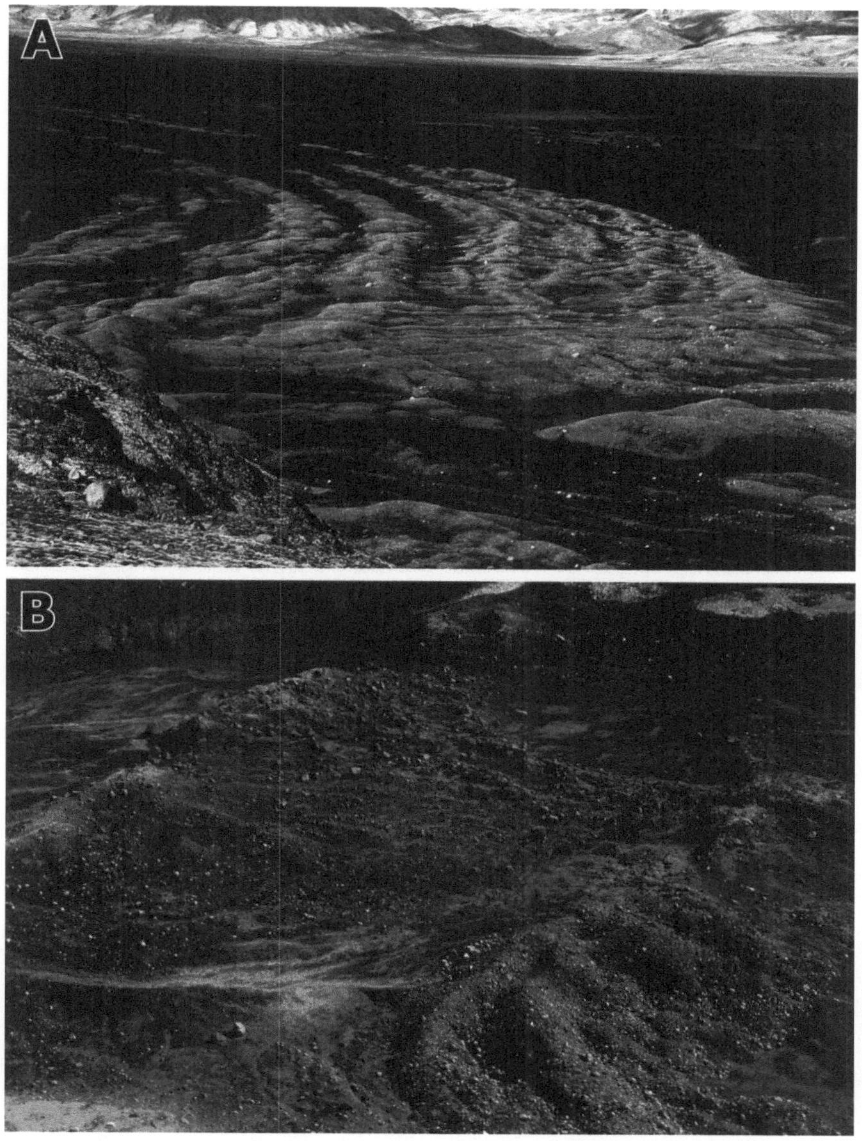

Fig. 6.31. (A) The massive series of six parallel ice-marginal moraine ridges about 3 km west of Mælifell, the innermost three ridges, however, have fluted surfaces and are overridden ridges probably produced around 1750. (B) A mossgrown marginal moraine probably from the mid-eighteenth century extending 50 m beyond the marginal moraine from around 1890 in the Bláfjöll area some 10 km west of Mælifell.

advances (Sigurðsson, 2005). The forefield consists of fluted and drumlinized ground moraine with marginal moraine ridges, some produced annually, and others in periods with a stationary or re-advancing ice margin (Fig. 6.32). The glacial river Jökulsá á Sólheimasandi dissects the forefield in the centre of the valley, providing a view into the glacial deposits in a number of natural sedimentary sections.

A prominent marginal moraine ridge from the LIA is located approximately 1.1 km from the present margin (Fig. 6.33). The moraine ridge has a moderately well-developed sawtooth shape indicating splaying/radial crevasses due to lateral extension in an advancing ice margin (Benn and Evans, 1998; Burki *et al.*, 2009). Several depressions occur in the area proximal to the moraine ridge, giving the moraine a hill-hole appearance in some places (Fig. 6.33). The structure and sedimentary composition of the moraine is visible from a river cut section in the centre of the valley. The sedimentology is characterized by diamicts with interbeds or lenses of sand. A few small outcrops and a section on the opposite side of the river indicate that glaciofluvial sandur dominates the stratigraphy below the moraine. The lowermost sediment layer properly exposed is, however, a >50 cm thick, massive, fine-to-medium-grained, matrix-supported, clast-rich, fissile and firm diamict interpreted as basal till (Benn and Evans, 1998; Krüger and Kjær, 1999). The basal till, which lies horizontally below the moraine ridge and on top of sandur deposits, can be traced undisturbed in the forefield beyond the moraine ridge, thus predating the formation of the moraine ridge. The till is cut at base of the backslope of the moraine and towards a depression on the upglacier side (Fig. 6.34). Slightly below the backslope surface, there is an upglacier-dipping slab of diamict showing the same characteristics as the basal till below. Small-scale folds in thin layers and lenses of sand and silt were observed at the base of the dipping till slab. A coarse-grained, matrix-supported, loose diamict with

Fig. 6.32. Marginal moraine ridge formed in 1995 by a sustained advance of Sólheimajökull up through the 1970s and early 1990s. May 2009.

Fig. 6.33. Little Ice Age marginal moraine ridge in the forefield of Sólheimajökull. The lower panel illustrates the composition of the moraine and a simple model of formation.

moderate amount of clasts forms the top-layer of the backslope and the distal part of the moraine ridge, and also occupies the space beneath the dipping till slab. Clasts were mostly angular to sub-rounded and many of them were striated, but with no preferred orientation of striations. Owing to its properties, this upper diamict is interpreted as a re-deposited diamict (Paul and Eyles, 1990; Krüger and Kjær, 1999).

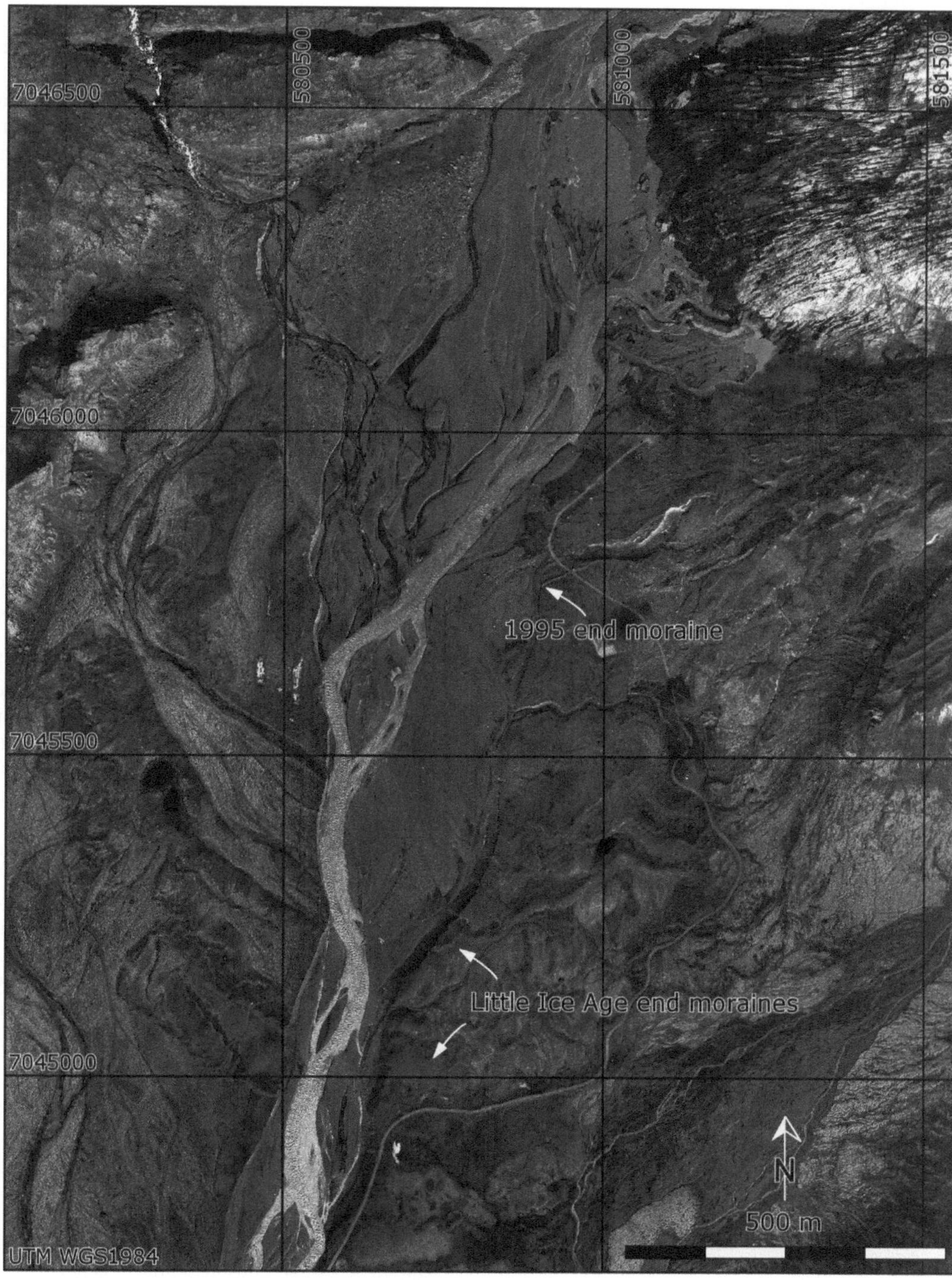

Fig. 6.34. Aerial orthophotograph showing the Sólheimajökull forefield in 2007.

The morphology and architecture of the moraine ridge implies that it was formed along an advancing ice margin by a combination of thrusting and dumping – a process also described from Kötlujökull (Krüger *et al.*, 2002; Krüger *et al.*, this volume). The hill-hole morphology suggests that the advancing ice shoved up pre-existing sediments to form the moraine ridge. The river cut section shows that by the end of the advance, the glacier detached a sheet of the pre-existing basal till covering the underlying sandur deposits and thrust it up to produce the moraine ridge. During the forward movement of the till sheet, sands and silt were slightly folded below its frontal foot. As the ice advance stopped, debris melted out of the glacier snout and dumped down-slope the ice front onto the marginal moraine ridge.

6.5. Conclusions

On the basis of the studies reported here from the margin of the principal outlet glaciers Kötlujökull, Sléttjökull and Sólheimajökull on respectively the east, north and south flank of the Mýrdalsjökull ice cap, the following conclusions may be drawn about the development of ice-marginal moraines.

(1) In front of Kötlujökull, a single-crested dump moraine combined of front cliff-fall sediments and generally without evidence of thrusting marks the maximum extent of a sustained glacier advance in 1980s. During this advance, however, the basic mechanisms involved in the frontal moraine

formation were pushing and thrusting. Without having observed the active mechanism of ridge formation, we could mistakenly have interpreted this ridge as a frontal moraine produced in association with a pause in a general ice recession.

(2) The annual-moraine ridges in front of Sléttjökull result from a mechanism that couples the ridge-forming process to limited basal freezing during winter, transportation of a frozen-on sediment slab as an integral part of the thin glacier foot during limited winter re-advances and superposition of the slab on the frontal ground. This type of annual-moraine ridges is likely to be characteristic of retreating temperate glaciers where the glacier toe is thin and winters are cold.

(3) During the retreat phases of Sléttjökull, such as before 1981 and after 1991, ridges were usually produced annually even when retreats were most rapid; the speed of glacier retreat is finely balanced with changes in summer temperature suggesting that the north part of the Mýrdalsjökull ice cap is not a surging glacier, but that the ice-front fluctuations are climatically induced.

(4) During the stillstand of Sléttjökull in the 1980s, the same basic mechanisms were involved in the moraine-ridge production. But instead of creating a series of discrete moraine ridges, the annual cycle of ice front activity produced a composite ridge composed of imbricately stacked slabs of frozen basal till, or sorted deposits, by recurrent superposition of slabs during small winter re-advances; one slab of sediments was added to the ridge each winter. Consequently, some of the so-called push-moraine ridges in former glaciated areas may mark pauses in glacier retreat, or limits of glacier recession, and not always ice limits of glacier advance as generally accepted in literature.

(5) The girdle of ice-marginal moraine ridges extending 1.5–2 km beyond the present front of Sléttjökull indicates the two most prominent extensions of the glacier in historical times, around 1750 and 1890, and these ice-front positions were almost identical.

(6) Sólheimajökull reached its Neoglacial maximum around 1705. Since then, the glacier has generally been retreating, although punctuated by short intervals of advances, the last sustained advance produced a well-defined marginal moraine in 1995.

Acknowledgements

We are grateful for financial support from The Danish Natural Science Research Council and the Carlsberg Foundation for research at Kötlujökull and Sléttjökull and from The University of Iceland research fund and Nordvulk for research at Sólheimajökull.

References

Aber, J.S. & Ber, A., 2007. Glaciotectonism. Developments in Quaternary Science 6, 1–246. Elsevier, Amsterdam.

Aber, J.S., Croot, D.G., & Fenton, M.M., 1989. Glaciotectonic landforms and structures. Dordrecht, Kluwer Academic Publishers.

Benediktsson, Í.Ö., Möller, P., Ingólfsson, Ó., van der Meer, J.J.M., Kjær, K.H., & Krüger, J., 2008. Instantaneous end moraine and sediment wedge formation during the 1890 surge of Brúarjökull, Iceland. Quaternary Science Reviews 27, 209–234.

Benn, D.I., 1995. Fabric signature of subglacial till deformation, Breidamerkurjökull, Iceland. Sedimentology 42, 735–747.

Benn, D.I. & Evans, D.J.A., 1998. Glaciers and Glaciation. Arnold, London, 734 pp.

Bennett, M.R., 2001. The morphology, structural evolution and significance of push moraines. Earth-Science Reviews 53, 197–236.

Bennett, M.R., Huddart, D., Hambrey, M.J., & Ghienne, J.F., 1996. Moraine development at the high-Arctic valley glacier Pedersensbreen, Svalbard. Geografiska Annaler 78A, 209–222.

Bennett, M.R., Huddart, D., Waller, R.I., Cassidy, N., Tomio, A., Zukowskyj, P., Midgley, N.G., Cook, S.J., Gonzalez, S., & Glasser, N.F., 2004. Sedimentary and tectonic architecture of a large push moraine: a case study from Hagafellsjökull-Eystri, Iceland. Sedimentary Geology 172, 269–292.

Boothroyd, J.C. and Ashley, G.M., 1975. Process, bar morphology and sedimentary structures on braided outwash fans, northeastern Gulf of Alaska. In: Jopling, A.V., McDonald, B.C. (Eds.), Glaciofluvial and glaciolacustrine sedimentation, Society of Economic Paleontologists and Mineralogists, Special Publication, 23, 193–222.

Boulton, G.S., 1972. Modern arctic glaciers as depositional models for former ice sheets. Journal of the Geological Society of London 128, 361–393.

Boulton, G.S., 1986. Push-moraines and glacier-contact fans in marine and terrestrial environments. Sedimentology 33, 677–698.

Boulton, G.S. & Eyles, N., 1979. Sedimentation in valley glaciers, a model and genetic classification. In: Schlüchter, Ch. (Ed), Moraines and varves. Rotterdam, A.A. Balkema, 11–25.

Boulton, G.S., van der Meer, J.J.M., Hart, J., Beets, D., Ruegg, G.H.J., van der Wateren, F.M., & Jarvis, J., 1996. Till and moraine emplacement in a deforming bed surge – an example from a marine environment. Quaternary Science Reviews 15, 961–987.

Boulton, G.S., van der Meer, J.J.M., Beets, D.J., Hart, J.K., & Ruegg, G.H.J., 1999. The sedimentary and structural evolution of a recent push moraine complex: Holmströmbreen, Spitsbergen. Quaternary Science Reviews 18, 339–371.

Burki, V., Larsen, E., Fredin, O., & Margreth, A., 2009. The formation of sawtooth moraine ridges in Bødalen, western Norway. Geomorphology 105, 182–192.

Croot, D.G., 1987. Glacio-tectonic structures: a mesoscale model of thin-skinned thrust sheets? Journal of Structural Geology 9, 797–808.

Croot, D.G., 1988. Morphological, structural and mechanical analysis of neoglacial ice-pushed ridges in Iceland. In: Croot, D.G. (Ed), Glaciotectonics – forms and processes. Rotterdam, A.A. Balkema, 33–47.

Dreimanis, A., 1989. Tills: their genetic terminology and classification. In: Goldwait, R.P., Matsch, C.L. (Eds), Genetic classification of glacigenic deposits. Rotterdam, A.A. Balkema, 17–83.

Evans, D.J.A., 1989. The nature of glaciotectonic structures and sediments at sub-polar glacier margins, Northwest Ellesmere Island, Canada. Geografiska Annaler 71A, 113–123.

Eybergen, F.A., 1987. Glacier snout dynamics and contemporary push moraine formation at the Tutmannglacier, Wallis, Switzerland. In: van der Meer, J.J.M. (Ed), Tills and glaciotectonics. Rotterdam, A.A. Balkema, 217–231.

Gripp, K., 1975. Hochsander-Satzmoräne-Endmoränenvertreter. Zeitschrift für Geomorphologie Neue Folge 19, 490–496.

Grove, J.M., 1988. The little ice age. London, Methuen.

Hambrey, M.J. & Huddart, D., 1995. Englacial and proglacial glaciotectonic processes at the snout of a thermally complex glacier in Svalbard. Journal of Quaternary Science 10, 313–326.

Harris, C. & Bothamley, K., 1984. Englacial deltaic sediments as evidence for basal freezing and marginal shearing, Leirbreen, southern Norway. Journal of Glaciology 30, 30–34.

Hart, J.K. & Watts, R.J., 1997. A comparison of the styles of deformation associated with two recent push moraines, south van Keulenfjorden, Svalbard. Earth Surface Processes and Landforms 22, 1089–1107.

Heim, D., 1983. Glaziare Entwasserung und Sanderbildung am Kötlujökull, Südisland. Polarforschung 53, 17–29.

Heim, D., 1984. Stauchmoränengenese durch die Entwicklung eines ''Gletscherfusses'' am Kötlujökull, Südisland. Polarforschung 54, 21–36.

Huddart, D. & Hambrey, M.J., 1996. Sedimentary and tectonic development of a high-Arctic thrust-moraine complex: Comfortlessbreen, Svalbard. Boreas 25, 227–243.

Humlum, O., 1981. Observations on debris in the basal transport zone of Mýrdalsjökull, Iceland. Annals of Glaciology 2, 71–77.

Humlum, O., 1985. Genesis of an imbricate push moraine, Höfdabrekkujökull, Iceland. Journal of Geology 93, 185–195.

Kälin, M., 1971. The active push moraine of the Thompson Glacier, Axel Heiberg Island, Canadian Arctic Archipelago. Research Reports of McGill University, Montreal, Glaciology 4, 1–64.

Kjær, K.H. & Krüger, J., 1998. Does clast size influence fabric strength? Journal of Sedimentary Research 68, 746–749.

Kjær, K.H., Krüger, J., & van der Meer, J.J.M., 2003. What causes till thickness to change over distance? Answers from Mýrdalsjökull, Iceland. Quaternary Science Reviews 22, 1687–1700.

Kjær, K.H., Sultan, L., Krüger, J., & Schomacker, A., 2004. Architecture and sedimentation of fan-shaped outwash in front of the Mýrdalsjökull ice cap, Iceland. Sedimentary Geology 172, 139–163.

Krüger, J., 1984. Clasts with stoss-lee form in lodgement tills: a discussion. Journal of Glaciology 30, 241–243.

Krüger, J., 1985. Formation of a push moraine at the margin of Höfdabrekkujökull, South Iceland. Geografiska Annaler 67A, 199–212.

Krüger, J., 1988. Glacial geomorphological research at Mýrdalsjökull, Iceland, 1977–86. I: the northern margin. Geografisk Tidsskrift 88, 68–77.

Krüger, J., 1993. Moraine-ridge formation along a stationary ice front in Iceland. Boreas 22, 101–109.

Krüger, J., 1994. Glacial processes, sediments, landforms and stratigraphy in the terminus region of Mýrdalsjökull, Iceland. Folia Geographica Danica 21, 1–233.

Krüger, J., 1995. Origin, chronology and climatological significance of annual-moraine ridges at Mýrdalsjökull, Iceland. The Holocene 5, 420–427.

Krüger, J., 1996. Moraine ridges formed from subglacial frozen-on sediment slabs and their differentiation from push moraines. Boreas 25, 57–63.

Krüger, J., 1997. Development of minor outwash fans at Kötlujökull, Iceland. Quaternary Science Reviews 16, 649–659.

Krüger, J. & Kjær, K.H., 1999. A data chart for field description and genetic interpretation of glacial diamicts and associated sediments – with examples from Greenland, Iceland, and Denmark. Boreas 28, 386–402.

Krüger, J., Kjær, K.H., & van der Meer, J.J.M., 2002. From push moraine to single-crested dump moraine during a sustained glacier advance. Norsk Geografisk Tidsskrift 56, 87–95.

Lawson, D.E., 1979. Sedimentological analysis of the western terminus region of the Matanuska Glacier, Alaska. U.S. Army Cold Regions Research and Engineering Laboratory, Report 79-9, 1–112.

Lehmann, R., 1992. Arctic push moraines, a case study of the Thompson Glacier moraine, Axel Heiberg Island, N.W.T., Canada. Zeitschrift für Geomorphologie SB 86, 161–171.

Lyså, A. & Lønne, I., 2001. Moraine development at a small High-Arctic valley glacier: Rieperbreen, Svalbard. Journal of Quaternary Science 16, 519–529.

Lønne, I. & Lauritsen, T., 1996. The architecture of a modern push-moraine at Svalbard as inferred from ground-penetrating radar measurements. Arctic and Alpine Research 28, 488–495.

Maizels, J.K., 1995. Sediments and landforms of modern proglacial terrestrial environments. In: Menzies, J. (Ed), Modern glacial environments: processes, dynamics and sediments. Oxford, Butterworth-Heinemann, 365–416.

Matthews, J.A., McCarroll, D., & Shakesby, R.A., 1995. Contemporary terminal-moraine ridge formation at a temperate glacier: Styggedalsbreen, Jotunheimen, southern Norway. Boreas 24, 129–139.

van der Meer, J.J.M. (Ed), 1987. Tills and glaciotectonics. Rotterdam, A.A. Balkema.

Miall, A.D., 1977. A review of the braided river depositional environment. Earth-Science Reviews 13, 1–62.

Miall, A.D., 1983. Principles of sedimentary basin analysis. New York, Springer.

Möller, P., 1995. Subrecent moraine ridge formation on Cuff Cape, Victoria Land, Antarctica. Geografiska Annaler 77A, 83–94.

Paul, M.A. & Eyles, N., 1990. Constraints on the preservation of diamict facies (melt-out tills) at the margins of stagnant glaciers. Quaternary Science Reviews 9, 51–69.

Prest, V.K., 1983. Canada's heritage of glacial features. Geological Survey Canada, Miscellaneous Report 28, 1–119.

Price, R.J., 1970. Moraines at Fjallsjökull, Iceland. Arctic and Alpine Research 2, 27–42.

Ravn, L., 2006. Interaktionen mellem klima og landskab-sediment-jordsystemet i et periglacialt miljø foran Brúarjökull, Island. Cand. scient. thesis, University of Copenhagen, 57 pp.

Roberts, D.H., Yde, J.C., Knudsen, N.T., Long, A.J., & Lloyd, J.M., 2009. Ice marginal dynamics during surge activity, Kuannersuit Glacier, Disko Island, West Greenland. Quaternary Science Reviews 28, 209–222.

de Ruyter de Wildt, M.S., Klok, E.J., & Oerlemans, J., 2003. Reconstruction of the mean specific mass balance of Vatnajökull (Iceland) with a seasonal sensitivity characteristic. Geografiska Annaler 85A, 57–72.

Sapper, K., 1909. Bemerkungen über einige südisländische Gletscher. Zeitschrift für Gletscherkunde 3, 297–305.

Shakesby, R.A., 1989. Variability in neoglacial moraine morphology and composition, Storbreen, Jotunheimen, Norway: within-moraine patterns and their implications. Geografiska Annaler 71A, 17–29.

Sharp, M.J., 1984. Annual moraine ridges at Skálafellsjökull, South-east Iceland. Journal of Glaciology 30, 82–93.

Sigurðsson, O., 2005. Variations of termini of glaciers in Iceland in recent centuries and their connection with climate. Developments in Quaternary Science 5, 241–255. Elsevier, Amsterdam.

Specht, A., 2003. Vintermoræner ved Sléttjökull, Island. MSc thesis, Department of Geography, University of Copenhagen, 1–73.

Thorarinsson, S., 1943. Vatnajökull. Scientific results of the Swedish-Icelandic investigations 1936-37-38. Geografiska Annaler 1-2, 1–54.

Winkler, S. & Nesje, A., 1999. Moraine formation at an advancing temperate glacier: Brigdalsbreen, western Norway. Geografiska Annaler 81A, 17–30.

Worsley, P., 1974. Recent ''annual'' moraine ridges at Austre Okstindbreen, Okstindan, north Norway. Journal of Glaciology 13, 265–277.

7

Dead-Ice Environments: A Landsystems Model for a Debris-Charged, Stagnant Lowland Glacier Margin, Kötlujökull

Johannes Krüger[1,*], Kurt H. Kjær[2] and Anders Schomacker[3]

[1]*Department of Geography and Geology, University of Copenhagen, Geocenter Copenhagen, Øster Voldgade 10, DK-1350 Copenhagen K., Denmark*
[2]*Natural History Museum of Denmark, University of Copenhagen, Øster Voldgade 5-7, DK-1350 Copenhagen K., Denmark*
[3]*Institute of Earth Sciences, University of Iceland, Askja, Sturlugata 7, IS-101 Reykjavík, Iceland*
**Correspondence and requests for materials should be addressed to Johannes Krüger (e-mail: jk@geo.ku.dk)*

7.1. Introduction

Dead-ice is stagnant glacier ice where movement by glacier flow has ceased (Benn and Evans, 1998; Evans, 2003). Whenever glaciers experience negative mass balance and have significant debris cover on their terminal part, dead-ice may form. Therefore, dead-ice most commonly appears as part of stagnant, debris-mantled glacier snouts (Sharp, 1949; Clayton, 1964; Boulton, 1972; Eyles, 1979; Kirkbride, 1993; Krüger, 1994; Lønne and Lyså, 2005; Schomacker and Kjær, 2007), or in various types of ice-cored landforms (Pickard, 1984; Kjær and Krüger, 2001; Evans and Rea, 2003; Schomacker *et al.*, 2006). Dead-ice also appears as remnants of buried Pleistocene glacier ice preserved in permafrost areas (Astakhov and Isayeva, 1988; Ingólfsson and Lokrantz, 2003).

Sediment accumulations on the surface of lowland glaciers originate from different debris sources such as melt-out of subglacial sediments, which have been thrust or squeezed up into the ice (Johnson, 1975; Sharp, 1985; Huddart and Hambrey, 1996; Hambrey *et al.*, 1999; Krüger and Aber, 1999; Glasser and Hambrey, 2002), or as melt-out of englacial debris bands (Sharp, 1949; Boulton, 1970; Kirkbride and Spedding, 1996; Hambrey *et al.*, 1999; Krüger and Aber, 1999), or melt-out of channel- or tunnel-fill material (Fitzsimons, 1991; Kirkbride and Spedding, 1996; Krüger and Aber, 1999; Vatne, 2001). Supraglacial sediment accumulations may also result from transportation of material directly to the glacier surface by meltwater bursts through the crevasse and conduit system (Näslund and Hassinen, 1996; Krüger and Aber, 1999; Roberts *et al.*, 2000; Russell and Knudsen, 2002), or they may result from deposition of aeolian material on the glacier surface (Kirkbride, 1995; Krüger and Aber, 1999; Kjær *et al.*, 2004). The type and amount of sediment cover on dead-ice strongly influence the melting rates, the reworking of the sediment cover and thereby the final sedimentary and topographic product left in the geological record. Basically, the processes working in dead-ice environments are, however, controlled by the climate, most importantly, by temperature and precipitation (Lawson, 1979; Krüger, 1994; Krüger and Kjær, 2000; Schomacker and Kjær, 2007; Schomacker, 2008).

Models for dead-ice melting and its imprint on sediment reworking on top of ice-cored moraines are well established from qualitative studies on a large number of glaciers all over the world, such as Svalbard, Iceland, Greenland, Arctic Canada, Alaska and Antarctica (Clayton, 1964; Boulton, 1968, 1971, 1972; Eyles, 1979, 1983; Lawson, 1979, 1989; Fitzsimons, 1990; Krüger, 1994; Bennett *et al.*, 1996, 2000; Hambrey *et al.*, 1997; Sletten *et al.*, 2001; Lyså and Lønne, 2001; Glasser and Hambrey, 2003). Such models have served as analogs for development of dead-ice moraines in former glaciated areas in northern Europe and North America (Sharp, 1949; Boulton, 1972; Clayton and Moran, 1974; Shaw, 1979; Krüger, 1983; Brodzikowski and van Loon, 1991; Benn, 1992; Johnsson *et al.*, 1995; Ham and Attig, 1996; Eyles *et al.*, 1999; Boone and Eyles, 2001; Johnson and Clayton, 2003).

A number of workers have attempted to quantify specific de-icing processes and sediment reworking processes (Østrem, 1959; McKenzie, 1969; Carrara, 1975; Lawson, 1979; Driscoll, 1980; Nakawo and Young, 1981, 1982; Pickard, 1983, 1984; McKenzie and Goodwin, 1987; Johnson, 1992; Bennett *et al.*, 2000; Etzelmüller, 2000; Nicholson and Benn, 2006; Schomacker and Kjær, 2007). The data sets, however, were obtained chiefly from consideration of cold polar or subpolar environments of dry continental nature, and few data are available for melting rates and sedimentary processes taking place in ice-cored moraine fields under humid, maritime conditions (Krüger, 1994; Syverson and Mickelson, 1995; Krüger and Kjær, 2000; Kjær and Krüger, 2001). The impact of multiple glacier surges on dead-ice melting and landscape formation has most recently been studied in the forefield of Brúarjökull, East Iceland (Schomacker *et al.*, 2006; Schomacker and Kjær, 2007).

Kötlujökull, the south-eastern outlet of the Mýrdalsjökull ice cap, has produced considerable amounts of

DEVELOPMENTS IN QUATERNARY SCIENCES
VOLUME 13 ISSN 1571-0866

supraglacial debris in its outer snout zone, also compared to nearby outlet glaciers from the same ice cap (Fig. 7.1). Most probably, this specific nature of the outer snout of Kötlujökull is a result of complex frontal oscillations with frequent ice advances with glacial thrusting followed by periods with ice-margin wastage and stagnation. Aerial photographic documentation of Kötlujökull extends back to 1945, and several scientific field expeditions have studied the glacier forefield, providing important 'time slices' of different stages in the dead-ice melting and landscape formation. Therefore, Kötlujökull provides suitable conditions of studying dead-ice melting and its imprint on sedimentary processes and landscape formation in a debris-charged lowland glacier margin environment. The scientific challenges and research rationale are to answer the following key questions. What are the specific conditions for dead-ice formation in the Kötlujökull area? From which sources does the sediment cover originate on the terminal part of the glacier? Which melting and reworking processes are acting in the ice-cored moraine terrain? What is the rate of de-icing the ice-cored moraine under humid, sub-polar conditions? Does

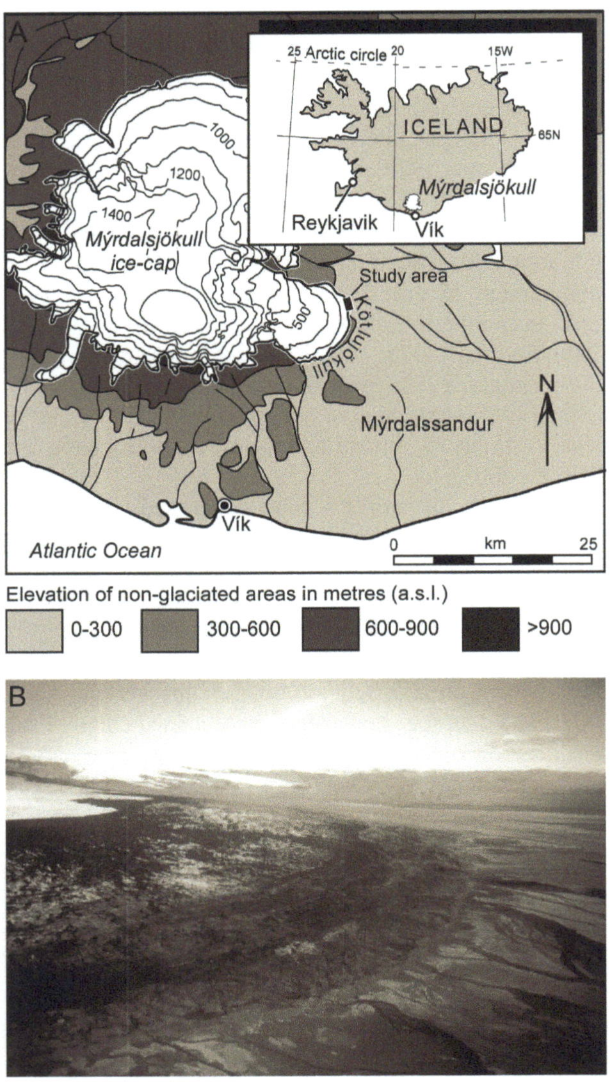

Fig. 7.1. (A) Location map showing Kötlujökull and the study site. (B) Oblique aerial view of the debris-covered terminus of Kötlujökull. August 2001.

this rate differ from that reported from cold polar environments of dry continental nature? How will the sedimentary architecture appear in the geological record? How will the final landsystem appear?

We intend to answer the above key questions by reviewing the research carried out and the conclusions drawn concerning dead-ice melting and landscape formation in the terminus region of Kötlujökull and link the identified processes and specific landform-sediment association to the current climate and glacier–volcano interaction.

7.2. Glacier Dynamics and Sources of Supraglacial Sediments

Kötlujökull transports huge quantities of debris, which melt out on the glacier surface from numerous debris bands and thrust planes in the terminal part of the glacier (Fig. 7.1B). Aerial photographs show that the upper limit of almost coherent dirt cover has migrated down-glacier through time and demonstrate a drastic decrease in surface velocity with distance down-glacier (Krüger, 1994; Krüger and Aber, 1999; Krüger and Kjær, 2000). This has given rise to an intense upward-directed thrusting of the ice mass and repetition of the debris band sequences.

The supraglacial sediment accumulations originate from different sources. Sorted sediments black in colour, chiefly of sand size and fine gravel, emerge from numerous debris bands all over the debris-covered portion of the glacier snout (Fig. 7.2A,B). Some of these bands consist of material originating from volcanic eruptions, but many of the thin debris bands containing finer materials may also represent dust storm events or old ablation surfaces. Snow diagenesis and ice flow have incorporated the individual accumulation layers and moved them from the catchment area to the ablation area. The ice may also contain lenses and discontinuous layers of sorted sandy material, 0.5–1.5 m in thickness, with prominent, but heavily deformed sedimentary structures. Observations of fans and sheets of similar sediments deposited in depressions on the glacier surface suggest that overriding by higher ice due to longitudinal compression may incorporate such sediment bodies in the subsurface ice.

Diamict sediments emerge mainly from debris-loaded thrust planes (Fig. 7.3). They are gravel-silt-sand or silty sand (according to the classification of Lawson, 1979) with subangular to subrounded stones and boulders, many of which are well striated, suggesting that this material represents outcrops of basal traction-zone debris and basal till which has been thrust up into the ice (Fig. 7.4). When the glacier was advancing in the mid-1980s (Krüger *et al.*, this volume), the terminal ice was strongly affected by dynamic activity related to upward-directed thrusts, so that overriding by higher ice had produced an overhanging ice front cliff (Fig. 7.5A). Major thrust planes heavily loaded with 10–50-cm-thick debris including striated stones and boulders – outsized boulders of 1 m in diameter occurred – could be traced several hundred metres along the overhanging ice wall (Fig. 7.5B). They thinned and thickened laterally and were often slightly deformed. Similar thrust planes were very common

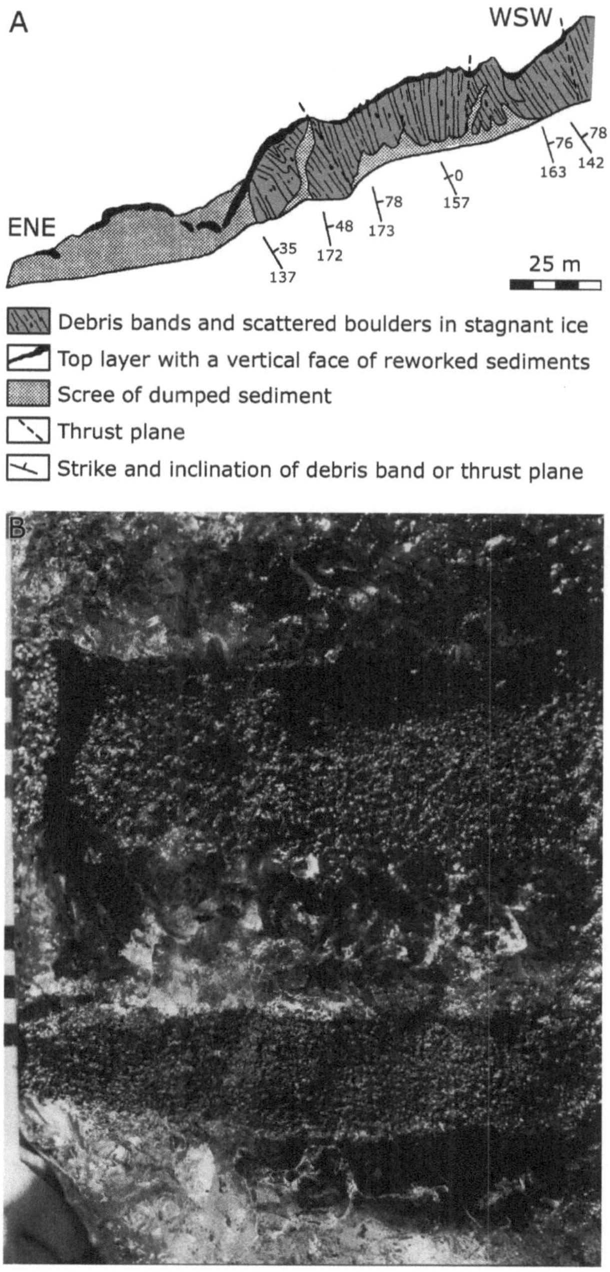

Fig. 7.2. (A) Ice wall section flanking a meltwater stream up-glacier from terminus. (B) Two debris bands, 5–10 cm thick, of sand alternating with bands of relatively clean ice.

within the terminal part of the glacier where they occurred on the glacier surface as curved bands, or ice-cored ridges, light greyish-brown in colour due to the outcropping gravel-silt-sand sediment.

In addition to the debris bands of diamict and sorted sediments, some supraglacial sediment accumulations comprising water-worked, poorly sorted gravely sand with non-striated subangular to rounded clasts emerge higher up on the glacier. A large accumulation of this type of sediment was seen in the 1980s in a restricted zone 600–800 m up-glacier from the ice margin as a slightly curved girdle, 1.2 km long and 200–300 m wide, comprising two to four parallel, sharp-crested, ice-cored ridges, 2–15 m high, with intervening ice-floored troughs (Fig. 7.6). The geometry of the ridges was expressed as cropping-out debris bands, or debris-loaded thrust planes. In the mid-1990s, however, this large sediment accumulation had reached the terminus region because of ice flow and developed into a ridge-shaped ice-disintegration field of irregular topography (Näslund and Hassinen, 1996; Krüger and Aber, 1999). Similar water-worked material was also seen cropping out as circular accumulations higher up on the glacier. The shape of these cropping-out sediment accumulations simply reflects the geometry caused by the intersection of the ice surface and abandoned conduit fills. Much of the englacial water-worked debris load may have originated from the catastrophic Katla eruption in 1918, which triggered a large jökulhlaup (Jonsson, 1983; Tómasson, 1996; Russell *et al.*, this volume). During this event, meltwater transported enormous amounts of material both across the surface of the glacier and englacially through conduits, shear planes and crevasses. The water-worked sediments have subsequently been transported and rearranged by ice movement, deformation and ablation.

7.3. Stages of Dead-Ice Moraine Development

The fluctuations of Kötlujökull up through the 20th century have resulted in the formation of a series of ice-cored moraines due to intermittent detachment of dead-ice fields of irregular topography covered by large quantities of coarse-grained diamict (Fig. 7.7A). In 1945, after an ice advance around 1940, the northeast ice-front section terminated in a zone, 100–400 m wide, of debris-covered dead-ice, bounded by an end-moraine ridge 700–1,000 m behind the outermost end-moraine produced around 1900 (Krüger and Kjær, 2000). In 1960, the glacier was retreating after another advance around 1955 (Þórarinsson, 1957), and the northeast glacier section terminated in a debris-covered dead-ice zone, 200–500 m wide, bounded by an end-moraine ridge, which was located 250–400 m behind the 1940 moraine, except in the northern ice-front section where the glacier had advanced 80–90 m beyond its 1940 position. From the end of the 1970s and until around 1987, the northeast front of Kötlujökull advanced across a forefield of ground moraine, ice-cored moraine, and outwash fans producing a frontal moraine ridge (Fig. 7.8A) (Krüger *et al.*, this volume). The terminal part of the advancing glacier was capped by a debris-mantled ice-disintegration field very much similar to that found on dead-ice. In the southern part of the northeast ice-front section, the glacier advanced to a position 100–200 m beyond its 1955 stage; in the central part, it almost equalled this stage, but more northerly, the glacier advanced to a position just 80–120 m behind its 1955 position. Since then, the glacier has been retreating, and the most terminal part of the debris-mantled glacier has developed into a zone, 300–500 m wide, of dead-ice (Fig. 7.8B).

The three advances of Kötlujökull, around 1940, 1955 and in the 1980s, and the following intermittent detachment of terminal dead-ice fields of irregular topography covered by large quantities of sediments have resulted in the formation of three different stages of dead-ice moraine development in the present-day Kötlujökull

Fig. 7.3. View across irregular glacier surface of dirt-covered ice peaks high up on Kötlujökull. The crevasse wall in the foreground shows relatively low-angle debris bands cut out by a higher angle diamict-laden thrust plane.

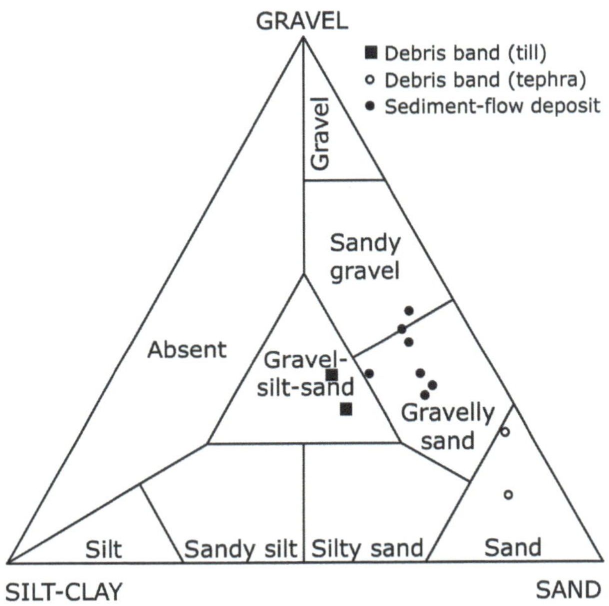

Fig. 7.4. Textural characteristics of various types of englacial debris and sediment flow deposits. Classification slightly modified after Lawson (1979).

terminus region, which are distinct from each other and demonstrate the evolutionary series of changes that the wastage of ice-cored moraines passes through (Krüger and Kjær, 2000) (Fig. 7.7B): (1) a fully ice-cored dead-ice field initiated in the late 1980s and now in a mature stage of collapse; (2) a partially ice-cored dead-ice moraine initiated in the late 1950s and now in a final stage of collapse and (3) a post-melt hummocky moraine landscape de-iced during the past 60–70 years. In addition, a young stage represented by parallel, ice-cored ridges occurs on the lower part of the active glacier. The rapidity with which the dead-ice field passes through this evolutionary series of changes is controlled by the current climate, thickness of sediment cover and hydrological processes.

7.3.1. The Fully Ice-Cored Dead-Ice Field

In the part of the terminal dead-ice terrain that is fully ice-cored, the thickness of dead-ice is between 10 and 40 m. The morphology of the terrain is typical of thermokarst on dead-ice (cf. Clayton, 1964): hummocky and very irregular with numerous mounds, ridges, collapsing edges and ice walls (Fig. 7.9). Steep-sided niches extend backwards into the flanks of broad-crested ridges or table-topped mounds to form cirque-like ice-degradation features with ice-cored backwalls or with ice exposed in them. The major steep-sloping free faces of ice result from mass movement processes, or because of over-steepening of slopes due to fluvial erosion along their base, but ice is also exposed by widening of crevasses or collapse of tunnel roofs. In walls where ice is visible, the mantle of heterogeneous sediment covering the ice core is 0.2–1.0 m thick and locally up to 2 m. Elevated patches of fluvial and lacustrine sediments show the former presence of streams and ponds and document relief inversion. In other parts of the terrain, ice-walled fluvio-glacial outwash with braided streams occurs. Downstream some of the watercourses terminate in an ice cave, and further drainage is through the ice core. The occurrence of dry stream beds and dry caves illustrate that drainage experienced frequent changes due to ice degradation and gradual lowering of the terrain.

7.3.1.1. De-icing processes

In the fully ice-cored terrain, the ice-melt processes can be divided into backwasting, defined as the lateral retreat of near-vertical free ice walls, or steep ice-cored slopes, and

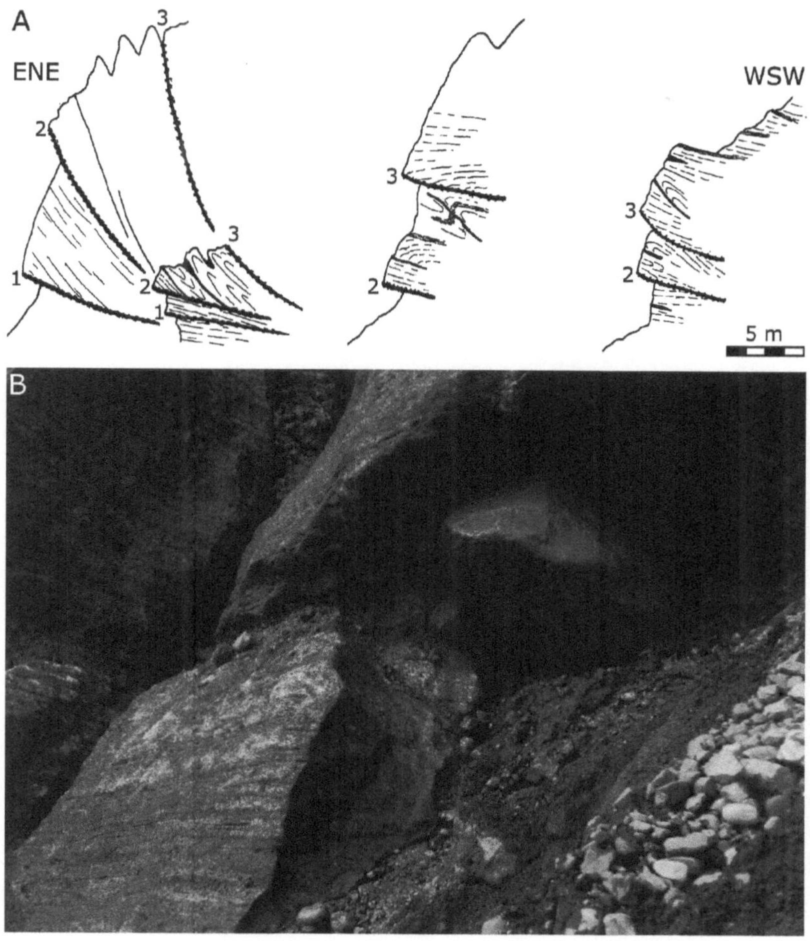

Fig. 7.5. (A) Ice wall sections transverse to the overhanging ice-front cliff showing debris-banded ice disrupted along debris-loaded thrust planes (1–3). (B) Stoss-lee shaped boulder, about 1 m in size, melting out from diamict-laden thrust plane.

downwasting, defined as the thinning of ice core by melting along the top and bottom surfaces (Fig. 7.10) (Eyles, 1979; Lawson, 1979; Krüger, 1994; Krüger and Kjær, 2000; Schomacker and Kjær, 2007).

In the field, backwasting of near-vertical ice walls or ice-cored slopes with sediment cover of 0.2–0.5 m was measured with tape as distance from the retreating edge above each wall or slope to a fixed benchmark (Pickard, 1984; Krüger, 1994). An average daily rate of melt of 6.7 cm was recorded during the field seasons; southerly exposed ice walls, however, retreated about 1.5 times faster than north-facing walls (Fig. 7.11). An average daily rate of backwasting of ice-cored slopes amounted 2.3 cm. Recordings of backwasting of steep ice walls and ice-cored slopes with a sediment cover of 0.1–0.7 m over 1-year periods gave an annual backwasting rate of respectively 7.4 and 1.5 m.

The ice-face retreat across the fully ice-cored terrain consumes high ground, causing surface lowering and flattening of the ice-cored slopes. The result is seen in some parts of the area where the terrain is relatively smooth and dominated by shallow niches with low-angled backslopes. Collapse due to karstic destruction of the ice core gradually break the sediment cover and expose ice leading to development of new ice walls and a new cycle of ice-degradation.

Downwasting of ice covered by sediment was determined by measuring the rate of lowering of terrain surface using repeated stadia surveys. The sites chosen were located on top of table-topped mounds or broad-crested ridges (Krüger and Kjær, 2000). All stations had a mantle of gravelly sandy diamict between 0.4 and 1.3 m thick. During the period of monitoring, sites were lost when rapid backwasting of nearby slopes overtook the sites, but new sites were then established. The average annual downwasting rate of ice core amounts 63 cm. The relative contributions of top melt (melt at the ice–sediment cover interface) and bottom melt (melt at the ice–bed interface) to the total downwasting have also been recorded. The field procedure is described in Krüger and Kjær (2000). The annual top melt was about 10 cm in average, while the mean annual bottom melt was 49 cm. In each case, the bottom melt was significantly greater than the top melt. The bottom melt mainly results from through-flow of water at the ice–bed interface, as the annual geothermal heat flow contributed to the ice melt with only 2 cm (Krüger and Kjær, 2000).

The relative contribution of the de-icing processes to the total rate of ice degradation in the fully ice-cored dead-ice field is presented in Table 7.1. Within a mapped area covering about 3,000 m^2 of the fully ice-cored terrain 250–300 m up-glacier from the end-moraine ridge, the calculated annual volume of ice lost through ablation of ice walls and ice-cored slopes was respectively approximately 1,600 and 600 m^3. If these volumes of ice were uniformly distributed, then backwasting of exposed ice faces and ice-cored slopes contributed to the surface lowering with respectively 0.5 and 0.3 m/yr. According to Table 7.1, top melt and bottom melt

Fig. 7.6. (A) Water-worked sediments cropping out in the ridge/trough zone high up on Kötlujökull in 1984. (B) Cover of water-worked sediment in the ridge/trough zone.

contributed separately to the surface lowering with respectively 0.1 and 0.5 m/yr. Thus, an annual surface lowering of about 1.4 m due to all melting processes is representative for the central part of the fully ice-cored dead-ice field.

Repeated stadia survey along fixed transects across the fully ice-cored dead-ice field gives supplementary information about the annual surface lowering. A transect running from a stable benchmark in the glacier forefield and across the terminal 200–300 m of the glacier documents the gradual stagnation and surface lowering of the terminal part of the glacier up through the 1990s (Fig. 7.12). During the two time periods 1986–1995 and 1995–2001, the surface of the developed dead-ice field has on average been reduced in altitude by respectively 2.5 and 1.6 m/yr as a result of ice disintegration. The main reasons for the declined rate of lowering of the ice-cored terrain are a thicker sediment cover and much less exposed ice.

7.3.1.2. Reworking of sediments

In the dead-ice environment, the debris cover is exposed to cycles of resedimentation due to melting of buried ice before a final product is left in the geological record. By far the most important processes are mass movements governed by gravity, such as fall, slumping, sliding and flowing of sediments (Figs. 7.13 and 7.14).

Sediment reworking is also due to fluvial and aeolian action. Backwasting of free ice walls results in undermining of the top layer above and subsequent fall and sliding of sediment from the working edge to the base of source slope. The recorded average daily rate of sediment cover erosion by fall and sliding along with recession of free ice walls is about 7 cm (Fig. 7.15). Apparently, rain is unimportant as a cause of sediment erosion associated with fall; no marked increase in sediment erosion along working edges accompanies the peaks of rainfall, probably because the near-vertical faces of relatively

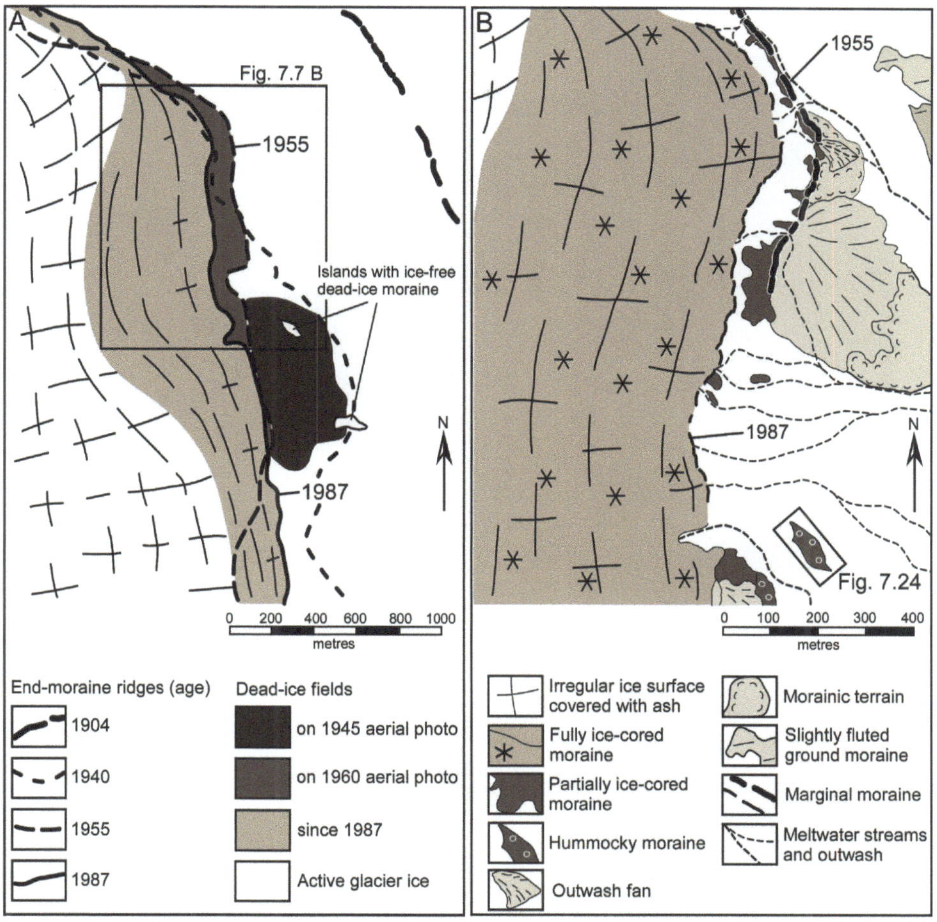

Fig. 7.7. (A) Marginal position of Kötlujökull in 1904, 1940, 1955 and 1987 and associated dead-ice fields. (B) Geomorphological map of the study area in 1996 indicating the location of areas with, respectively, fully ice-cored moraine, partially ice-cored moraine and ice-free dead-ice moraine. The position of the glacier margin in 1955 and 1987 is shown. Modified after Kjær and Krüger (2001).

coarse-grained sediment are well-drained, so that the material remains unsaturated and resists flow, even during periods of heavy rain (Krüger, 1994).

At the base of ice walls, the increasing accumulation of sediment with repeated input of material from above results in sliding of sediment cover as a relatively coherent mass on the sloping ice. Sliding of sediment masses occurs sporadically throughout the fully ice-cored field, where they leave niches and ice-floored depressions bounded by nearly vertical faces of reworked sediments. Due to excess of water, the sediments often become water-soaked, so that the majority of sliding is combined with sediment flow behaviour. Mass movement of unsaturated sediments by backslumping along discrete planes is also very common and originates from recurrent slope-fracturing upslope as the support is removed by erosion due to sliding, flow or water activity. Backslumping may also result from over-steepening of slopes due to irregularity of melting of the surface of buried ice. Stones and boulders are frequently forced to roll, or slide, forming distinct fall-sorting patterns of the material with the direction of coarsening downslope with accumulations of stones and boulders at the base of the slope (Fig. 7.16).

Remobilization by sediment gravity flows, where the water content is the most important parameter controlling the behaviour of the flow, occurs, but due to their coarse-grained nature, most sediment is not of a very liquid consistency during movement (cf. Lawson, 1979). In the highly humid climate in the Kötlujökull region, the mechanism of sediment cover saturation by intense precipitation is critical for initiation of slope processes and maintenance of ice walls, whereas the mechanism of ice–sediment interface saturation by melting of the ice core is of less importance (cf. Eyles, 1983); sunny days, or extended dry, cloudy periods, mostly restore the stability and strength of the covering sediment (Krüger, 1994).

The movement of four selected sediment flows were measured by placing lines of painted wooden blocks across flow half-way upslope (Table 7.2). It appears that periods of heavy rain generally start sediment flow of varying magnitude. Furthermore, interesting points emerge from consideration of the flow activity and the daily rainfall rates. For example, the 13 July 1986 case of flow followed a period when sediments were so slow moving as to be imperceptible, accompanying a period of general moisture deficit. On 12 July, there was 26 mm of rain, but no instability occurred. Obviously, it was the higher daily peak of 35 mm of rain the following day that initiated the surge-like event involving different materials, slopes and flow thicknesses at the same point of time. The average surface movement amounted approximately 1–3 m. The preceding period of sunny weather had

Fig. 7.8. (A) Steep overhanging front of the advancing glacier Kötlujökull in July 1986. In the foreground, a meltwater gap cuts through a terminal moraine ridge from a glacier advance in 1955. (B) The same locality in 2003. In the background, the debris-covered stagnant margin of Kötlujökull is seen.

reduced the moisture content of the flow material and increased its shear strength to a degree, which obviously reduced the total effectiveness of rainfall. Therefore, flow only took place after a *preparation day*, when moderate influx of rainwater had increased the moisture of the sediment followed by a day of heavy rain, which further increased its shear stress and reduced its strength to the critical value where failure occur. There is a supporting line of evidence from the flow behaviour of flow 4 after the surge-like event (Table 7.2). During the following three weeks, there were peaks of rainfall, but they never totalled more then 29 mm over a period of 1–2 days and occurred after stability had been restored during the preceding days of sunny weather. No perceptible flow along the reestablished block lines accompanied the peaks of rain, apparently because the recorded precipitation was insufficient to saturate the sediment and initiate flow. Thus, extremely high daily rainfalls with a preceding *preparation day* of moderate precipitation usually initiate pronounced accelerations of sediment movement, whereas warm, sunny days mostly increase the time between flow events and restore sediment stability and strength (Krüger, 1994).

The drainage of surface water is another significant process of erosion and transportation operating in the fully ice-cored terrain. On sunny days, meltwater containing fines flows off the ablating ice faces in thin sheets and numerous small rills. Streamlets erode into the base of slopes and removed fine components and deposited fans of sorted silt and sand in local depressions, or they met with major streams. Numerous small ponds intermittently filled during periods of heavy rainfall developed on the sediment cover in isolated depressions formed by irregular ice melt and collapse. The major streams were typically located in ice-walled gorges where gravely and stony materials had accumulated along the courses, or in depositional trough regions where undermining of stream banks of deposits provided material for fluvial transport, or they were located in broad basins where gradients were low and the stream became braided with widespread deposition of sand and gravel (Fig. 7.17). Many surface streams continued directly to the ice margin along courses deeply incised into the

Fig. 7.9. Oblique aerial view of the most recently detached dead-ice field in the terminus region of Kötlujökull showing a mature phase of dead-ice moraine development with a typical terrain of thermokarst.

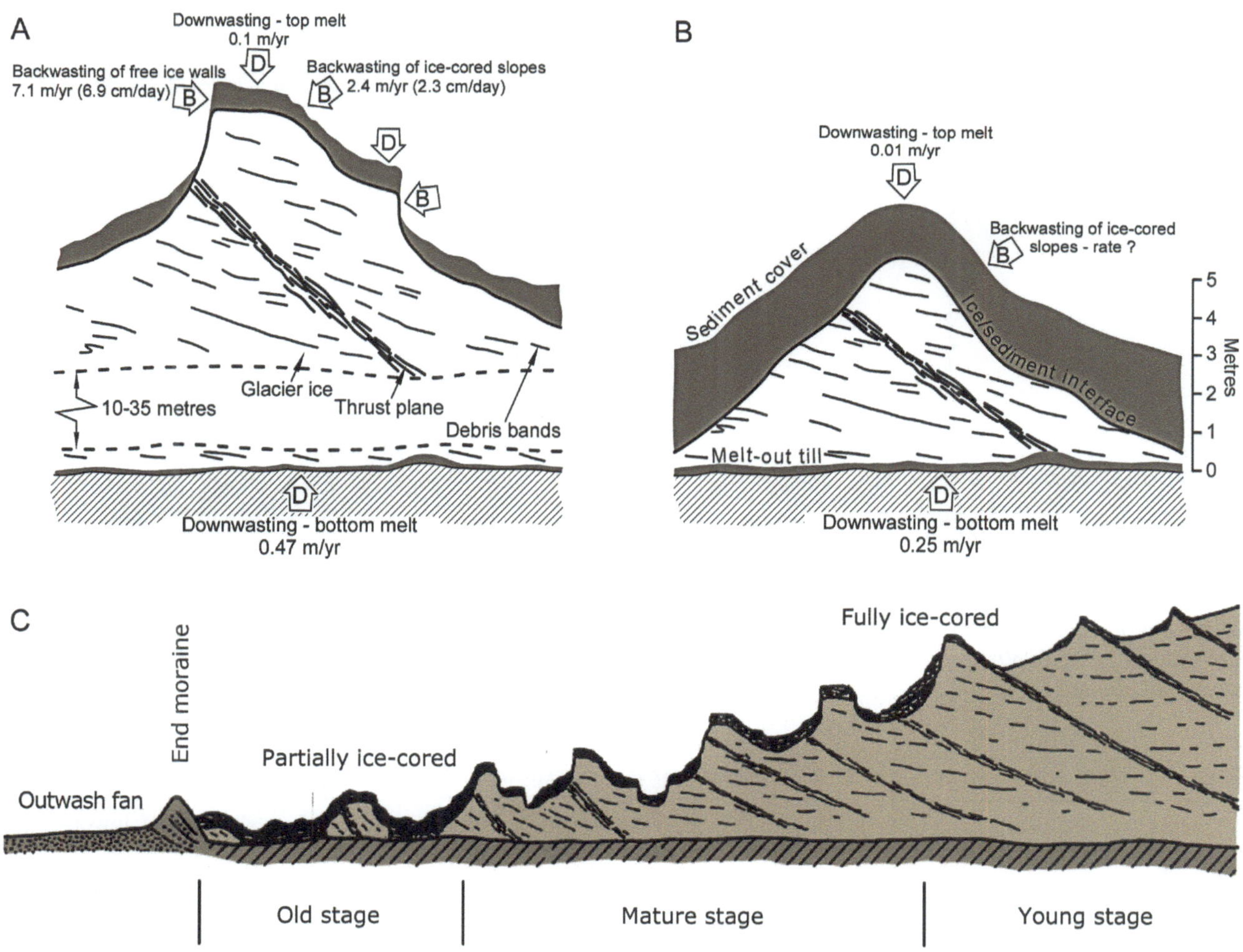

Fig. 7.10. Modes and rates of de-icing progression in the fully ice-cored dead-ice field (A) and in the partially ice-cored dead-ice field (B) in the terminus region of Kötlujökull. (C) Idealized profile across a dead-ice terrain from the terminal moraine to the active glacier.

terminal dead-ice. Others found lake basins, or englacial tunnels, many of which were in a process of karstic enlargement and collapse. The majority of field evidence shows that sediment concentration in supraglacial streams fluctuates with changing weather. On days of strong precipitation, streams were heavily loaded with sand and gravel, because large quantities of rain generally initiated surge-like movement of water-soaked sediment downslope to the water courses providing material for fluvial transport. On the contrary, extended sunny periods were essential to maintain a water supply by ablation, but only small amounts of sediments were entrained in streams due to the temporary stability of many ice-cored slopes. Extended dry, cloudy periods reduce meltwater production, so that many water courses became temporarily dry. One must conclude that when the gradients of surface streams are relatively high, large-scale sediment transportation mainly takes place in very restricted periods of rain. It is, furthermore, evident that stony material is generally left on the ice or in supraglacial stream channels, and only silt, sand, and some gravel are washed away in the water and deposited as ice lake or trough deposits.

All these basic sediment-forming mass movement processes were progressively recycled along with ice degradation. Breakage of the sediment cover as a result of development of a crevasse, or collapse of a tunnel roof due to karstic enlargement of englacial or subglacial

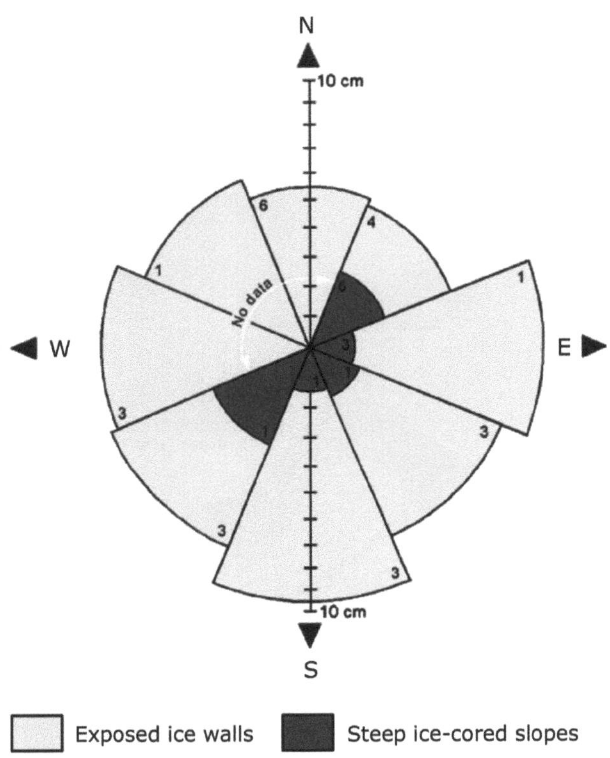

Fig. 7.11. Average daily rates of backwasting of free ice walls and ice-cored slopes relative to orientation in July–August in the 1980s and 1990s. The figures indicate number of slopes.

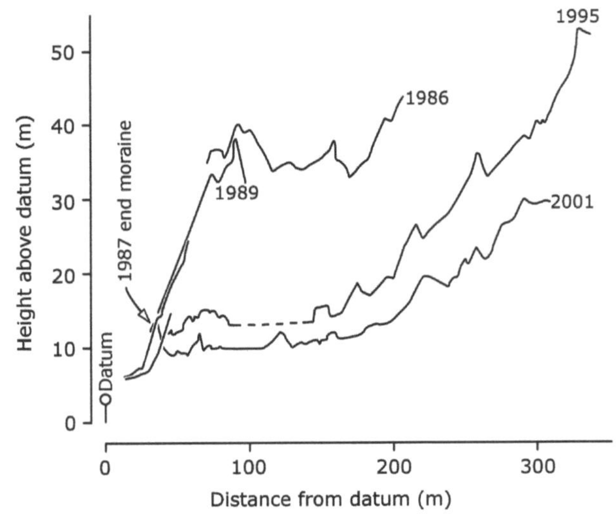

Fig. 7.12. Surface lowering along a transect illustrating the de-icing progression in the fully ice-cored dead-ice terrain during the time period 1986–2001 (after Krüger et al., 2002).

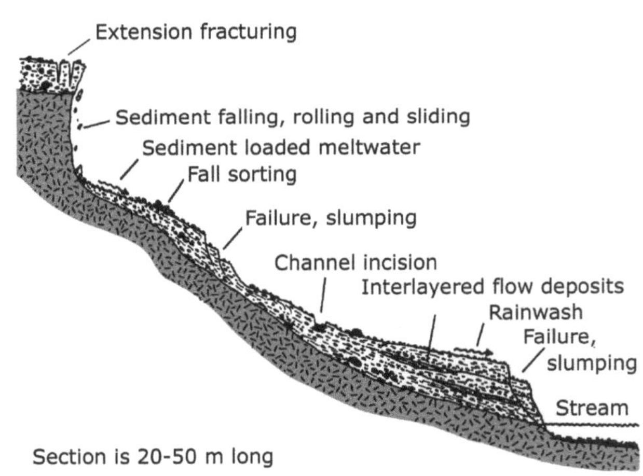

Fig. 7.13. Associations of resedimentation processes acting in a single cycle of reworking.

Table 7.1. Total annual melting due to all processes for a surveyed area of 3,000 m², 1995–1998.

Processes	Ice volume (m^3)	Surface lowering (m)
Top melt	300	0.1
Bottom melt	1,500	0.5
Ice wall retreat	1,500	0.5
Ice-cored slope retreat	1,000	0.3
Total	4,300	1.4

Source: After Krüger and Kjær (2000).

Fig. 7.14. View across fully ice-cored field showing various resedimentation processes. (1) Cover of multiple redeposited sediments. (2) Dump of sediment from the top edge of backwasting ice cliff. (3) Sliding and flowing of debris on ice-cored slope. (4) Progressive backslumping. (5) Group of stony material as a result of gravity sorting. (6) Gully formed water erosion. (7) Sorted sediment deposited by water. (8) Beginning subsidence of sediment flow deposits.

stream channels, typically exposes ice and initiates large-scale backwasting. This in turn initiates a new resedimentation cycle implying a repetition of the sequence of the many resedimentation processes such as falling, rolling, sliding, slumping and flowing of material downslope into a new depositional trough region with development of overlapping debris-flow deposits (Fig. 7.18). Because this multi-cycle event of reworking is controlled by backwasting rather than by downwasting, it leads to a stepwise lowering of the fully ice-cored terrain. A general lack of topographical inversion is also a consequence of the high rates of backwasting relative to downwasting. When free ice walls or ice-cored slopes are attacked by backwasting, the sediment cover will slide or flow downslope into the adjacent depressions. Ideally, these depressions are the potential new topographic highs. Generally, however, these areas will not become topographical highs, because backwasting of ice-cored slopes across the ice-cored terrain consumes the high ground causing flattening of the terrain; sediments move from one level to a lower one as backwasting repeatedly overtakes downwasting (Kjær and Krüger, 2001). Field experiments have documented that the combined effects of the basic mass movement processes may rework the sediment cover one to two times during an ablation period (Krüger, 1994). Observations also show, however, that large masses of the sediment are reworked more frequently, while the remainder of the sediment cover is relatively stable (cf. Lawson, 1979).

7.3.2. The Partially Ice-Cored Dead-Ice Field

Beyond the fully ice-cored terrain is the area that was covered by masses of dirty stagnant ice in 1960. Since then, the ice has thinned considerably, and therefore, the former coherent ice mass is disintegrated into isolated dead-ice blocks, 5–10 m thick, capped by 1–3 m of multi-cycle redeposited sediments with an almost complete vegetation cover dominated by mosses (Figs. 7.19 and 7.20). Ice is not visible, but still remains as cores, especially beneath topographic high areas. The presence of ice cores, however, is evident from the many backslumping features associated with ice-cored slopes and from the many sinkholes that reflect the space vacated by melting ice. The intervening mainly ice-free areas consist of low-relief hummocky moraine. A significant feature is formed by the numerous groups of stones and boulders scattered throughout the terrain. They mainly result from gravitational sorting during the former stage of collapse, where they accumulated at the foot of free ice faces or ice-cored slopes, but since then, many of them have been converted to hummock cappings (Krüger, 1994; Krüger and Kjær, 2000).

Within the partially ice-cored terrain, de-icing is dominated by downwasting processes and backwasting of ice-cored slopes, because of the relatively thick, complete sediment cover that prevents exposure of ice (Fig. 7.10B,C). At places where ice-degradation is at a more advanced stage, however, the contribution from backwasting of ice-cored slopes is considered to be very limited, because of the thicker sediment cover and the reduced relief. Figure 7.21 displays the magnitude of the surface lowering for selected observation points in the partially ice-cored terrain. It appears that the annual reduction in altitude due to downwasting amounted to 28 cm (Kjær and Krüger, 2001). The data do not allow a differentiation of the relative contribution of top melt and bottom melt, but data from the fully ice-cored terrain show that below more than 1 m of sediment, the annual

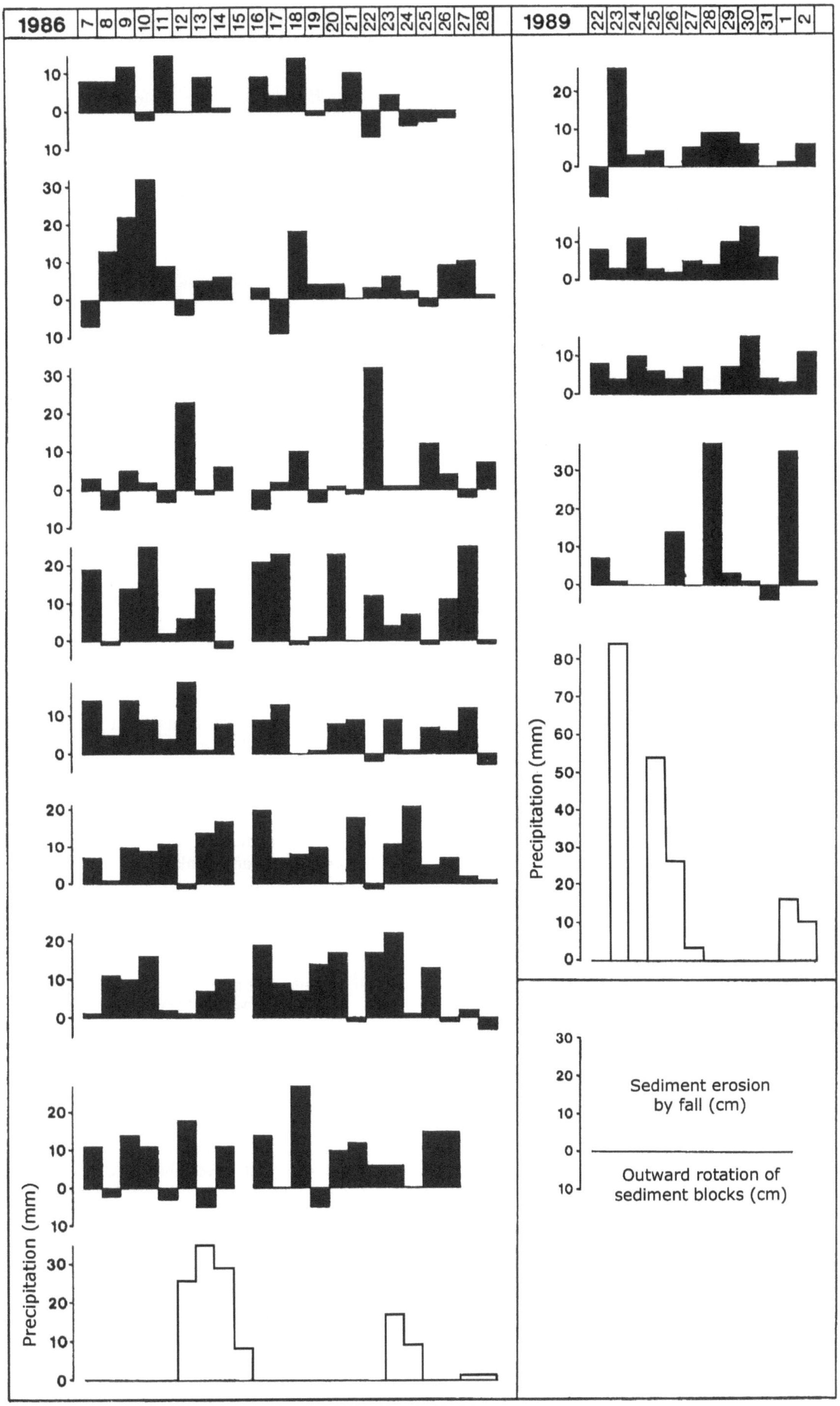

Fig. 7.15. Erosion of sediment cover by fall from above backwasting ice walls during the periods 7–28 July, 1986, and 22 July–2 August, 1989, compared with daily mean precipitation during the same periods (after Krüger, 1994).

Fig. 7.16. Fall-sorting patterns of stones and boulders at the base of a steep ice-cored slope.

contribution from top melt is approximately 1 cm (Krüger and Kjær, 2000). In addition, the annual contribution to bottom melt from the geothermal heat flux is only 2 cm. It means that the annual bottom melt due to other factors averaged 25 cm, which is almost half of that found in the fully ice-cored dead-ice field (see Fig. 7.10A,B). Factors contributing to this are insulation by an increased amount of vegetation and decreased through-flow of subglacial water; wherever meltwater is flowing from the fully ice-cored dead-ice field, drainage patterns of stream channels are cut deeply into the fronting field of partially ice-cored moraine, so that ice cores are relatively dry-based (Krüger, 1994).

The typical surface features and structures related to the de-icing progression of isolated dead-ice blocks are shown in Fig. 7.22. Reworking of the sediment cover is initiated by sinkhole formation at the base of slopes as a result of melting of the toe of buried dead-ice blocks. The formation of sinkholes causes over-steepening of the adjacent slopes and introduces backslumping, which in turn produces niches into the slopes. Thus, the final phase of dead-ice moraine development at Kötlujökull is characterized by a gradual lowering of the ice-cored moraine combined with a lateral movement of the sediment cover (Kjær and Krüger, 2001). Consequently, the most recent resedimentation resulting from sinkhole formation and backslumping will overprint the sedimentary architecture of previous events (Krüger, 1994; Kjær and Krüger, 2001). The many groups of boulders, however, is a sedimentary characteristic that might be traced back to processes working in the fully ice-cored terrain. They represent accumulation of boulders in topographic lows during an earlier phase of de-icing the dead-ice terrain.

According to the de-icing progression model shown in Fig. 7.23, it takes around 50–60 years to melt down 40 m of stagnant, dirty glacier ice in the terminus region of Kötlujökull; about 20 years are required to disintegrate the ice into isolated dead-ice blocks and another 30–40 years to entirely melt out the ice cores (Krüger and Kjær, 2000). However, if a glacier advance set in, the melt-out of the ice in the partially ice-cored terrain in front of the advancing glacier will proceed very much faster. In front of the advancing glacier in the 1980s, the predominating processes of de-icing the partially ice-cored dead-ice field between the glacier and the 1955 end-moraine ridge were fluvio-thermal erosion of buried ice from beneath followed by rapid collapsing of the sediment cover. The fluvio-thermal erosion was favoured by the through-flow of groundwater and running water, the level being raised along with the glacier advance. Annual rates of surface lowering of fronting ice-cored moraines of 1.0–2.1 m were recorded (Krüger, 1994; Krüger and Kjær, 2000).

7.3.3. The Post-Melt Landscape

Beyond the partially ice-cored terrain, a low-relief hummocky moraine terrain that represents a relict from the ice stagnation around 1940 does not show any evidence of active ice melt and is therefore qualified to demonstrate the nature of the fully de-iced terrain (Krüger, 1994; Kjær and Krüger, 2001). This terrain is irregular, but without the steep-sided mounds that characterise the partially ice-cored moraine (Fig. 7.24). Its supraglacial origin, however, is clearly illustrated by the occurrence of numerous small mounds and crescent-shaped ridges dotted by groups of stony and bouldery material. The low relief of 0.5–3 m reflects the thickness of the supraglacial sediment association that drapes a clast-paved basal-till bed (Krüger and Kjær, 1999).

Table 7.2. Comparison of movement patterns of four sediment flows with rainfall data from 6 July to 7 August, 1986.

Date, 1986	Movement of sediment flows (cm)				Rainfall (mm)
	1	2	3	4	
July					
6	0	0	0	0	2
7	0	0	0	0	0
8	0	6	0	0	0
9	0	0	0	0	0
10	0	0	0	0	0
11	0	0	0	0	0
12	3	0	0	0	26
13	300	276	243	117	35
14					29
15					8
16					0
17				0	0
18				0	0
19				0	0
20				0	0
21				0	0
22				0	0
23				0	17
24				0	9
25				0	0
26				0	0
27				0	1
28				0	1
29				0	25
30				0	4
31				0	0
August					
1				0	0
2				0	3
3				0	24
4				0	0
5				0	0
6				0	2
7				0	2

Well-exposed sections along meltwater streams display sediments associated with an advance of Kötlujökull and subsequent stagnation and de-icing of the ice-cored terrain. A datachart comprising a detailed description of the sedimentary succession recognized in the dead-ice moraine landscape is given in Fig. 7.25 (cf. Kjær and Krüger, 2001). The lowermost part of the succession consists of two diamict units (1 and 3) separated by a thin unit (2) of fines and gravel. The lower diamict (unit 1) is interpreted as a slightly deformed lodgement till deposited in response to an advance of Kötlujökull. The upper diamict unit (unit 3) is interpreted as deposited in a subglacial environment by *in situ* melt-out of debris from stagnant ice, while the sorted sediments separating the two beds of diamict are interpreted as water-laid cavity fill as a result of meltwater activity beneath the stagnant ice (Krüger and Kjær, 1999). Above unit 3 follows a succession (units 4–7) of both sorted and diamict sediments, which show slump structures associated with collapse and normal faults with prominent offset related to rotational sliding and, therefore, constitutes a supraglacial sediment association. They are interpreted as trough fillings caused by meltwater activity and with entry of diamict sediments from the final phase of de-icing (Kjær and Krüger, 2001). Because of the absence of collapse structures, the gravel and laminated silt and sand (unit 8) that caps the supraglacial sediments is interpreted as trough-filling, washed down from the hummock side slope after the melting of underlying ice.

7.4. A Landsystems Model

An idealized sequential model for landscape formation along an advancing and finally stagnating debris-charged lowland glacier margin is shown in Fig. 7.26. The model is based on field studies in the terminus region of Kötlujökull during the time period 1977–2003 spanning 16 field seasons:

(A) *The phase of glacier advance*: The terminal part of the glacier is capped by a debris-mantled ice-disintegration field very much similar to that found on dead-ice. Along the steep, advancing ice-front, a series of minor ice-contact fans, so-called hochsander fans, are produced as a consequence of the glacier transport of large quantities of supraglacial and englacial debris. Along with the glacier advance, the glacier toe thrusts and bulldozes the fan sediments into an end-moraine ridge, and from source areas on and above the frontal ice cliff, the moraine ridge is, furthermore, continuously fed by large quantities of material by dumping (Krüger *et al.*, this volume).

(B) *The initial, or young, phase of dead-ice melting*: The glacier advance has stopped and the terminal part of the glacier has stagnated. One factor contributing significantly to the ice-degradation during the first years of ice stagnation is the large-scale backwasting of the many major free faces of ice related to the steep frontal ice slope and the meltwater gorges deeply incised into the terminal ice. Many streams from the glacier surface still transport fine material from the source area to the hochsander fans.

(C) *The mature phase of dead-ice melting*: The terminal part of the glacier has decreased significantly in altitude due to ice disintegration. Cave enlargement and collapse of tunnel roofs have produced numerous mounds, ice-degradation niches, collapsing edges, steep ice walls and sinkholes. The debris cover is exposed to cycles of resedimentation processes such as fall or sliding of sediments from the top of ice cores and

Fig. 7.17. Basin in the fully ice-cored zone with widespread deposition of sand and gravel by braided streams.

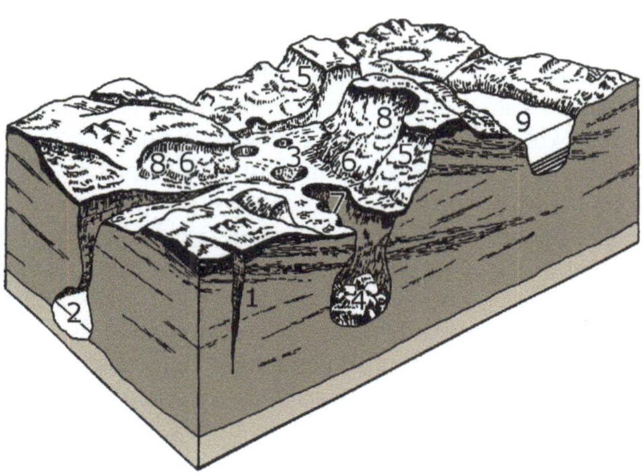

Fig. 7.18. Idealized characteristic of ice degradation and reworking of supraglacial sediments in the fully ice-cored terrain in the terminus region of Kötlujökull. (1) Enlargement of crevasse by ice melt. (2) Subglacial tunnel. (3) Sinkhole. (4) Collapsed tunnel roof. (5) Backwasting ice slope. (6) Lobate sediment flows. (7) Enlargement of sinkhole by ice melt and collapse. (8) Ice degradation niche. (9) Lake (after Krüger, 1994).

subsequent remobilization by sediment gravity flows. The decrease in altitude has reduced the gradient of the supraglacial streams and thereby their energy. Therefore, many streams find small lake basins, or englacial tunnels, many of which are in a process of karstic enlargement and collapse and deposit most of their sediment load in broad depressions within the dead-ice field and not at and beyond the end-moraine ridge. As a consequence, the production of hochsander fans has stopped.

(D) *The final, or old, phase of dead-ice melting*: The ice has thinned considerably, and therefore, the former coherent ice mass is disintegrated into isolated dead-ice blocks capped by a thick cover of multiple resedimented deposits. Due to this insulating layer, ice is not visible and ice melt and slope retreat is reduced at this stage. Reworking of the sediment cover is initiated by sinkhole formation at the base of slopes, which leads to collapse and slumping of adjacent sediment, which in turn dissects most of the sedimentary structures produced in the fully ice-cored terrain. Here and there, table-topped and steep-sided kame hills composed of layered fine gravel, sand and silt appear representing former depressions in the dead-ice field in which streams have deposited their sediment load.

(E) *The post-melt landscape*: The advance of the lowland glacier with a debris-charged margin, the following glacier stagnation with development and disintegration of dead-ice has produced a characteristic series of glacial and glaciofluvial landforms: an end-moraine ridge, (1) which separates a large-scale proglacial outwash plain (2) of sand, gravel and stones overlain by minor hochsander fans (3) of sand and laminated silt and clay from the hummocky dead-ice morainic landscape (4) of multiple resedimented deposits and with table-topped, steep-sided kame hills (5) of layered fine gravel, sand and silt.

7.5. Melting Rates Under Other Climate Conditions

In the humid, subpolar climate at Kötlujökull, the degradation of ice-cored terrain may take a few tens of years and is aided by subglacial meltwater drainage and precipitation. The huge annual precipitation accelerates the de-icing progression and favours backwasting over top melt beneath the sediment cover as the dominant ablation process (cf. Eyles, 1983; Johnson, 1992). Under more continental conditions, where exposed ice faces are short lived, it takes much longer time

Fig. 7.19. View across partially ice-cored terrain showing slow backwasting of steep, ice-cored mounds.

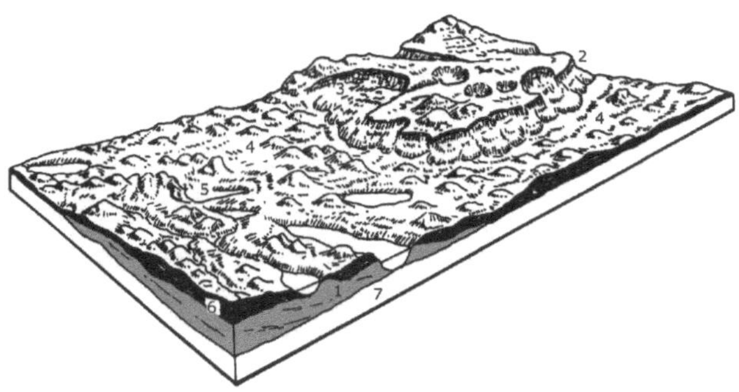

Fig. 7.20. Idealized characteristics of the old stage of dead-ice moraine development in front of Kötlujökull. (1) Dead ice. (2) Backwasting ice-cored slope. (3) Ice degradation niche with lobate sediment flows. (4) Ice-free low-relief hummocks. (5) Lake. (6) Deposits of supraglacial environments. (7) Deposits of subglacial environments (after Krüger, 1994).

(Shaw, 1977), although the rates of backwasting of ice walls might be of the same order of magnitude as obtained in the Kötlujökull terminus region (Driscoll, 1980; Pickard, 1984). Thus, at the surge-type glacier Brúarjökull, the north part of Vatnajökull, where the climate is at the limit of permafrost and the precipitation is moderate, ice cores below thick debris covers persist or melt only at very low rates; this is due to the lack of extensive ice cliff areas where effective backwasting could have occurred (Schomacker, 2007). At Holmströmbreen, Svalbard, where the environment is characterized by semi-arid conditions and continuous permafrost, the rates of lowering of ice-cored terrain are strikingly at the same order of magnitude as dead-ice melt rates in the humid, subpolar climate in South Iceland. The permafrost prevents percolation of meltwater into the ice and therefore also the formation of glacier karst. It means

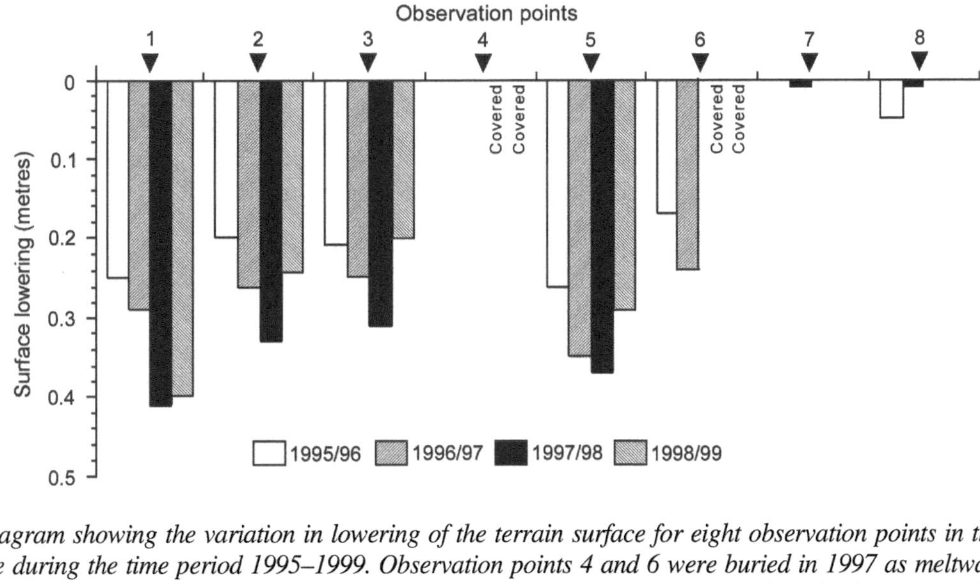

Fig. 7.21. Diagram showing the variation in lowering of the terrain surface for eight observation points in the partially ice-cored moraine during the time period 1995–1999. Observation points 4 and 6 were buried in 1997 as meltwater invaded the low-lying part of the study area and deposited 2–3 m of sorted sediment, mostly gravel and sand (after Kjær and Krüger, 2001).

Fig. 7.22. Surface features related to the de-icing progression of isolated dead-ice blocks. Sinkholes develop at the base of slopes and backslumping upslope. Stones and boulders are mainly located in depressions or the crest of hills as a result of fall-sorting processes in the preceding phases of de-icing.

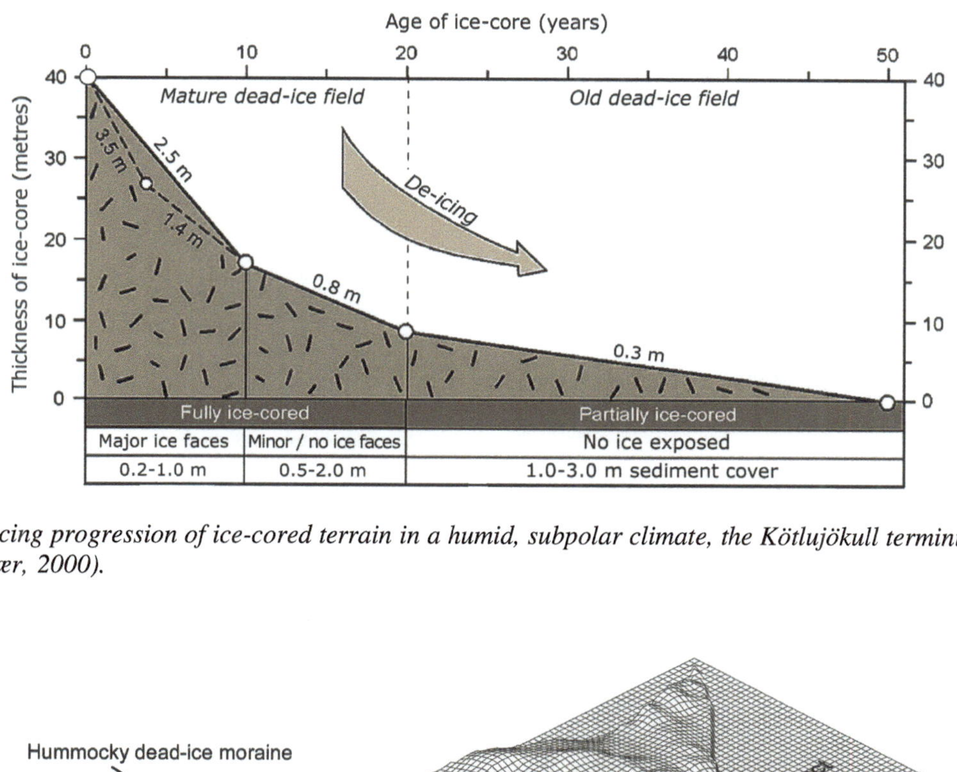

Fig. 7.23. De-icing progression of ice-cored terrain in a humid, subpolar climate, the Kötlujökull terminus region (after Krüger and Kjær, 2000).

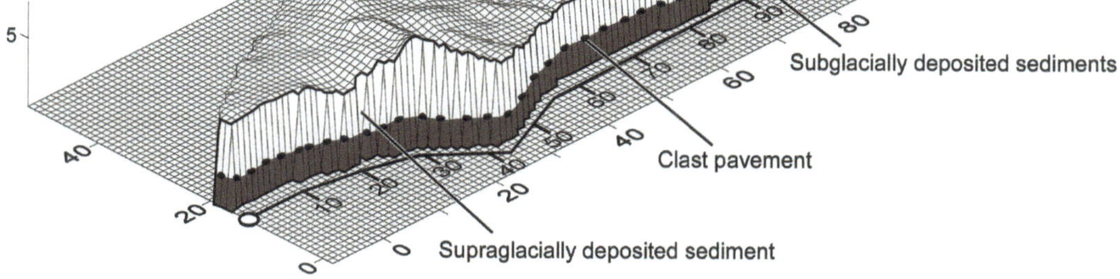

Fig. 7.24. Three-dimensional terrain model illustrating the morphology and stratigraphy of the ice-free dead-ice moraine. Subglacially and supraglacially deposited sediments along a well-exposed geological section, 110 m long, are shown (after Kjær and Krüger, 2001).

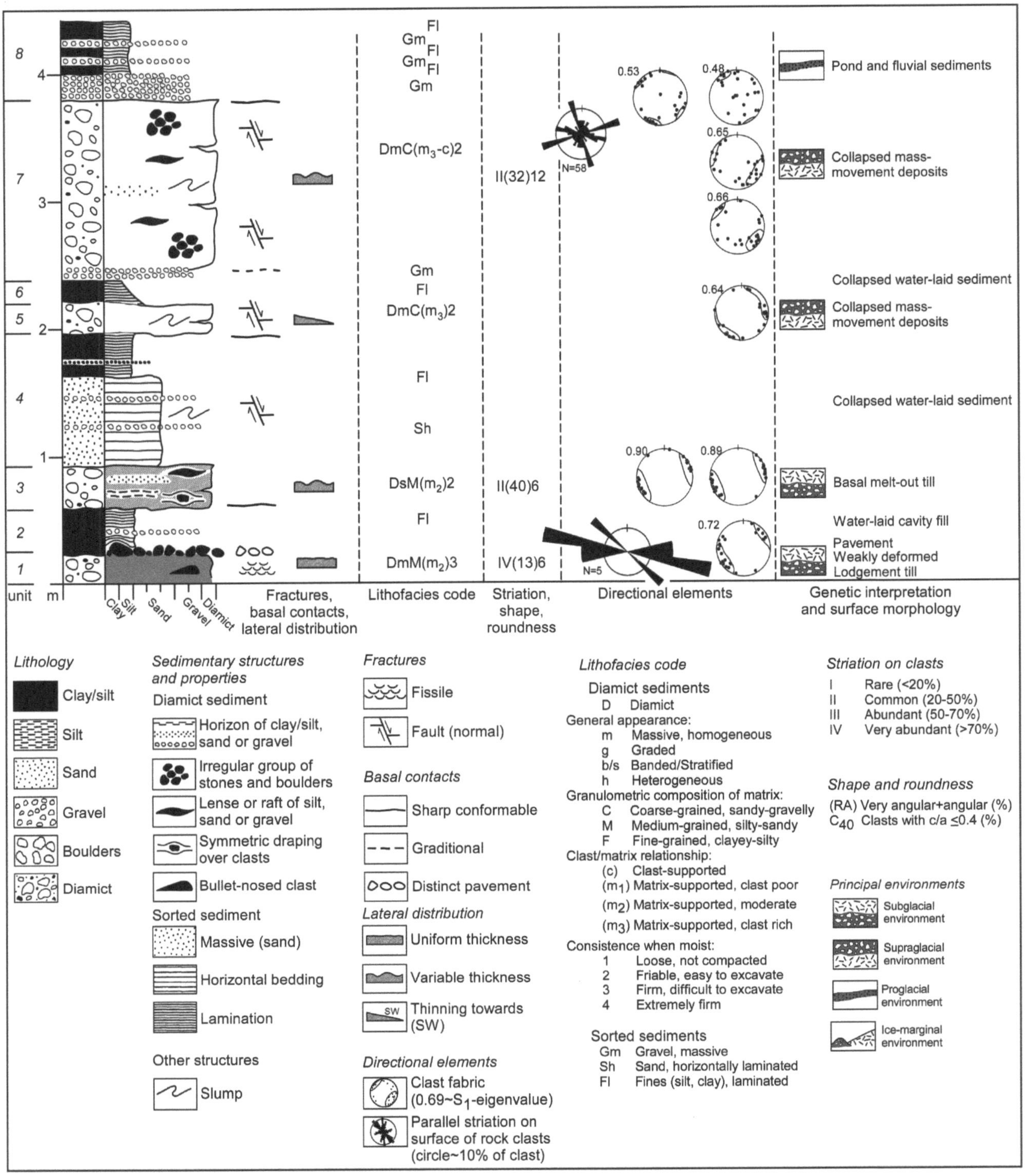

Fig. 7.25. Data chart comprising a detailed description of the sedimentary succession recognized in the dead-ice moraine landscape. Partly after data chart developed by Krüger and Kjær (1999).

that meltwater stays on the extensive and very gently sloping ice surface and keeps sediment gravity flows with very high water content and exposure of new dead-ice and ice-cliff formation active (Fig. 7.27). This may explain the relatively high melt rates (Schomacker, 2007, 2008). As long as backwasting and mass movement processes prevent build-up of an insulating debris cover and expose ice cores to melting, the de-icing will continue even though the Holmströmbreen area is within the zone of continuous permafrost (Schomacker, 2007). However, if the debris overburden under permafrost conditions exceeds the active layer, the buried glacier ice will be protected against melting. In Siberia, Arctic Canada and northeast Greenland, remnants of Pleistocene glaciers which stagnated for more than 50 ka BP have been identified in dead-ice morainic terrain (Astakhov and Isayeva, 1988; French and Harry, 1990).

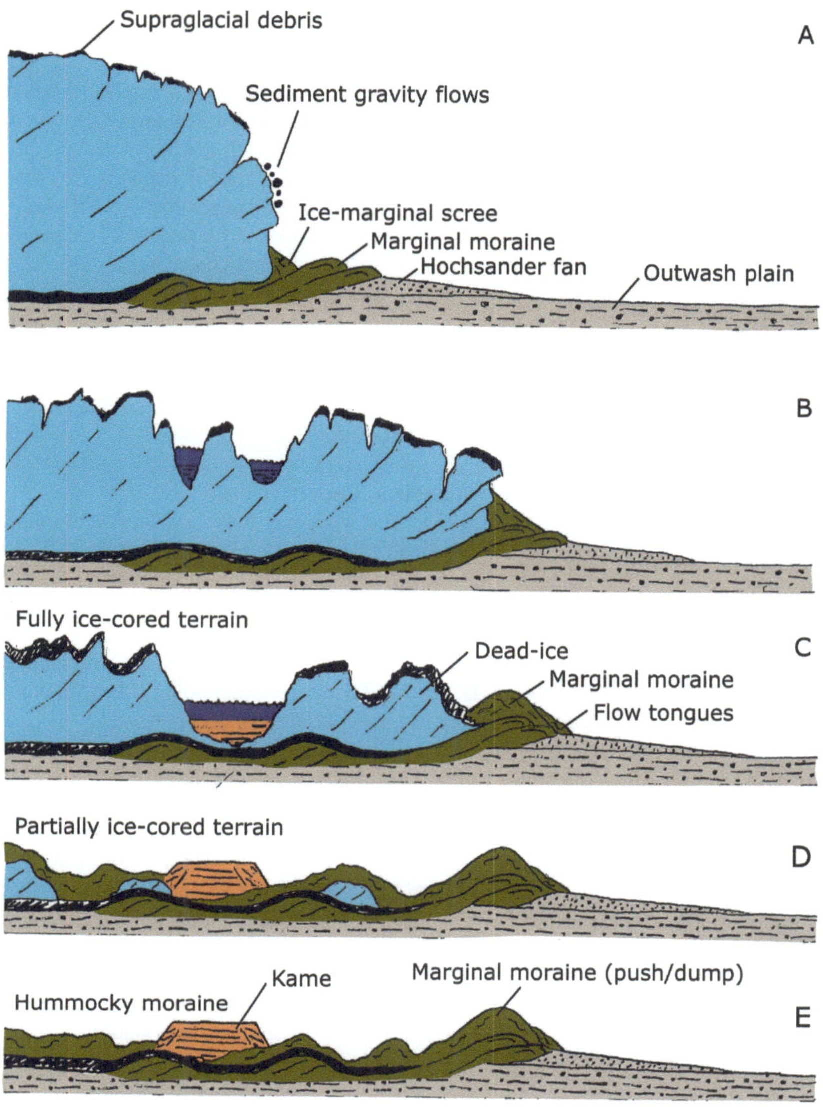

Fig. 7.26. Sequential model for landscape formation along a debris-charged lowland glacier margin. (A) The glacier advance phase with production of minor ice-contact fans fed by supraglacial streams and formation of an end-moraine. (B) The ice stagnation and initial, or young, dead-ice phase with continued development of hochsander fans and beginning decrease in altitude of the frontal part of the glacier because of large-scale backwasting of the major free faces of ice related to the steep frontal ice slope. (C) The mature phase of dead-ice melting where the terminal part of the glacier has decreased significantly in altitude due to ice disintegration. Cave enlargement and collapse of tunnel roofs have produced numerous mounds, ice-degradation niches, collapsing edges, steep ice walls and sinkholes. The production of hochsander fans has stopped. (D) The final, or old, phase of dead-ice melting where the ice has thinned considerably, and therefore, the ice mass is disintegrated into isolated dead-ice blocks capped by a thick cover of multiple resedimented deposits. (E) The post-melt landscape with a characteristic series of glacial and glaciofluvial landforms.

Fig. 7.27. View across fully ice-cored moraine in front of Holmströmbreen, Svalbard. Note the extensive sediment gravity flow transporting material downslope from the backwasting free ice face seen to the right.

7.6. Conclusions

- The supraglacial debris at Kötlujökull originates from two main sources: volcanic ash deposited on the ice cap (black sand and fine gravel) and marginal thrusting-up of subglacial sediments (diamict material).
- Three different stages of dead-ice development are currently observed in the forefield of Kötlujökull. (1) The fully ice-cored stage with a coherent dead-ice mass originating from an advance in the late 1980s. (2) The partially ice-cored stage with dead-ice in its final phase of collapse and melt-out. This dead-ice was initiated in the 1950s. (3) The post-melt hummocky moraines produced after a complete de-icing during the past 60–70 years.
- In the fully ice-cored stage, melting takes place as backwasting, that is, lateral retreat of near-vertical ice cliffs or slopes, and downwasting, defined as melting at the top and bottom surfaces of dead-ice. In this stage, the total mean annual melting corresponds to a surface lowering of 1.4 m. Backwasting is most prominent in the fully ice-cored stage, but as the debris cover develops and the number of ice cliffs decreases in the partially ice-cored moraines, backwasting also becomes less important. A model for the de-icing progression indicates that it takes about 50 years to completely melt down approximately 40 m of debris-covered dead-ice under the present climatic conditions.
- The debris cover in the dead-ice environment is exposed to resedimentation. The most important processes are mass movements governed by gravity, that is, fall, slumping, sliding and flowing of sediments.
- Using Kötlujökull as an example, we present a conceptual landsystems model for de-icing of a debris-charged, temperate lowland glacier margin. The model is based on field monitoring of dead-ice melting in the Kötlujökull forefield since 1977 and suggests five main steps in the landscape development: a phase of glacier advance, the young phase of dead-ice melting, a mature phase of dead-ice melting, a final phase of dead-ice melting and the post-melt landscape after a complete de-icing.

References

Astakhov, V.I. & Isayeva, L.L., 1988. The ‘Ice Hill’: an example of ‘retarded deglaciation’ in Sibiria. Quaternary Science Reviews 7, 29–40.

Benn, D.I., 1992. The genesis and significance of ‘hummocky moraine’: evidence from the Isle of Skye, Scotland. Quaternary Science Reviews 11, 781–799.

Benn, D.I. & Evans, D.J.A., 1998. Glaciers and glaciation. London, Arnold.

Bennett, M.R., Huddart, D., Glasser, N.F., & Hambrey, M.J., 2000. Resedimentation of debris on an ice-cored lateral moraine in the high-Arctic (Kongsvegen, Svalbard). Geomorphology 35, 21–40.

Bennett, M.R., Huddart, D., Hambrey, M.J., & Ghienne, J.F., 1996. Moraine development at the high-Arctic valley glacier Pedersensbreen, Svalbard. Geografiska Annaler 78A, 209–222.

Boone, S.J. & Eyles, N., 2001. Geotechnical model for great plains hummocky moraine formed by till deformation below stagnant ice. Geomorphology 38, 109–124.

Boulton, G.S., 1968. Flow tills and related deposits on some Vestspitsbergen glaciers. Journal of Glaciology 7, 391–412.

Boulton, G.S., 1970. On the origin and transport of englacial debris in Svalbard glaciers. Journal of Glaciology 9, 213–229.

Boulton, G.S., 1971. Till genesis and fabric in Svalbard, Spitsbergen. In: Goldthwait, R.P. (Ed), Till: a symposium. Colombus, Ohio State University Press, 41–72.

Boulton, G.S., 1972. Modern Arctic glaciers as depositional models for former ice sheets. Journal of the Geological Society of London 128, 361–393.

Brodzikowski, K. & van Loon, A.J., 1991. Glacigenic sediments. Amsterdam, Elsevier.

Carrara, P.E., 1975. The ice-cored moraines of Akudnirmuit Glacier Cumberland Peninsula, Baffin Island, N.W.T., Canada. Arctic and Alpine Research 7, 461–468.

Clayton, L., 1964. Karst topography on stagnant glaciers. Journal of Glaciology 5, 107–112.

Clayton, L. & Moran, S.L., 1974. A glacial process-form model. In: Coates, D.R. (Ed), Glacial geomorphology. Binghamton, State University of New York, 89–119.

Driscoll, F.G., Jr., 1980. Wastage of the Klutlan ice-cored moraines, Yukon Territory, Canada. Quaternary Research 14, 31–49.

Etzelmüller, B., 2000. Quantification of thermo-erosion in proglacial areas: examples from Svalbard. Zeitschrift für Geomorphologie N.F. 44, 343–361.

Evans, D.J.A., 2003. Ice-marginal terrestrial landsystems: active temperate glacier margins. In: Evans, D.J.A. (Ed), Glacial landsystems. London, Arnold, 12–43.

Evans, D.J.A. & Rea, B.R., 2003. Surging glacier landsystem. In: Evans, D.J.A. (Ed), Glacial landsystems. London, Arnold, 259–288.

Eyles, N., 1979. Facies of supraglacial sedimentation on Icelandic and Alpine temperate glaciers. Canadian Journal of Earth Sciences 16, 1341–1361.

Eyles, N., 1983. Modern Icelandic glaciers as depositional models for ‘hummocky moraine’ in the Scottish Highlands. In: Evenson, E.B., Schlüchter, Ch., Rabassa, J. (Eds), Tills and related deposits. Rotterdam, Balkema, 47–60.

Eyles, N., Boyce, J.I., & Barendregt, R.W., 1999. Hummocky moraine: sedimentary record of stagnant Laurentide Ice Sheet lobes resting on soft beds. Sedimentary Geology 123, 163–174.

Fitzsimons, S.J., 1990. Ice-marginal depositional processes in a polar maritime environment, Vestfold Hills, Antarctica. Journal of Glaciology 36, 279–286.

Fitzsimons, S.J., 1991. Supraglacial eskers in Antarctica. Geomorphology 4, 293–299.

French, H.M. & Harry, D.G., 1990. Observations on buried glacier ice and massive segregated ice, western arctic coast, Canada. Permafrost and Periglacial Processes 1, 31–43.

Glasser, N.F. & Hambrey, M.J., 2002. Sedimentary facies and landform genesis at a temperate outlet glacier: Soler Glacier, North Patagonian Icefield. Sedimentology 49, 43–64.

Glasser, N.F. & Hambrey, M.J., 2003. Ice-marginal terrestrial landsystems: Svalbard polythermal glaciers. In: Evans, D.J.A. (Ed), Glacial Landsystems. London, Arnold, 65–88.

Ham, N.R. & Attig, J.W., 1996. Ice wastage and landscape evolution along the southern margin of the Laurentide Ice Sheet, north-central Wisconsin. Boreas 25, 171–186.

Hambrey, M.J., Bennett, M.R., Dowdeswell, J.A., Glasser, N.F., & Huddart, D., 1999. Debris entrainment and transfer in polythermal valley glaciers. Journal of Glaciology 45, 69–86.

Hambrey, M.J., Huddart, D., Bennett, M.R., & Glasser, N.F., 1997. Genesis of 'hummocky moraines' by thrusting in glacier ice: evidence from Svalbard and Britain. Journal of the Geological Society, London 154, 623–632.

Huddart, D. & Hambrey, M.J., 1996. Sedimentary and tectonic development of a high-Arctic thrust-moraine complex: Comfortlessbreen, Svalbard. Boreas 25, 227–243.

Ingólfsson, Ó. & Lokrantz, H., 2003. Massive ground ice body of glacial origin at Yugorski Peninsula, Arctic Russia. Permafrosat and Periglacial Processes 14, 199–215.

Johnson, M.D. & Clayton, L., 2003. Supraglacial landsystems in lowland terrain. In: Evans, D.J.A. (Ed), Glacial landsystems. London, Arnold, 228–258.

Johnson, P.G., 1975. Recent crevasse fillings at the terminus of the Donjek Glacier, St. Elias Mountains, Yukon Territory. Quaestiones Geographicae 2, 53–59.

Johnson, P.G., 1992. Stagnant glacier ice, St. Elias Mountains, Yukon. Geografiska Annaler 74A, 13–19.

Johnsson, M.D., Mickelson, D.M., Clayton, L., & Attig, J.W., 1995. Composition and genesis of glacial hummocks, western Wisconsin, USA. Boreas 24, 97–116.

Jonsson, J., 1983. Notes on the Katla volcanoglacial debris flows. Jökull 32, 61–68.

Kirkbride, M.P., 1993. The temporal significance of transitions from melting to calving termini at glaciers in the central Southern Alps of New Zealand. The Holocene 3, 232–240.

Kirkbride, M.P., 1995. Ice flow vectors on the debris-mantled Tasman Glacier, 1957–1986. Geografiska Annaler 77A, 147–157.

Kirkbride, M.P. & Spedding, N., 1996. The influence of englacial drainage on sediment-transport pathways and till texture of temperate valley glaciers. Annals of Glaciology 22, 160–166.

Kjær, K.H. & Krüger, J., 2001. The final phase of dead-ice moraine development: processes and sediment architecture, Kötlujökull, Iceland. Sedimentology 48, 935–952.

Kjær, K.H., Sultan, L., Krüger, J., & Schomacker, A., 2004. Architecture and sedimentation of fan-shaped outwash in front of the Mýrdalsjökull ice cap, Iceland. Sedimentary Geology 172, 139–163.

Krüger, J., 1983. Glacial morphology and deposits in Denmark. In: Ehlers, J. (Ed), Glacial deposits in North-west Europe. Rotterdam, Balkema, 181–191.

Krüger, J., 1994. Glacial processes, sediments, landforms and stratigraphy in the terminus region of Mýrdalsjökull, Iceland. Folia Geographica Danica 21, 1–233.

Krüger, J. & Aber, J.S., 1999. Formation of supraglacial sediment accumulations on Kötlujökull, Iceland. Journal of Glaciology 45, 400–402.

Krüger, J. & Kjær, K.H., 1999. A data chart for field description and genetic interpretation of glacial diamicts and associated sediments – with examples from Greenland, Iceland, and Denmark. Boreas 28, 386–402.

Krüger, J. & Kjær, K.H., 2000. De-icing progression of ice-cored moraines in a humid, subpolar climate, Kötlujökull, Iceland. The Holocene 10, 737–747.

Krüger, J., Kjær, K.H., & van der Meer, J.J.M., 2002. From push moraine to single-crested dump moraine during a sustained glacier advance. Norsk Geografisk Tidsskrift 56, 87–95.

Lawson, D.E., 1979. Sedimentological analysis of the western terminus region of the Matanuska Glacier, Alaska. CRREL Report 79-9, 1–111.

Lawson, D.E., 1989. Glacigenic resedimentation: classification concepts and application to mass-movement processes and deposits. In: Goldthwait, R.P., Matsch, C.L. (Eds), Genetic classification of glacigenic deposits. Rotterdam, Balkema, 147–169.

Lønne, I. & Lyså, A., 2005. Deglaciation dynamics following the Little Ice Age on Svalbard: implications for shaping of landscapes at high latitudes. Geomorphology 72, 300–319.

Lyså, A. & Lønne, I., 2001. Moraine development at a small High-Arctic valley glacier: Rieperbreen, Svalbard. Journal of Quaternary Science 16, 519–529.

McKenzie, G.D., 1969. Observations on a collapsing kame terrace in Glacier Bay National Monument, south-eastern Alaska. Journal of Glaciology 8, 413–425.

McKenzie, G.D. & Goodwin, R.G., 1987. Development of collapsed glacial topography in the Adams Inlet area, Alaska, USA. Journal of Glaciology 33, 55–59.

Nakawo, M. & Young, G.J., 1981. Field experiments to determine the effect of a debris layer on ablation of glacier ice. Annals of Glaciology 2, 85–91.

Nakawo, M. & Young, G.J., 1982. Estimate of glacier ablation under a debris layer from surface temperature and meteorological variables. Journal of Glaciology 28, 29–34.

Näslund, J.-O. & Hassinen, S., 1996. Supraglacial sediment accumulations and large englacial water conduits at high elevations in Mýrdalsjökull, Iceland. Journal of Glaciology 42, 190–192.

Nicholson, L. & Benn, D.I., 2006. Calculating ice melt beneath a debris layer using meteorological data. Journal of Glaciology 52(178), 463–470.

Østrem, G., 1959. Ice melting under a thin layer of moraine and the existence of ice cores in moraine ridges. Geografiska Annaler 41, 228–230.

Pickard, J., 1983. Surface lowering of ice-cored moraine by wandering lakes. Journal of Glaciology 29, 338–342.

Pickard, J., 1984. Retreat of ice scarps on an ice-cored moraine, Vestfold Hills, Antarctica. Zeitschrift für Geomorphologie 28, 443–453.

Þórarinsson, S., 1957. The jökulhlaup from the Katla area in 1955 compared with other jökulhlaups in Iceland. Jökull 7, 21–25.

Roberts, M.J., Russell, A.J., Tweed, F.S., & Knudsen, Ó., 2000. Controls on englacial sediment deposition during the November 1996 jökulhlaup, Skeidarárjökull, Iceland. Earth Surface Processes and Landforms 26, 935–952.

Russell, A.J., Knudsen, Ó., 2002. The effects of glacier-outburst flood flow dynamics on ice-contact deposits: November 1996 jökulhlaup. Skeiðarársandur, Iceland. Special Publications, International Association of Sedimentologists, 32, 67–83.

Schomacker, A., 2007. Dead-ice under different climate conditions: processes, landforms, sediments and melt rates in Iceland and Svalbard. LUNDQUA Thesis 59.

Schomacker, A., 2008. What controls dead-ice melting under different climate conditions? A discussion. Earth-Science Reviews 90, 103–113.

Schomacker, A. & Kjær, K.H., 2007. Origin and de-icing of multiple generations of ice-cored moraines at Brúarjökull, Iceland. Boreas 36, 411–425.

Schomacker, A., Krüger, J., & Kjær, K.H., 2006. Ice-cored drumlins at the surge-type glacier Brúarjökull, Iceland: a transitional-state landform. Journal of Quaternary Science 21, 85–93.

Sharp, M.J., 1985. 'Crevasse-fill' ridges – a landform type characteristic of surging glaciers? Geografiska Annaler 67A, 213–220.

Sharp, R.S., 1949. Studies of supraglacial debris on valley glaciers. American Journal of Science 247, 289–315.

Shaw, J., 1977. Tills deposited in arid polar environments. Canadian Journal of Earth Sciences 14, 1239–1245.

Shaw, J., 1979. Genesis of the Sveg tills and Rogen moraines of central Sweden: a model of basal melt-out. Boreas 8, 409–426.

Sletten, K., Lyså, A., & Lønne, I., 2001. Formation and disintegration of a high-arctic ice-cored moraine complex, Scott Turnerbreen, Svalbard. Boreas 30, 272–284.

Syverson, K.M. & Mickelson, D.M., 1995. Ablation of debris-covered ice and the formation of pitted outwash plains at Burroughs Glacier, south-eastern Alaska. Proceedings of the Third Glacier Bay Science Symposium 1993, Anchorage, National Park Service, 66–74.

Tómasson, H., 1996. The jökulhlaup from Katla in 1918. Annals of Glaciology 22, 249–254.

Vatne, G., 2001. Geometry of englacial meltwater conduits, Austre Brøggerbreen, Svalbard. Norsk Geografisk Tidsskrift 55, 85–93.

8

Subglacial Environments, Sediments and Landforms at the Margins of Mýrdalsjökull

Anders Schomacker[1,*], Kurt H. Kjær[2] and Johannes Krüger[3]

[1]*Institute of Earth Sciences, University of Iceland, Askja, Sturlugata 7, IS-101 Reykjavík, Iceland*
[2]*Natural History Museum of Denmark, University of Copenhagen, Øster Voldgade 5-7, DK-1350 Copenhagen K, Denmark*
[3]*Department of Geography and Geology, University of Copenhagen, Øster Voldgade 10, DK-1350 Copenhagen K, Denmark*
**Correspondence and requests for materials should be addressed to Anders Schomacker (e-mail: anders@hi.is)*

8.1. Introduction

In Iceland, subglacial landscapes are well known from areas glaciated during the Pleistocene and from modern glacier forefields. In these forefields, subglacial research has focused on till and landform genesis as well as dynamics at the interface between glaciers and their beds. Significant parts of our current view on subglacial till formation owe to investigations in Icelandic glacier forefields (Evans *et al.*, 2006). A tunnel below Breiðamerkurjökull, for instance, served as an experimental site for measurements of subglacial/submarginal till deformation in the classical study by Boulton and Hindmarsh (1987). Later, the two-tiered deformation tills at Breiðamerkurjökull were used in studies of clast fabric signatures and characteristics of glacially deformed sediments (Benn, 1995; Benn and Evans, 1996). Utilizing the high frequency of glacier surges in Iceland, several studies of subglacial tills deposited during surges have been undertaken (e.g. Fuller and Murray, 2000, 2002; Nelson *et al.*, 2005; Kjær *et al.*, 2006; Carr and Goddard, 2007). Together with numerous other studies, the Mýrdalsjökull forefield has contributed with sedimentological investigations of subglacial landform genesis. Particularly, the genesis of drumlins and flutes has attracted glacial geologists (e.g. Krüger and Thomsen, 1984; Krüger, 1987b; Hart, 1995; Evans *et al.*, 1999; Evans and Twigg, 2002; Evans and Rea, 2003; Schomacker *et al.*, 2006). Many studies have also proposed subglacial landsystem models on the basis of research in Iceland (e.g. Krüger, 1987a, 1994; Evans *et al.*, 1999; Evans and Twigg, 2002; Evans, 2003). The scope of many of these investigations was to use observations from modern (sub)glacial processes and sediments as tools to decipher the depositional environments of ancient (sub)glacial sediments.

The aim of this chapter is to provide a review of research on the subglacial environments, sediments and landforms at Mýrdalsjökull and its outlet glaciers. The effects of volcanic activity on the glacial processes are given special attention.

8.2. Study Areas

The north margin of Mýrdalsjökull is known as Sléttjökull, and terminates as a broad lobe (Figs. 8.1 and 8.2). The clean margin retreats frontally, thereby exposing a subglacial landscape characterized by flutes and drumlins. Meltwater rivers have dissected the landscape, thus providing an insight into the subglacial sediments. Research in this glacier forefield has added substantial progress to our understanding of subglacial sediments, processes and landforms. Focus has been directed towards the understanding of modern glacial landforming processes as an analogue to ancient glacial landscapes and sediments.

In contrast, Kötlujökull, the major eastern outlet glacier, is characterized by its debris-covered, stagnant margin (Fig. 8.3). Kötlujökull terminates as an expanded piedmont lobe, and is therefore also considered a good modern analogue to Pleistocene lowland glaciers. Currently, the glacier is retreating by melting of the stagnant debris-covered snout. Although the main research focus has been supraglacial processes and sediments, subglacial till surfaces occur in the glacier forefield (Krüger, 1994; Wisniewski *et al.*, 1999). In many locations, the subglacial tills are, however, covered by supraglacial sediments originating from dead-ice melting (Krüger and Kjær, 2000; Kjær and Krüger, 2001). Because Kötlujökull does not retreat frontally, subglacial melt-out tills are preserved below dead-ice sediments (Krüger and Kjær, 1999).

The 15 km long Sólheimajökull is a valley glacier draining the southwestern part of the Mýrdalsjökull ice cap through a 1–2 km wide valley (Figs. 8.1 and 8.4). Studies at Sólheimajökull have focused on subglacial water-escape structures (WES), meltwater chemistry and jökulhlaups (e.g. Lawler *et al.*, 1996; Sigurðsson *et al.*, 2000; Le Heron and Etienne, 2005).

DEVELOPMENTS IN QUATERNARY SCIENCES
VOLUME 13 ISSN 1571-0866

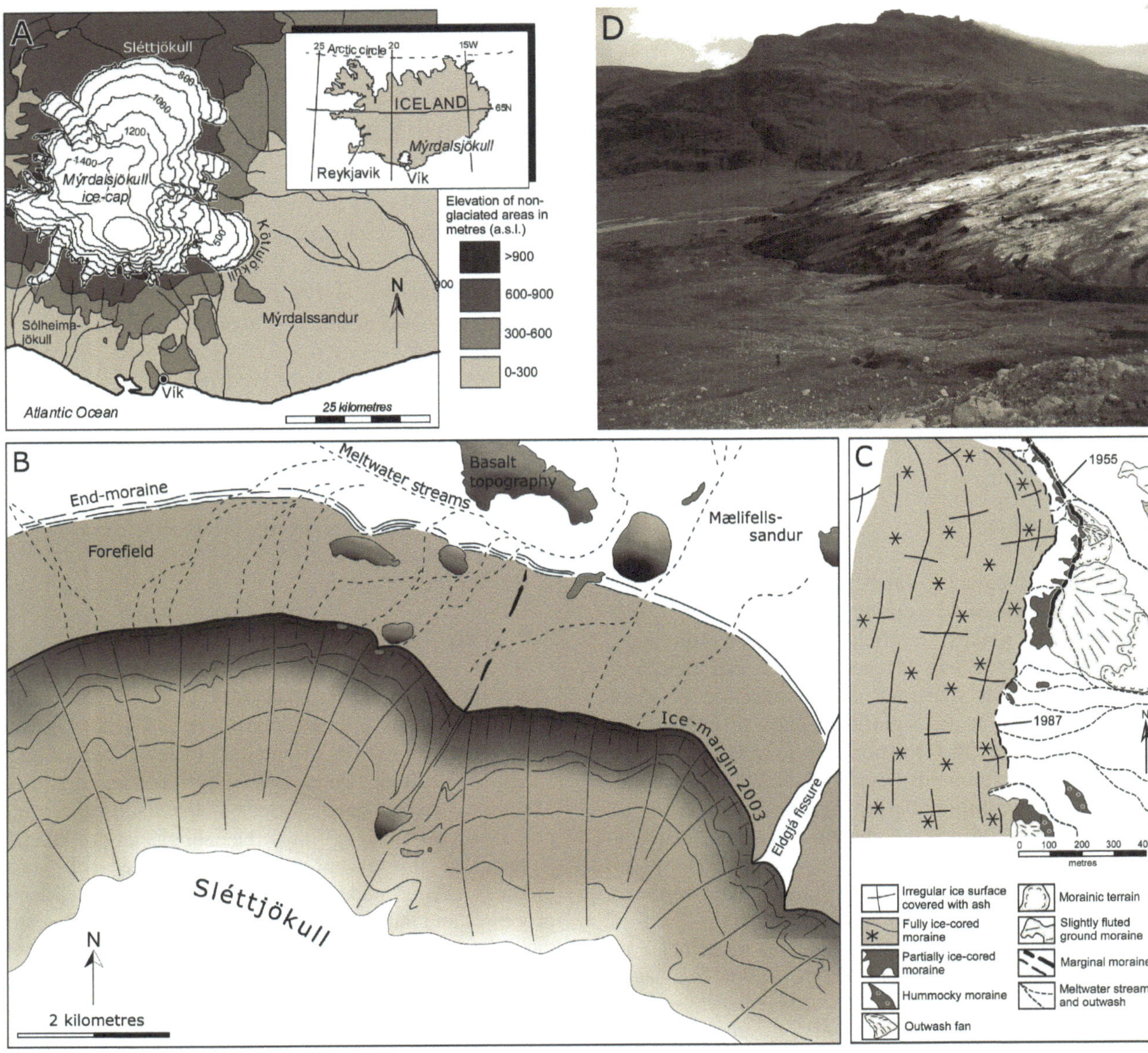

Fig. 8.1. (A) Location of the study areas at Mýrdalsjökull, south Iceland. (B) Sléttjökull, the northern margin of Mýrdalsjökull, terminates as a 20 km wide glacier lobe. Most of the data presented here are gathered in the eastern part of this glacier forefield. The margin is currently at an elevation of c. 600 m a.s.l. and has retreated more than 1,500 m from its maximum position during the Little Ice Age (Björnsson et al., 2000; Magnússon et al., 2005). Meltwater drains northwards to Mælifellssandur. Glacier fluctuations since the Little Ice Age have been shown to be climatically controlled, and the eastern part of Sléttjökull has no records of surges (Krüger, 1995; Björnsson et al., 2003). (C) The debris-covered, stagnant margin of Kötlujökull (also known as Höfðabrekkujökull). The glacier is 15 km long and descends from an elevation of 1,200 m a.s.l. and terminates at 220 m a.s.l. Below the 600 m level the glacier spreads out to form an expanded piedmont lobe with a maximum width of 8 km terminating in low-relief terrain consisting of glacial and glaciofluvial sediments. Meltwater is drained southeastwards to the Atlantic Ocean by the braided rivers of Mýrdalssandur. The map shows the geomorphology in the study area in the northeastern part of the glacier snout. (D) Sólheimajökull descends from an elevation of 1,500 m a.s.l. and terminates at 150 m a.s.l. Meltwater is evacuated southwards to the Atlantic Ocean by the river Fulilækur (also known as Jökulsá á Sólheimasandi). Following two decades of advance from the early 1970s to the mid-1990s, Sólheimajökull continued its post-Little Ice Age retreat (Sigurðsson, 2005). The photograph shows the glacier margin in May 2005. Note persons for scale.

8.3. The Subglacial Environment

8.3.1. Large-Scale Subglacial Topography

The ice surface morphology of Mýrdalsjökull reveals that the ice cap is underlain by the Katla central volcano. On aerial photographs and Digital Elevation Models, the ice-covered caldera rim is clearly seen (Fig. 8.5). During the last 50 years, seismic and radio echo soundings have been undertaken to map the ice thickness. When the surface position and altitude at the point of ice thickness measurement is known, the altitude of the substratum may be calculated. Early single-point soundings on the central part of the ice cap indicated a maximum ice thickness of 370 m (Rist, 1967). In recent decades, several field campaigns have established the ice thickness and

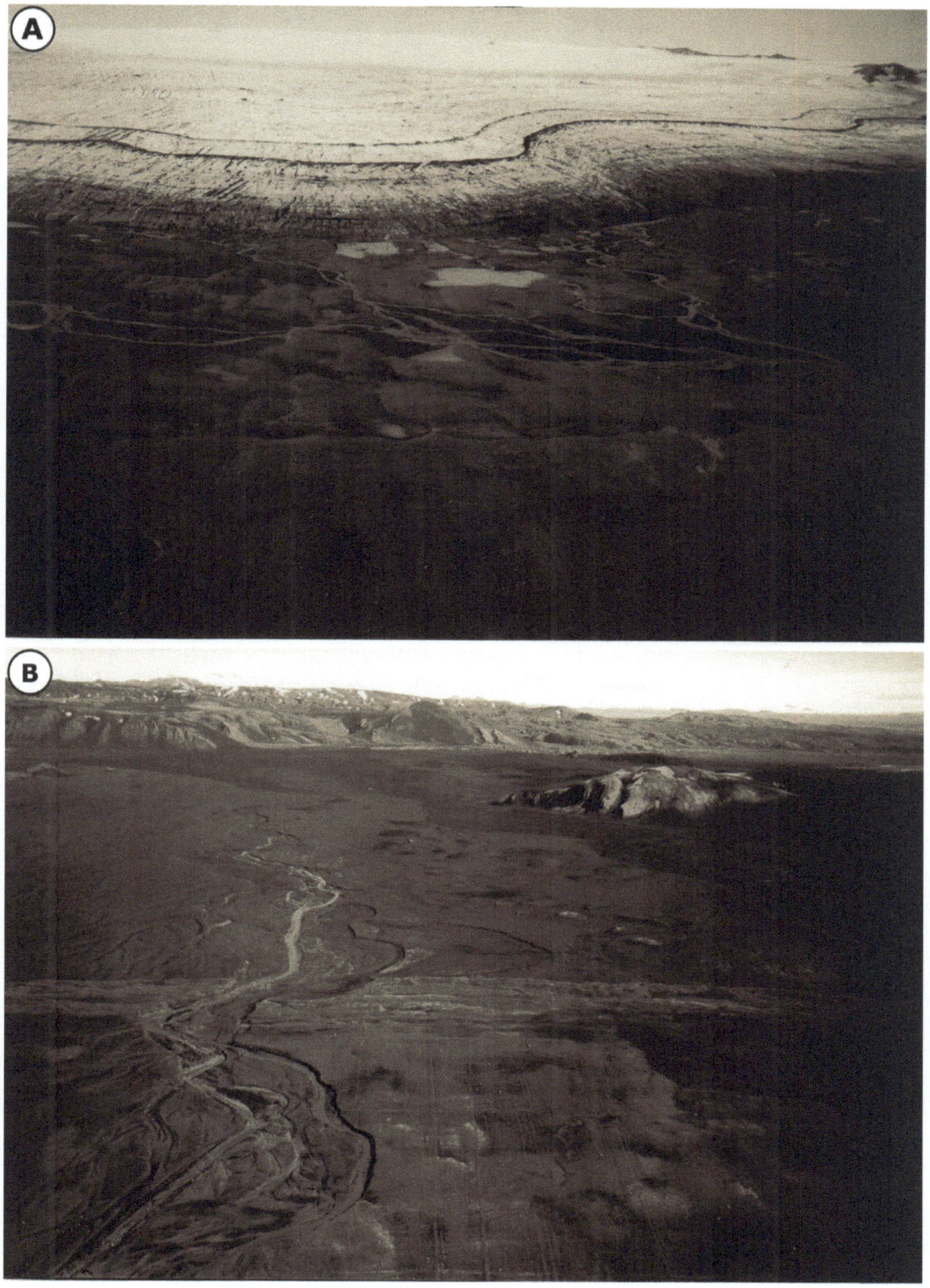

Fig. 8.2. Aerial photographs of the Sléttjökull forefield, August 2001. (A) Oblique view towards south of the glacier margin and the glacially streamlined till plain. (B) Oblique view towards north of the end moraine from c. AD 1900, the distalmost part of the till plain and the Mælifellssandur outwash plain.

subglacial topography of outlet glaciers and parts of the ice cap (e.g. Lawler *et al.*, 1996; Mackintosh *et al.*, 2000; Brandt *et al.*, 2005; Thorsteinsson *et al.*, 2005). Accurate maps of the ice cap surface and the subglacial topography were presented by Björnsson *et al.* (2000) (Fig. 8.5). The maps clearly display the Katla caldera with a depth of c. 700 m. A maximum ice thickness of approximately 740 m was measured in the central area of the caldera (Fig. 8.5). The caldera rim is, however, cut by several glacially eroded troughs where ice is currently drained from the ice cap to the outlet glaciers. Most prominent are the troughs hosting Kötlujökull and Sólheimajökull. To the north, the ice thickness in the Sléttjökull lobe is 200–300 m, gently decreasing towards the ice margin. Figure 8.5 also shows that the flow of Sléttjökull is nearly unconfined by subglacial topography, and that the glacier spreads out as the characteristically piedmont lobe north of the caldera.

8.3.2. Present-Day Subglacial Meltwater Drainage

The most extreme signs of subglacial meltwater activity are the volcano-glacial jökulhlaups from Mýrdalsjökull (Russell *et al.*, this volume). However, as at other

Fig. 8.3. Oblique view towards north of Kötlujökull, August 2001. Note the rim of ice-cored moraine along the glacier margin.

Fig. 8.4. The margin of Sólheimajökull, 2006. Photo: Ívar Örn Benediktsson.

temperate glaciers, subglacial meltwater also originates from percolating supraglacial meltwater, melting due to the 'normal' geothermal heat release at the glacier bed, pressure melting and melting due to frictional heat release. The meltwater activity from Mýrdalsjökull and Eyjafjallajökull is continuously monitored at several hydrological measuring stations in the glacial rivers. The hydrograph shown in Fig. 8.6 demonstrates the water level in the river Jökulsá Gigjökull over 3 years. It is evident that a base flow maintains a minimum water level of c. 60 cm, indicating that the Eyjafjallajökull ice cap is warm-based. This is because water flows throughout the winters, where the major input to the rivers comes from subglacial meltwater. Such sustained meltwater discharge, even during winters, is typical for rivers draining the warm-based ice caps of Iceland (Arnborg, 1955; Sigurðsson, 1990). The dramatic water level fluctuations, mainly during the spring and summer, reflect the snowmelt and heavy rainfall in the catchment.

Strong smells of volcanic gasses (H_2S) in the meltwater rivers Fúlilækur and Múlakvísl indicate subglacial interaction between the ice and geothermal areas. The smell from the meltwater rivers is not continuous but occurs frequently as distinct events. Lawler *et al.* (1996) studied the hydrochemistry of the river Fúlilækur during a melt season. They demonstrated a clear effect of subglacial geothermal events on the meltwater chemistry (Fig. 8.7). A few days after extraordinarily high subglacial seismic activity and enhanced geothermal fluid injection, the meltwater showed peaks in discharge, H_2S and total carbonate, decrease in pH and increase in electrical conductivity. This demonstrates the dynamic subglacial interplay between the volcanism and the ice at Mýrdalsjökull.

Localized subglacial heat release from eruptions or other geothermal activity even affects the surface morphology of the ice cap. Characteristic ice cauldrons appear on the surface of Mýrdalsjökull, bearing witness to

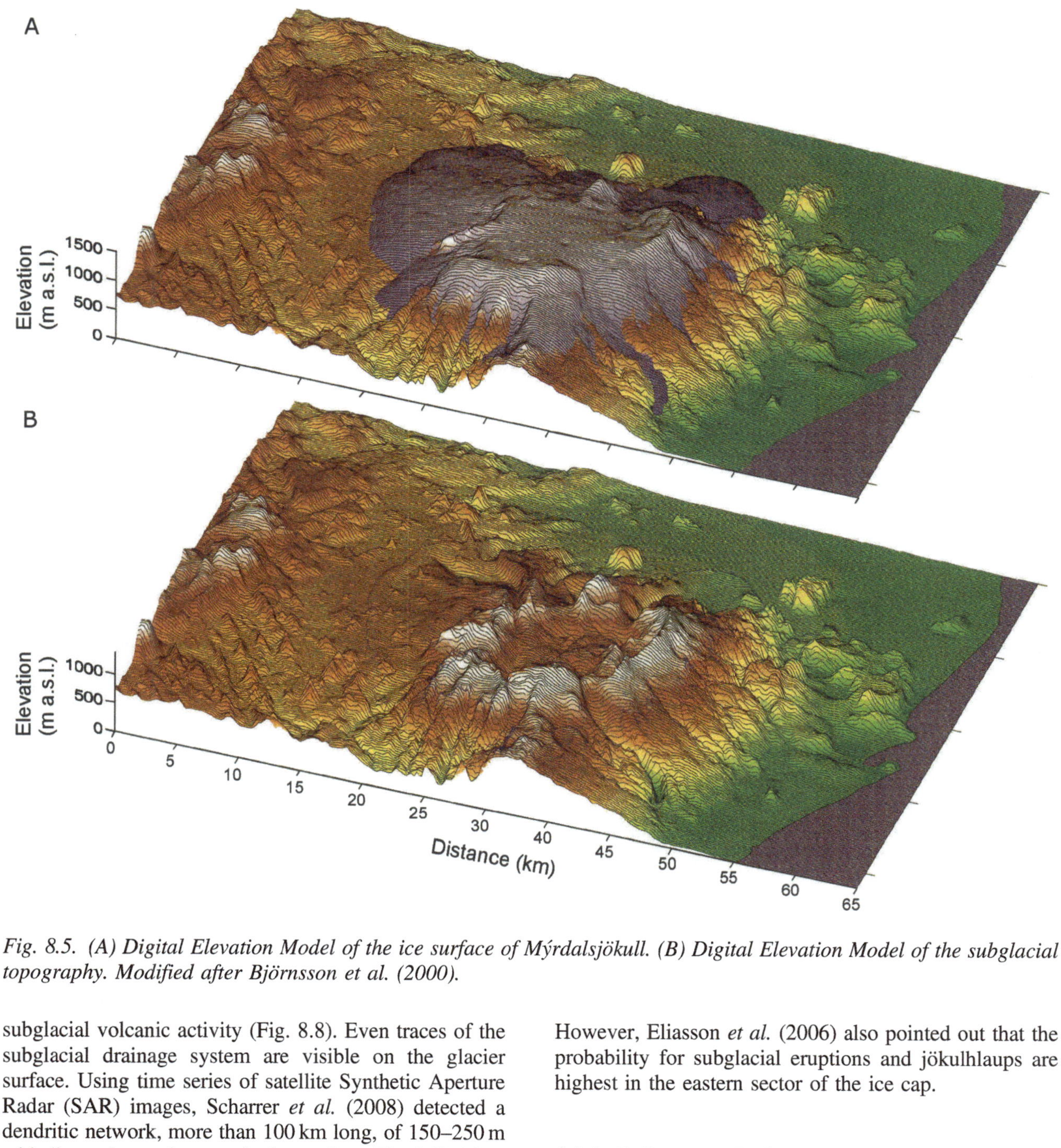

Fig. 8.5. (A) Digital Elevation Model of the ice surface of Mýrdalsjökull. (B) Digital Elevation Model of the subglacial topography. Modified after Björnsson et al. (2000).

subglacial volcanic activity (Fig. 8.8). Even traces of the subglacial drainage system are visible on the glacier surface. Using time series of satellite Synthetic Aperture Radar (SAR) images, Scharrer *et al.* (2008) detected a dendritic network, more than 100 km long, of 150–250 m wide linear depressions on the ice surface (Fig. 8.9). Most linear depressions originated at cauldrons and indicate meltwater drainage away from such cauldrons. The location and size of linear depressions was constant over the 1994 to 2006 period. Scharrer *et al.* (2008) interpreted the linear depressions as the surface expression of subglacial meltwater tunnels. It appears from Fig. 8.9 that the subglacial drainage system and water divides are more complex than suggested by Björnsson *et al.* (2000) because many subglacial tunnels cross the water divides. The discovery of subglacial tunnels also implies that the current view of jökulhlaup routing and flood prediction should be reassessed (Eliasson *et al.*, 2006). If the surface depressions mapped by Scharrer *et al.* (2008) reflect sub- or englacial tunnels, it indicates that jökulhlaups might escape through Sléttjökull – a routing that is not included in the present models for floods (Eliasson *et al.*, 2006). However, Eliasson *et al.* (2006) also pointed out that the probability for subglacial eruptions and jökulhlaups are highest in the eastern sector of the ice cap.

8.3.3. Sedimentary Evidence of Past Subglacial Meltwater Drainage

Laminated, wedge-shaped structures bifurcating downwards from the base of subglacial tills into the underlying sorted sediments have been described from Sléttjökull and Sólheimajökull (Fig. 8.10). van der Meer *et al.* (1999), working at Sléttjökull, interpreted the structures as subglacial WES. The wedges at Sléttjökull consist of clay, silt and sand, and have a width of 1–100 cm; they are commonly 1–2 m high, striking perpendicular to the ice-flow direction. The wedges record a water escape direction towards the glacier margin at the time of deposition and may be followed for tens of meters in length. Figure 8.11 shows an ice-flow parallel Ground Penetrating Radar (GPR) transect across sediments with WES. The radargram clearly shows the WES intruding

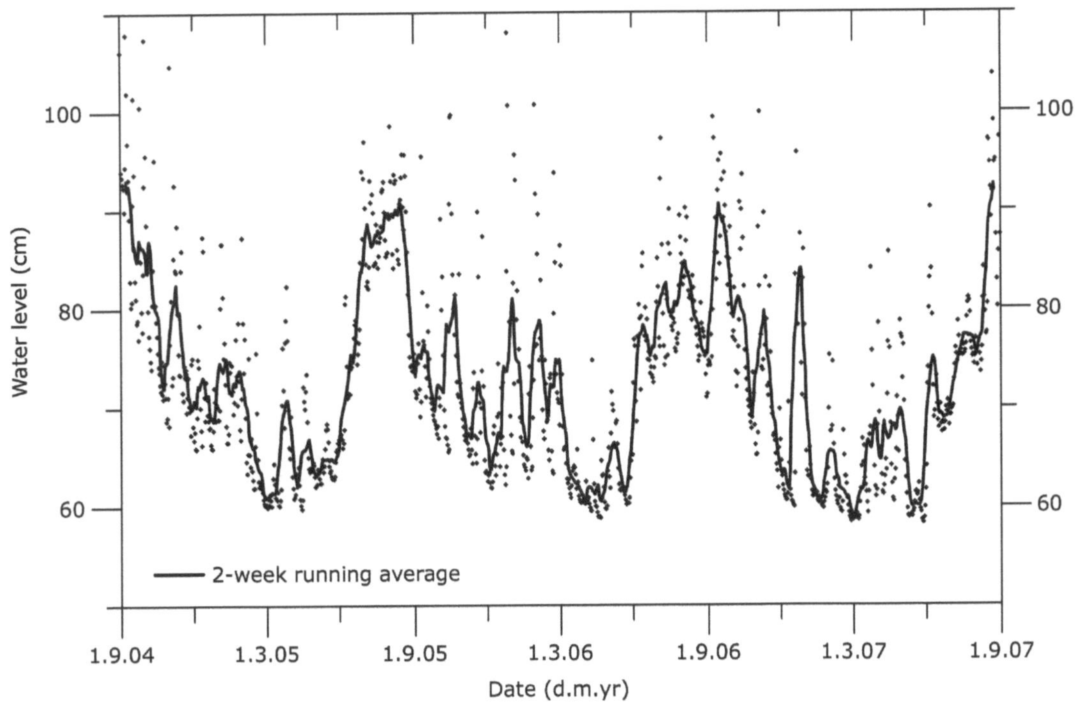

Fig. 8.6. Water level in the meltwater river Jökulsá Gigjökull from 2004 to 2007. The sustained base flow during the winters indicates that subglacial meltwater is produced throughout the year and that the ice cap is warm-based.

from the base of the tills into sorted sediments, confirming the results from the geological sections studied by van der Meer *et al.* (1999).

Because the WES are intruded from the glacier sole into the youngest till, they must post-date this till. It is clear that the WES indicate that at some point after deposition of the youngest till, the subglacial hydrological system has experienced an extreme event where it could not drain the meltwater in the normal mode. The most likely cause for such an event is the 1918 Katla eruption which produced extreme amounts of subglacial meltwater (e.g. Jónsson, 1982; Tómasson, 1996; Larsen, 2000). Even though most of the meltwater was evacuated as jökulhlaups from Kötlujökull, the subglacial water divides and pressure conditions do, indeed, allow meltwater from Katla eruptions to flow northwards (Björnsson *et al.*, 2000; Eliasson *et al.*, 2006). Furthermore, recent Synthetic Aperture Radar studies suggest that many subglacial tunnels drain the Katla caldera towards Sléttjökull (Scharrer *et al.*, 2008). According to van der Meer *et al.* (1999), the extraordinary amounts of meltwater were blocked to flow freely below Sléttjökull because the glacier toe was frozen to its bed at that time of the year. Therefore, meltwater escaped downwards through the till into the highly permeable outwash sediments where it produced the WES (Fig. 8.12).

Similar wedge-shaped intrusions occur at Sólheimajökull (Fig. 8.13), even though a different model of formation has been suggested (Le Heron and Etienne, 2005). The 'clastic dykes' at Sólheimajökull also consist of laminated sediments from clay to gravel and diamict in grain size, and they are injected from the base of a subglacial till. It appears from the sections in front of Sólheimajökull that the dykes extend from the till and at least to the present river level, i.e. more than 10 m. Their height is 1–10 m, and they reach a thickness of up to 0.5 m. Similar to the WES at Sléttjökull, they may be traced for tens of metres in the geological sections.

Le Heron and Etienne (2005) found that the clastic dykes at Sólheimajökull most likely formed due to hydraulic fracturing (water escape) during a Holocene glacier advance. In contrast to van der Meer *et al.* (1999), they suggest that dyke formation took place over a long period of time and was caused by the load from the overlying glacier.

The intrusions at Sléttjökull and Sólheimajökull need not necessarily have formed by the same processes. Although the details of dyke formation are still not understood in details, the escape of subglacial meltwater through clastic dykes most likely affect the glaciodynamics and landform/sediment genesis (van der Meer *et al.*, 2009). Further investigations of the dykes in the Mýrdalsjökull area should focus on assessing the exact relation between dykes and host sediments, the glacier thickness, depth to bedrock, hydraulic conductivity of involved sediments and the spatial relation between the dyke sites and the glacier extent (van der Meer *et al.*, 2009).

8.4. Subglacial Sediments and Landforms

8.4.1. Tills

Beneath the terrain surface at Sléttjökull, two subglacial till beds, Lower and Upper Till, are resting on outwash sediments and bedrock (Fig. 8.14; Kjær *et al.*, 2003,

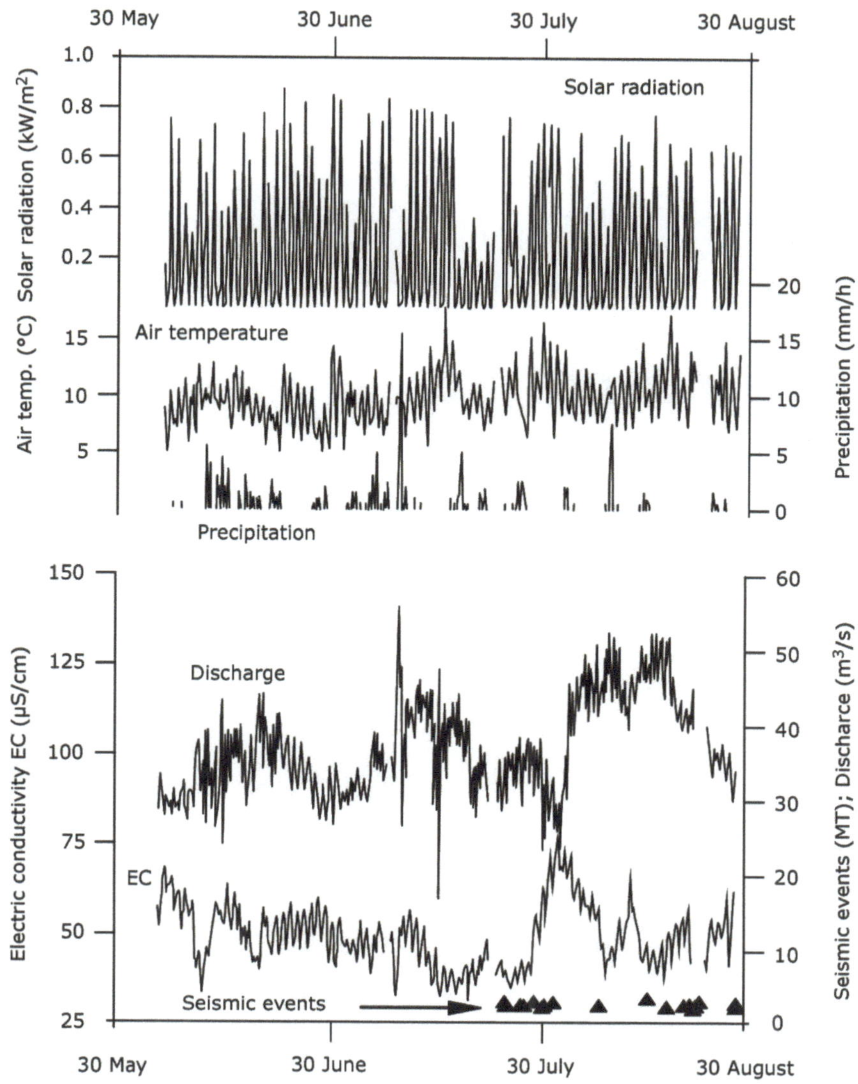

Fig. 8.7. Meltwater discharge and electric conductivity (EC) in Jökulsá á Sólheimasandi from May to August 1989. The timing and magnitude (M_T) of seismic events beneath SW Mýrdalsjökull are indicated. The upper panel shows meteorological parameters at Skógasandur south of Sólheimajökull. Modified after Lawler et al. (1996).

Fig. 8.8. Ice cauldron on the surface of Mýrdalsjökull, August 2001. Cauldrons form due to localized geothermal heat release at the base of the ice cap, causing very high melt rates and the formation of circular depressions on the ice surface. The cauldron has a diameter of c. 800 m.

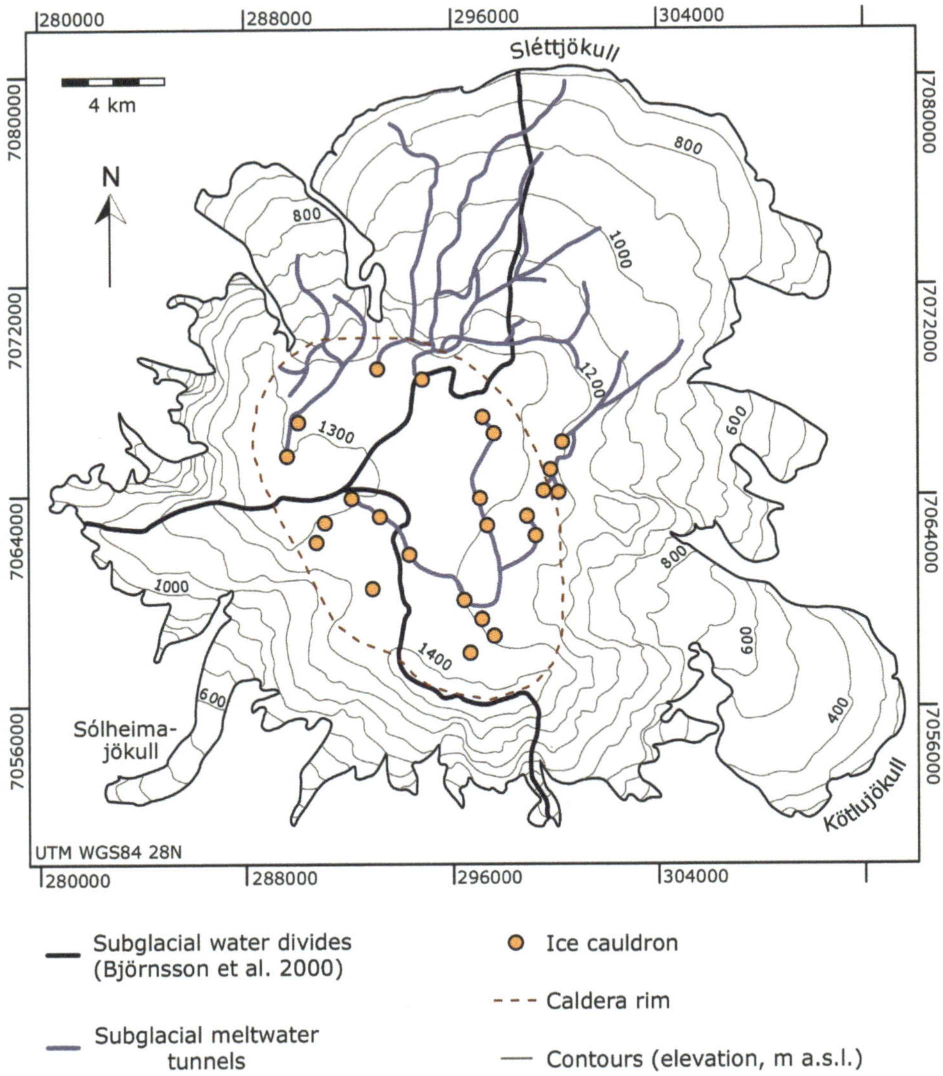

Fig. 8.9. Map of the surface of Mýrdalsjökull. The ice cauldrons and meltwater drainage system are extracted from time series of Satellite Aperture Radar (SAR) imagery from 1994 to 2006. Modified after Scharrer et al. (2008).

Fig. 8.10. Water-escape structures in the sediments in the Sléttjökull forefield. Photo: Jaap J.M. van der Meer (1997).

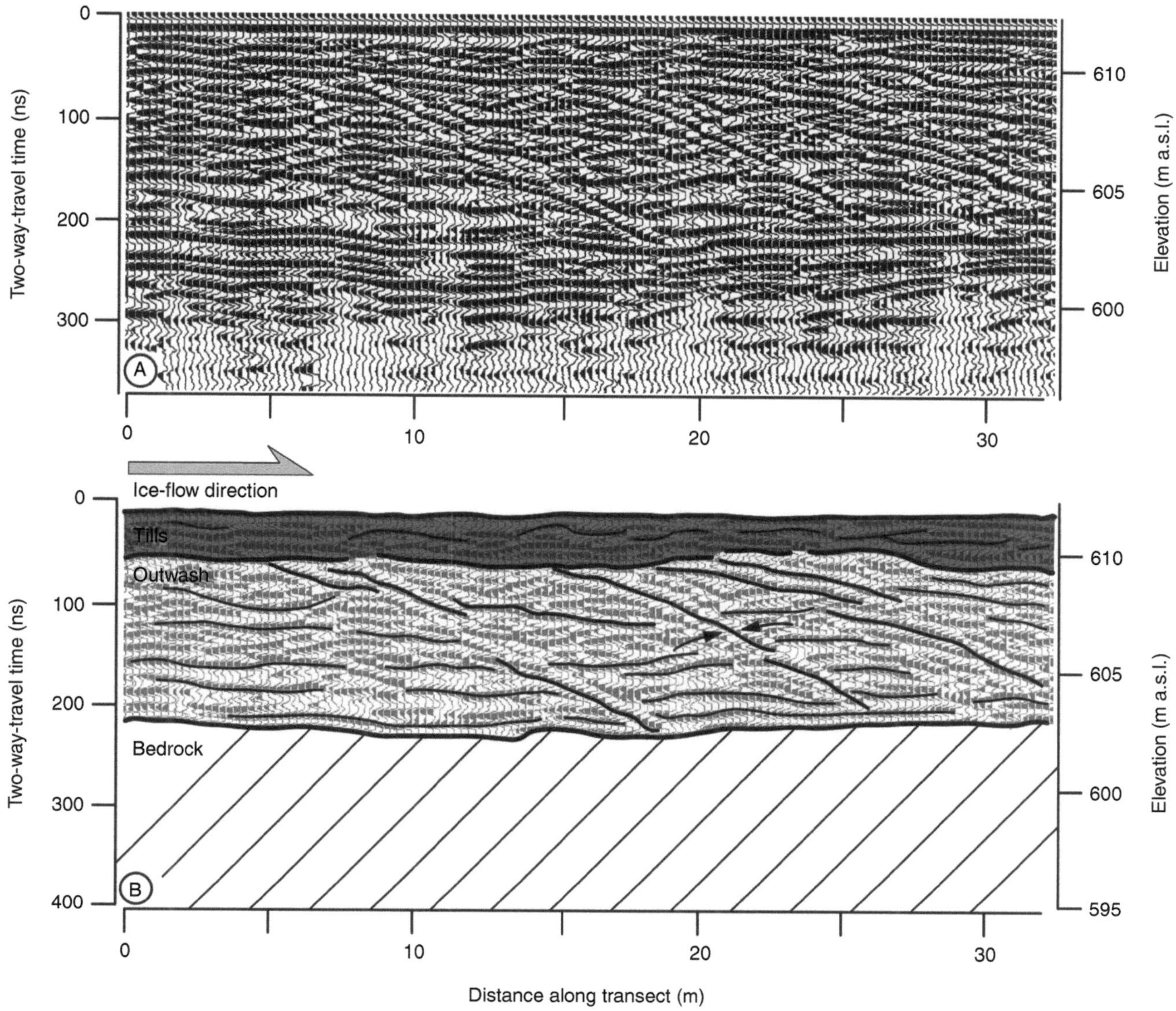

Fig. 8.11. (A) Ice-flow parallel Ground Penetrating Radar (GPR) transect over the Sléttjökull forefield. The radar survey was conducted with a PulseEKKO IV system with a pulser voltage of 400 V, using a centre frequency of 100 MHz and a step size of 0.25 m. For depth conversion, we used a velocity of 0.085 m/ns obtained by Common Mid-Point (CMP) analyses. (B) Interpretation of the GPR data. Water-escape structures are clearly seen extending from the base of the tills and down through the sorted sediments (arrows).

2004). A typical sediment succession appears in the geological section in Fig. 8.15.

The Lower Till does not occur as frequent as the Upper Till and has a discontinuous distribution with remnants up to 40 m long. It is a light grey, massive, silty-sandy, matrix-supported, extremely firm diamict. Generally, the thickness is 0.1–0.5 m with a high clast content, predominantly less than 10 cm long. The Lower Till bed is often broken into segments and dissected by WES or pods of outwash sediment. Dependent on local conditions the two till beds are separated by a clast pavement, re-deposited diamicts or sorted sediments (Krüger, 1987a, 1994; Kjær, 1999; van der Meer *et al.*, 1999; Kjær *et al.*, 2003).

It is often possible to distinguish two subunits within the Upper Till: a lower subunit rich in well-preserved structures and an upper homogeneous subunit. The structure-rich subunit is a dark grey, massive-to-banded, silty-sandy, matrix-supported, firm diamict, up to 80 cm thick and with a moderate clast content. Structures are related to numerous oriented lenses and streaks of blackish sandy material, clearly derived from the outwash below the till. Clasts of less competent lithologies located along the shear planes are crushed and stretched-out. The overlying homogeneous subunit consists of a dark grey, silty-sandy, matrix-supported, friable diamict, up to 60 cm thick with a fissile appearance and a moderate clast content. A few bullet-nosed clasts are present, their stoss side facing upglacier. A clast pavement covers the upper surface – many of these clasts are bullet-nosed and between 2 and 10 cm long.

The outwash below the two till beds consists of coarse tephra which has been shown to originate from the Eldgjá eruption in AD 934 (Krüger, 1994; Larsen, 2000; Sultan, 2002; Kjær *et al.*, 2004). This implies that the two till beds are younger than the Eldgjá event, and most likely

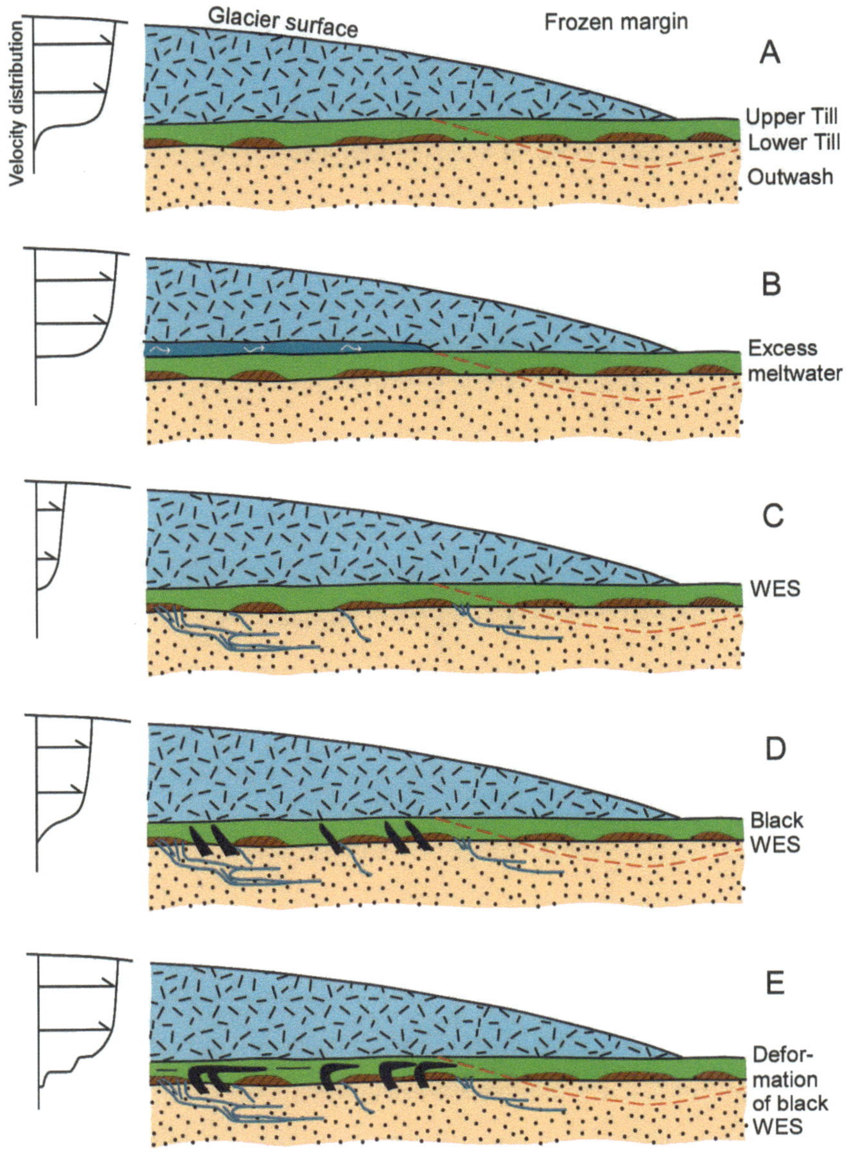

Fig. 8.12. Sequential model for the development of the water-escape structures (WES) and the till structure at Sléttjökull. (A) The situation c. AD 1900 before the overproduction of meltwater showing two till beds, Lower Till related to an older advance and Upper Till produced by the last ice advance, resting on outwash sediments. When the retreating ice-front approaches, the sediment freezes to a depth of about 2 m in the most terminal part of the glacier due to permafrost conditions in the glacier forefield. The velocity profile in this system (i.e. upglacier from the permafrost zone) illustrates creep in the ice, sliding at the ice–bed interface and decreasing velocity within the till. (B) The situation around the Katla eruption in 1918 with the appearance of an unusual amount of meltwater collecting at the ice–bed interface. This led to a higher velocity because of enhanced sliding over the ice–bed interface. As the glacier decoupled from the bed because of the meltwater layer, it is assumed that bed deformation ceased. (C) The increased pressure has forced meltwater to escape to the permeable outwash underneath the till and underneath the permafrost layer at the glacier front and further into the glacier forefield. The consequence is a slowing down of sliding. The opening up of the drainage paths in the shape of WES may well have changed the subglacial lubrication conditions completely meaning that the deformation of the bed was halted somewhat longer. As long as the WES were capable of carrying water, it is conceivable that all water available was drained, including the water that would be present also under normal conditions. (D) Stiffening as a result of drying out of the till bed combined with continuous foreward movement of the glacier caused the till bed to fracture followed by the movement of water and liquefied sand up into the till from the underlying outwash. Conceivably the fractures extended all the way to the ice–bed interface. The velocity profile changed once more, because the till bed became part of the system again, even though it must have started to move in a block-like jerky manner. (E) The situation shortly before deglaciation where the till bed has taken up moisture again after the WES has stopped to function, e.g. by sealing through clay laminae. By this time the till has developed three unit structures: the top unit is completely attenuated and becomes structureless and highly dilated; the middle unit shows deformation structures, because of the presence of the black WES; and finally, the lower unit is characterized by the presence of interweaving WES. Consequently, the velocity profile is composed of creep in the ice, sliding at the ice–bed interface, a high velocity in the upper till unit, sliding at shear planes at the upper unit–middle unit transition, and decreasing velocity with depth with occasional shearing just below the till base. Modified after van der Meer et al. (1999).

Fig. 8.13. (A) Overview of the geological sections along the banks of Jökulsá á Sólheimasandi. The clastic dykes are intruded from the base of a basal till in the upper part of the section. (B) Clastic dyke intruded into a coarse grained tephra layer. October 2007.

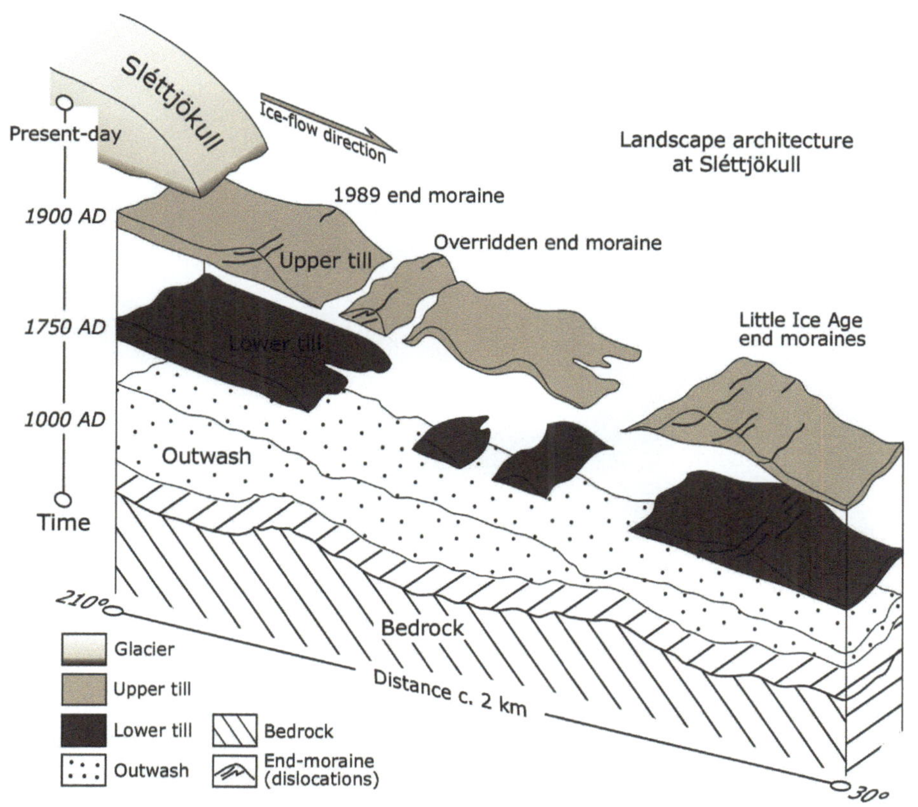

Fig. 8.14. General landscape architecture of the Sléttjökull forefield. Modified after Kjær et al. (2004).

they were deposited during the Little Ice Age. Krüger (1994) suggested that the Lower Till formed around AD 1750 and the Upper Till around AD 1890 – both periods where Icelandic glaciers generally advanced (e.g. Casely and Dugmore, 2004; Grove, 2004; McKinzey *et al.*, 2005a, 2005b; Flowers *et al.*, 2007).

8.4.2. Mode of Transport

Using a well-defined point source and a firm knowledge of ice-flow direction, transport distance, transport path and glacier regime, the mode of subglacial transport was reconstructed at the Sléttjökull margin (Kjær, 1999). Interestingly, within the first 250 m of transport distance significant changes occur in the grain size distribution, clast morphology and geochemical and mineral magnetic properties. Thus, only a very short transport distance is required for the different sedimentological properties to reach a mature state. Beyond 250 m the lateral development of till properties is likely to be a function of transport distance and type of substrate beneath the glacier and its deforming bed. The type of substrate controls the subglacial drainage conditions, thereby the rheological behaviour and subsequently the wear environment in the deforming bed. The transport distance partly controls the

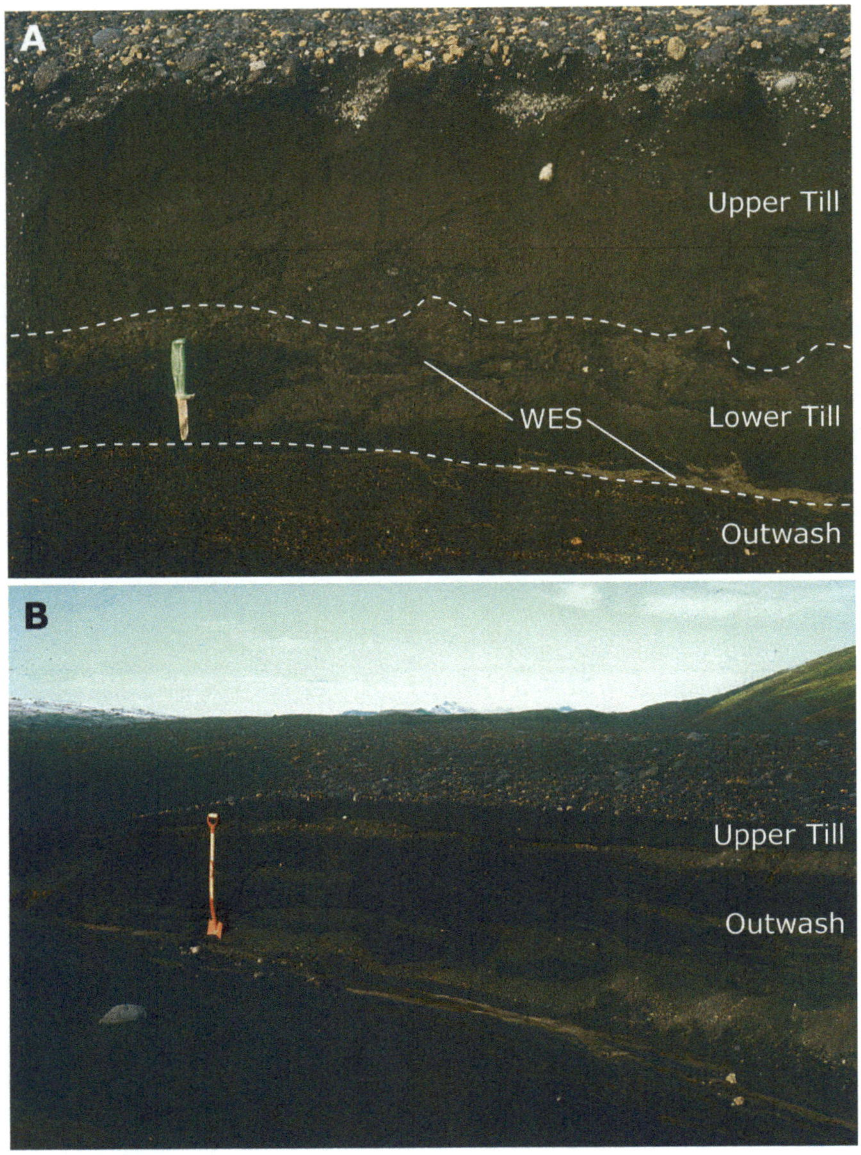

Fig. 8.15. Photographs of a geological section through a minor subglacial landform in the Sléttjökull forefield distinguishing between the Upper and Lower Tills. (A) Thicker succession with both Lower and Upper Tills above outwash. (B) Thin succession with only Upper Till present above outwash.

duration in which the sediment is in the deforming bed. It is inferred that the physical processes and conditions in a deformable bed are highly responsive to changes in type of substratum. As a consequence the till properties have a limited spatial continuity as they change rapidly in response to major changes in the basal conditions.

8.4.3. Till Sheet Architecture

As a part of a study of the architecture of a stratigraphically distinct till sheet (Upper Till) in front of Sléttjökull, Kjær *et al.* (2003) also investigated subglacial landforms. They found a strong correlation between till thickness and the type of substratum (Fig. 8.16). Where the substratum consists of sorted sediments, Upper Till never exceeds 40 cm in thickness. Conversely, where the substratum consists of Lower Till overlying outwash, the thickness of Upper Till varies between 20 and 125 cm. As a novel approach, Kjær *et al.* (2003) tied the till thickness to variable hydraulic conditions in a deforming subglacial bed. Transferred into the simplest scenario this involves high water content over areas with Lower Till and low water content in areas with outwash as a substrate. Thus, it was demonstrated that accumulation of Upper Till is facilitated in areas with a low-permeability substratum of Lower Till. This is because erosion and deformation of Lower Till nourished the accumulation of the new till. This implies that the distribution of till sheets is predetermined by the distribution of different substrata deposited subsequent to the ice advance.

8.4.4. Landform Interpretation and its Significance for Subglacial Processes

Many forefields of the Mýrdalsjökull outlet glaciers host drumlins and other subglacially streamlined landforms (Fig. 8.17). Drumlins are streamlined hills with a steep, blunt stoss side and a gentle, tapering lee side (e.g. Menzies, 1979; Krüger and Thomsen, 1984; Menzies and Rose, 1987; Patterson and Hooke, 1995; Jørgensen

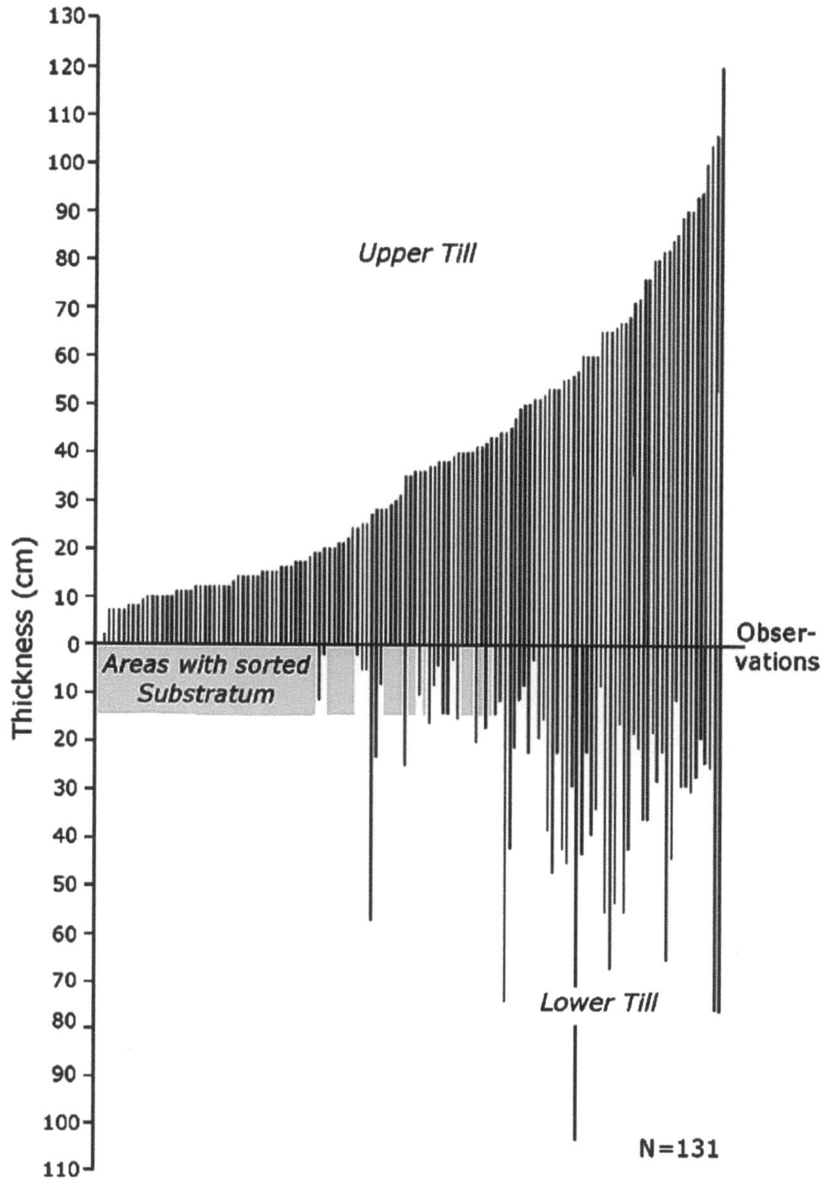

Fig. 8.16. Correspondence between the thickness of Upper Till – both subunits – and the underlying Lower Till unit. Where the Lower Till is absent the Upper Till rests on sorted outwash sediments only – grey shade. N *is number of observations.*

and Piotrowski, 2003; Rattas and Piotrowski, 2003; Schomacker *et al.*, 2006; Briner, 2007; Menzies and Brand, 2007). The long axis of a drumlin is parallel to the ice-flow direction. Even though alternative models of drumlin formation have been suggested, there is generally consensus that drumlins form at the bed of active, warm-based glaciers or ice streams (e.g. Krüger and Thomsen, 1984; Boulton, 1987; Krüger, 1987b; Kjær *et al.*, 2003; Briner, 2007; Menzies and Brand, 2007). Thus, drumlins reflect the processes acting in the subglacial environment.

Krüger and Thomsen (1984), Krüger (1987b, 1994) and Kjær *et al.* (2003) recognized three main types of drumlins at Sléttjökull (Fig. 8.18). In all cases, drumlins had formed around pre-existing landforms that had acted as obstacles to ice flow and favoured the initiation of drumlin formation. The obstacles were produced by glaciofluvial erosion in the forefield prior to the ice advance, and they consisted of either subglacial till or sorted sediments. Drumlins with a core of Lower Till were formed around pre-existing patches of an old till plain which had been heavily dissected by proglacial meltwater streams before the glacier advanced, shaped the drumlin and deposited a mantle of Upper Till (Fig. 8.18(A)). Where the glacier advanced over pre-existing obstacles of old glaciofluvial sediments, such as prominent interstream areas, bars or remnants of outwash fans, Upper Till mantled drumlins with a core of sorted sediments occur (Fig. 8.18(B)). The last type of drumlins was shown to have formed by drumlinization and deposition of Upper Till when the glacier overrode an end moraine (Fig. 8.18(C)). Thus, the glaciotectonic deformation in the drumlin core originates from the end moraine and has nothing to do with the drumlin formation. A similar distribution pattern with drumlins at an overridden end moraine was recognized at Brúarjökull at the north margin of the Vatnajökull ice cap, Iceland, by Kjær *et al.* (2008).

8.4.5. Interactions between Subglacial and Supraglacial Environments

Kötlujökull is characterized by its debris-covered margin of stagnant ice, and research has therefore focused on the transition from active ice to passive ice, and the interactions between the sub- and supraglacial environments (Krüger *et al.*, this volume). Debris-rich basal ice melts out at the

Fig. 8.17. (A) Overview of a section along the length axis of a drumlin in front of Sólheimajökull. (B) The drumlin consists of a stratified, extremely firm, grey basal till, and rests on coarse outwash gravels. October 2007.

Fig. 8.18. Three types of drumlins in the Sléttjökull forefield. (A) Drumlins formed around erosional remnants of Lower Till. (B) Drumlins formed around obstacles of sandy outwash, e.g. bars. (C) Drumlins formed around an overridden end moraine. Modified after Krüger (1994).

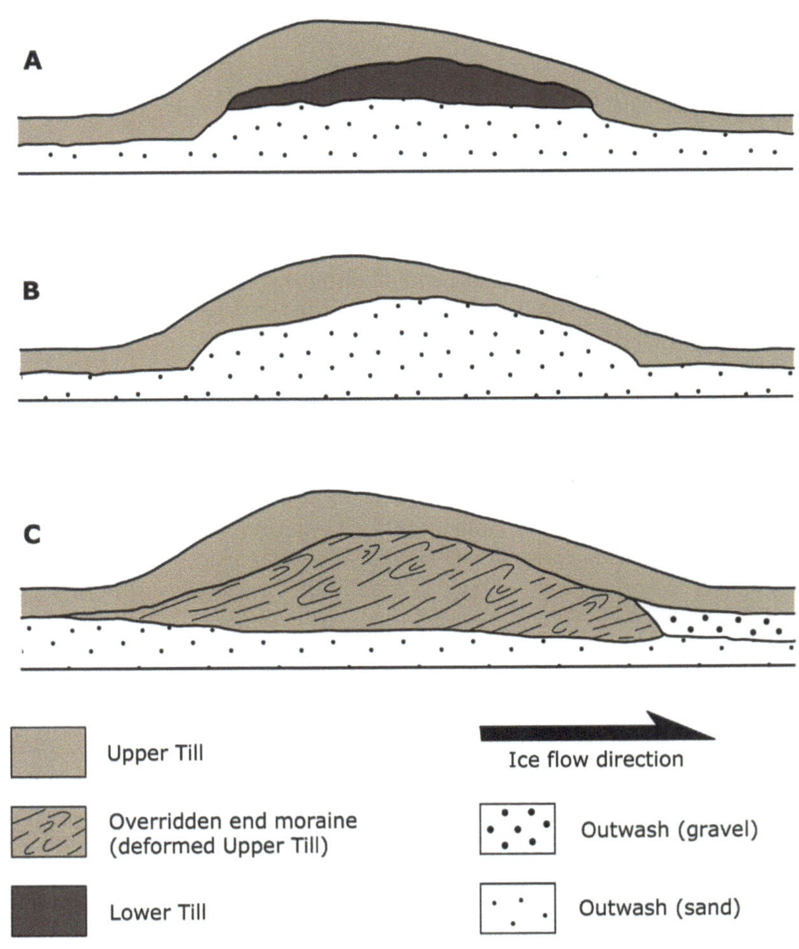

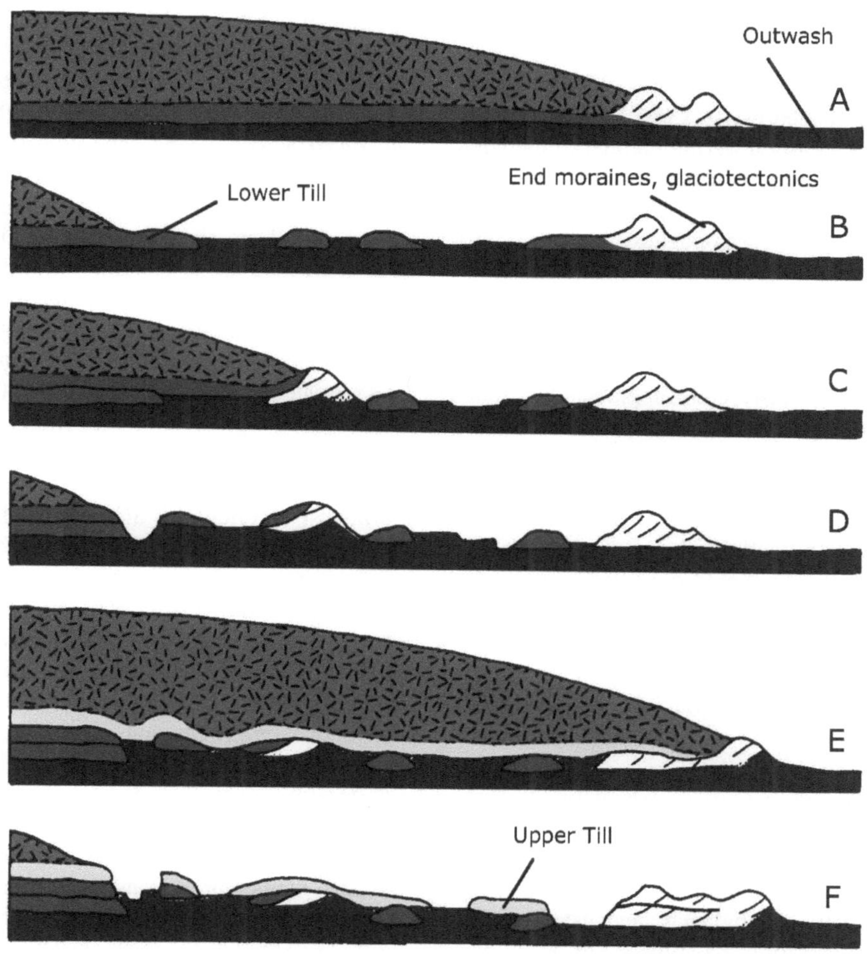

Fig. 8.19. Simplified sequential model for glacial landscape formation at the margin of Sléttjökull. The transects are about 2 km in length. (A) Around mid-eighteenth century. (B) At the end of the eighteenth century. (C) Around 1810. (D) Around mid-nineteenth century. (E) Around 1890. (F) Around 1970s. Partly after Krüger (1994).

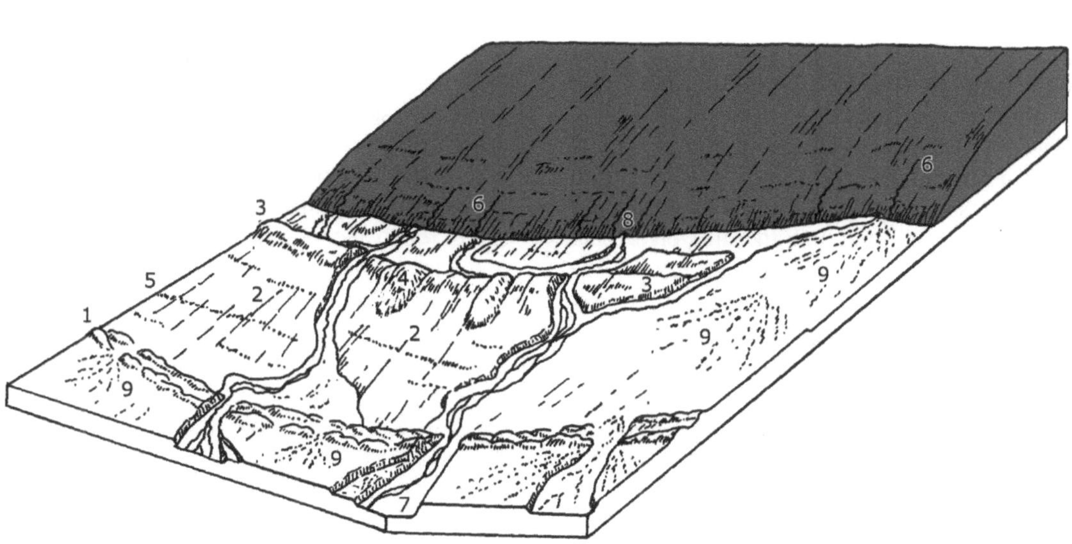

Fig. 8.20. Schematic model of the marginal–subglacial–proglacial landsystem in front of Sléttjökull. (1) Ice–marginal moraine. (2) Fluted till plain. (3) Overridden marginal moraine. (4) Drumlin. (5) Annual moraine. (6) Supraglacial stream. (7) Meltwater channel. (8) Ice cave (9) Outwash fan. After Krüger (1994).

base of dead-ice bodies, and the debris accumulates during de-icing progression. Krüger and Kjær (1999) described a subglacial melt-out till in ice-free sedimentary sections in the Kötlujökull forefield. The melt-out till was stratigraphically above a lodgement till and below supraglacial deposits, supporting the genetic interpretation.

The subglacial melt-out till is dark grey, crudely stratified, silty-sandy, matrix-supported and friable. Many clasts are draped by thin layers of silt and sand. It has a sharp, conformable basal contact. In contrast to the basal tills from Sléttjökull (Section 8.4.1), this diamict lacks fissility and stoss-lee shaped clasts. Clast fabrics display an extremely high degree of clustering parallel to the ice-flow direction (fabric strength $S_1 \sim 0.90$).

Particularly, the extremely clustered ice-flow parallel fabrics and the drapes over clasts suggest deposition by *in situ* melt-out in a subglacial environment (Lawson, 1979; Krüger, 1994). The coarse texture and lack of stoss-lee clasts supports this interpretation (Krüger and Kjær, 1999).

The preservation potential of undisturbed sediments in dead-ice environments is low, and subglacial melt-out tills have been investigated at relatively few localities in modern glacial environments (Lawson, 1979; Paul and Eyles, 1990; Evans *et al.*, 2006). Thus, the observed melt-out of basal dead-ice at Kötlujökull provides an important modern analogue to understand the depositional environments of Pleistocene subglacial melt-out tills (Klint *et al.*, this volume).

8.5. A Marginal–Subglacial–Proglacial landsystem in front of Sléttjökull

Non-surging glaciers without extensive debris cover respond rapidly to climate changes, so glacier margins tend to oscillate on an annual basis (Benn and Evans, 1998; Krüger *et al.*, this volume). During a long-term glacier retreat, however, a subglacial landsystem combined with subglacial landform assemblages of drumlins, flutes and overridden marginal moraines located between a marginal moraine indicating a sustained glacier advance and the present ice front with glacifuvial depo-centres is exposed (Evans *et al.*, 1999; Evans and Twigg, 2002; Evans, 2003). This is illustrated in front of Sléttjökull.

An idealized sequential model for development of the present stratigraphy relative to glacier dynamics and landscape formation at Sléttjökull is shown in Fig. 8.19 (Krüger, 1994). (A) Lower Till and an ice–marginal moraine system 1.5–2 km beyond the present glacier front are interpreted as responses to a glacier advance across a pre-existing outwash plain around the mid-eighteenth century. (B) During the following glacier retreat, the exposed subglacial landsystem became more or less dissected by meltwater streaming from the retreating glacier. The absence of Lower Till at several sites indicates that in the proglacial environment laterally migrating streams disintegrated parts of the subglacial landsystem and formed broad channels and outwash fans with intervening plateaux of resistant clast-paved plain of Lower Till like the scenery in the present glacier forefield. (C) An overridden, well-preserved frontal moraine ridge 0.5–1 km in front of the glacier originates from a subsequent limited re-advance, which probably dates back to around AD 1810. Beyond the frontal moraine ridge, proglacial streams deposited outwash fans, or continued to disintegrate the old till plain of Lower Till and the old ice–marginal moraine system in front of it. (D) During the following retreat a renewed subglacial landsystem became exposed to processes working in the proglacial environments. (E) This transformation of the subglacial landsystem in the direction of a proglacial landsystem of outwash deposits, however, was interrupted by a more extensive re-advance which dates back to the last decades of the nineteenth century. The ice advanced across a landscape consisting of a complexly branching network of outwash deposits with intervening remnants of clast-paved basal till plain, and a fragmentary ice–marginal moraine. From the glacier base, Upper Till was deposited as a continuous sheet. Along its most extended position the glacier disturbed the pre-existing ice–marginal moraine from the mid-eighteenth century and produced new moraine ridges. In 1906 the ice edge had retreated to a position only 50–140 m behind the most distal ice–marginal moraine ridge, i.e. the ice edge followed, in the main, the inmost boundary of the ice–marginal moraine ridges. In the Bláfjöll area in the northwest forefield of Sléttjökull, the ancient moraine ridges, probably from the mid-eighteenth century, are easily recognized as moss-overgrown ridges extending 50 m beyond the younger ones from around 1900 (Krüger *et al.*, this volume). (F) The following glacier retreat beginning in the late 1920s exposed different, superimposed patterns of clast-paved basal till plain. The most prominent terrain elements in the pre-existing landscape had offered good conditions for development of undulating basal till plain; they had acted as subglacial obstacles, leading to increased deposition of Upper Till and drumlin formation. On the other hand, distal parts of pre-existing outwash fans, or extensive remnants of till plains favoured formation of a rather smooth or terraced ground moraine. Thus, the actual subglacial landsystem is an integration of superimposed landforms derived from different development phases and glacial events, where some of the pre-existing landforms are reflected in the present topography. Along with the ice-front recession, which finely balanced the climatic changes (Krüger, 1995; Krüger *et al.*, this volume), the exposed subglacial landsystem underwent a gradual transition from subglacial landsystem through a landscape with poorly developed proglacial forms occasionally to a mature proglacial landsystem without evidence of the original subglacial landsystem (Fig. 8.20). During winter or early spring, small re-advances created annual moraine ridges, and ponding of water in depressions on the subglacial till plain resulted in deposition of fine-grained lacustrine deposits. Wherever meltwater was flowing from the glacier surface, channels were cut into the till plain, and in low-lying till areas were stripped and replaced by outwash fans. In the 1980s the ice front was stationary except for annual oscillations and a composite marginal moraine ridge was formed, but at present the ice edge is retreating again.

8.6. Conclusions

- The clean ice margin of Sléttjökull is currently retreating frontally, and thereby exposing a spectacular subglacial landscape with fluted and drumlinized till plains.
- Geophysical surveys have shown that Mýrdalsjökull rests on the Katla caldera, and that outlet glaciers

have carved out U-shaped troughs in the caldera rim. The maximum ice thickness over the caldera is about 740 m.

- There is a strong link between subglacial geothermal activity and meltwater chemistry and discharge. After some days with high geothermal activity, meltwater discharge and content of volcanogenic solutes peak. On the ice surface, ice cauldrons often form as a response to high, localized subglacial geothermal heat release. WES in glacial sediments are evidence of past events of extreme subglacial meltwater drainage.
- The properties and genesis of basal tills have been investigated in detail at Sléttjökull. Recently, it has been demonstrated that only 250 m transport distance after a point source (bedrock knob) in the subglacial environment is required for the sedimentological properties to reach a mature state. The Sléttjökull forefield also served as a field laboratory for studies of drumlin formation and till accumulation. The most recent model shows that drumlins and new till accumulate in areas with a low-permeability substratum of old till. Erosion and deformation of the old till nourished the accumulation of the new till.
- Subglacial landsystems models showing the formation of landscapes with fluted and drumlinized till plains have been proposed, based on research at the frontally retreating north margin of Mýrdalsjökull. Such models from modern glacial environments are important in reconstructions of the genesis of Pleistocene glacial landscapes.

Acknowledgements

The authors thank the Danish Natural Science Research Council, the Carlsberg Foundation and The University of Iceland Research Fund for financial support to the research at Mýrdalsjökull, and the Icelandic National Research Council (RANNÍS) for allowing us to work at the margins of Mýrdalsjökull. We extend our gratitude to all the members of the 1977–2008 field campaigns.

Ívar Örn Benediktsson, Jaap J.M. van der Meer and Kilian Scharrer kindly provided three figures. Óðinn Þórarinsson from the Icelandic National Energy Authority (Orkustofnun) is thanked for letting us use the hydrological data.

References

Arnborg, L., 1955. Hydrology of the glacial river Austurfljót. Geografiska Annaler 37, 185–201.

Benn, D.I., 1995. Fabric signature of subglacial till deformation, Breidamerkurjökull, Iceland. Sedimentology 42, 735–747.

Benn, D.I. & Evans, D.J.A., 1996. The interpretation and classification of subglacially-deformed materials. Quaternary Science Reviews 15, 23–52.

Benn, D.I. & Evans, D.J.A., 1998. Glaciers and glaciation. London, Arnold.

Björnsson, H., Pálsson, F., & Guðmundsson, M.T., 2000. Surface and bedrock topography of the Mýrdalsjökull ice cap, Iceland: the Katla caldera, eruption sites and routes of jökulhlaups. Jökull 49, 29–46.

Björnsson, H., Pálsson, F., Sigurðsson, O., & Flowers, G.E., 2003. Surges of glaciers in Iceland. Annals of Glaciology 36, 82–90.

Boulton, G.S., 1987. A theory of drumlin formation by subglacial sediment deformation. In: Menzies, J., Rose, J. (Eds), Drumlin Symposium. Rotterdam, Balkema, 25–80.

Boulton, G.S. & Hindmarsh, R.C.A., 1987. Sediment deformation beneath glaciers: rheology and geological consequences. Journal of Geophysical Research 92, 9059–9082.

Brandt, O., Björnsson, H., & Gjessing, Y., 2005. Mass–balance rates derived by mapping internal tephra layers in Mýrdalsjökull and Vatnajökull ice caps, Iceland. Annals of Glaciology 42, 284–290.

Briner, J., 2007. Supporting evidence from the New York drumlin field that elongate subglacial bedforms indicate fast ice flow. Boreas 36, 143–147.

Carr, S.J. & Goddard, M.A., 2007. Role of particle size in till-fabric characteristics: systematic variation in till fabric from Vestari-Hagafellsjökull, Iceland. Boreas 36, 1–15.

Casely, A.F. & Dugmore, A.J., 2004. Climate change and 'anomalous' glacier fluctuations: the southwest outlets of Mýrdalsjökull, Iceland. Boreas 33, 108–122.

Eliasson, J., Larsen, G., Gudmundsson, M.T., & Sigmundsson, F., 2006. Probabilistic model for eruptions and associated flood events in the Katla caldera, Iceland. Computational Geosciences 10, 179–200.

Evans, D.J.A. (Ed), 2003. Glacial landsystems. London, Arnold.

Evans, D.J.A. & Rea, B.R., 2003. Surging glacier landsystem. In: Evans, D.J.A. (Ed), Glacial landsystems. London, Arnold, 259–288.

Evans, D.J.A. & Twigg, D.R., 2002. The active temperate glacial landsystem: a model based on Breiðamerkurjökull and Fjallsjökull, Iceland. Quaternary Science Reviews 21, 2143–2177.

Evans, D.J.A., Lemmen, D.S., & Rea, B.R., 1999. Glacial landsystems of the southwest Laurentide Ice Sheet: modern Icelandic analogues. Journal of Quaternary Science 14, 673–691.

Evans, D.J.A., Phillips, E.R., Hiemstra, J.F., & Auton, C.A., 2006. Subglacial till: formation, sedimentary characteristics and classification. Earth-Science Reviews 78, 115–176.

Flowers, G.E., Björnsson, H., Geirsdóttir, Á., Miller, G.H., & Clarke, G.K.C., 2007. Glacier fluctuation and inferred climatology of the Langjökull ice cap through the Little Ice Age. Quaternary Science Reviews 26, 2337–2353.

Fuller, S. & Murray, T., 2000. Evidence against pervasive bed deformation during the surge of an Icelandic glacier. In: Maltman, A.J., Hubbard, B., Hambrey, M.J. (Eds), Deformation of glacial materials, Special Publications 176, London, Geological Society, 203–216.

Fuller, S. & Murray, T., 2002. Sedimentological investigations in the forefield of an Icelandic surge-type glacier: implications for the surge mechanism. Quaternary Science Reviews 21, 1503–1520.

Grove, J.M., 2004. Little Ice Ages: ancient and modern, 2nd ed., Routledge studies in physical geography and environment, London, Routledge, 718pp.

Hart, J.K., 1995. Recent drumlins, flutes and lineations at Vestari-Hagafellsjökull, Iceland. Journal of Glaciology 41, 596–606.

Jónsson, J., 1982. Notes on the Katla volcanoglacial debris flows. Jökull 32, 61–68.

Jørgensen, F. & Piotrowski, J.A., 2003. Signature of the Baltic ice stream on Funen Island, Denmark during the Weichselian glaciation. Boreas 32, 242–255.

Kjær, K.H., 1999. Mode of subglacial transport deduced from till properties, Mýrdalsjökull, Iceland. Sedimentary Geology 128, 271–292.

Kjær, K.H., Krüger, J., & van der Meer, J.J.M., 2003. What causes till thickness to change over distance? Answers from Mýrdalsjökull, Iceland. Quaternary Science Reviews 22, 1687–1700.

Kjær, K.H., Sultan, L., Krüger, J., & Schomacker, A., 2004. Architecture and sedimentation of outwash fans in front of the Mýrdalsjökull ice cap, Iceland. Sedimentary Geology 172, 139–163.

Kjær, K.H., Larsen, E., van der Meer, J., Ingólfsson, Ó., Krüger, J., Benediktsson, Í.Ö., Knudsen, C.G., & Schomacker, A., 2006. Subglacial decoupling at the sediment/bedrock interface: a new mechanism for rapid flowing ice. Quaternary Science Reviews 25, 2704–2712.

Kjær, K.H., Korsgaard, N.J., & Schomacker, A., 2008. Impact of multiple glacier surges – a geomorphological map from Brúarjökull, East Iceland. Journal of Maps 2008, 5–20.

Kjær, K.H. & Krüger, J., 2001. The final phase of dead-ice moraine development: processes and sediment architecture, Kötlujökull, Iceland. Sedimentology 48, 935–952.

Krüger, J., 1987a. Træk af et glaciallandskabs udvikling ved nordranden af Mýrdalsjökull, Island. Dansk Geologisk Forening Årsskrift 1986, 49–65.

Krüger, J., 1987b. Relation of drumlin shape and distribution to drumlin stratigraphy and glacial history, Mýrdalsjökull, Iceland. In: Menzies, J., Rose, J. (Eds), Drumlin Symposium. Rotterdam, Balkema, 257–266.

Krüger, J., 1994. Glacial processes, sediments, landforms, and stratigraphy in the terminus region of Mýrdalsjökull, Iceland. Folia Geographica Danica 21, 1–233.

Krüger, J., 1995. Origin, chronology and climatological significance of annual-moraine ridges at Mýrdalsjökull, Iceland. The Holocene 5, 420–427.

Krüger, J. & Kjær, K.H., 1999. A data chart for field description and genetic interpretation of glacial diamicts and associated sediments – with examples from Greenland, Iceland, and Denmark. Boreas 28, 386–402.

Krüger, J. & Kjær, K.H., 2000. De-icing progression of ice-cored moraines in a humid, subpolar climate, Kötlujökull, Iceland. The Holocene 10, 737–747.

Krüger, J. & Thomsen, H.H., 1984. Morphology, stratigraphy, and genesis of small drumlins in front of the glacier Mýrdalsjökull, South Iceland. Journal of Glaciology 30, 94–105.

Larsen, G., 2000. Holocene eruptions within the Katla volcanic system, south Iceland: characteristics and environmental impact. Jökull 49, 1–28.

Lawler, D.M., Björnsson, H., & Dolan, M., 1996. Impact of subglacial geothermal activity on meltwater quality in the Jökulsá á Sólheimasandi system, southern Iceland. Hydrological Processes 10, 557–578.

Lawson, D.E., 1979. Sedimentological analysis of the western terminus region of the Matanuska Glacier, Alaska. U.S. Army Cold Regions Research and Engineering Laboratory, Report 79-9, 1–112.

Le Heron, D.P. & Etienne, J.L., 2005. A complex subglacial clastic dyke swarm, Sólheimajökull, southern Iceland. Sedimentary Geology 181, 25–37.

Mackintosh, A.N., Dugmore, A.J., & Jacobsen, F.M., 2000. Ice-thickness measurements on Sólheimajökull, southern Iceland and their relevance to its recent behaviour. Jökull 48, 9–15.

Magnússon, E., Björnsson, H., Dall, J., & Pálsson, F., 2005. The 20th century retreat of ice caps in Iceland derived from airborne SAR: W-Vatnajökull and N-Mýrdalsjökull. Earth and Planetary Science Letters 237, 508–515.

McKinzey, K.M., Orvin, J.F., & Bradwell, T., 2005a. A revised chronology of key Vatnajökull (Iceland) outlet glaciers during the Little Ice Age. Annals of Glaciology 42, 171–179.

McKinzey, K.M., Ólafsdóttir, R., & Dugmore, A., 2005b. Perception, history, and science: coherence or disparity in the timing of the Little Ice Age maximum in southeast Iceland. Polar Record 41, 319–334.

Menzies, J., 1979. A review of the literature on the formation and location of drumlins. Earth-Science Reviews 14, 315–359.

Menzies, J. & Brand, U., 2007. The internal sediment architecture of a drumlin, Port Byron, New York State, USA. Quaternary Science Reviews 26, 322–335.

Menzies, J. & Rose, J. (Eds), 1987. Drumlin Symposium. Rotterdam, Balkema.

Nelson, A.E., Willis, I.C., & Ó Cofaigh, C., 2005. Till genesis and glacier motion inferred from sedimentological evidence associated with the surge-type glacier, Brúarjökull, Iceland. Annals of Glaciology 42, 14–22.

Patterson, C.J. & Hooke, R.L., 1995. Physical environment of drumlin formation. Journal of Glaciology 41, 30–38.

Paul, M.A. & Eyles, N., 1990. Constraints on the preservation of diamict facies (melt-out tills) at the margins of stagnant glaciers. Quaternary Science Reviews 9, 51–69.

Rattas, M. & Piotrowski, J.A., 2003. Influence of bedrock permeability and till grain size on the formation of the Saadjärve drumlin field, Estonia, under an east-Baltic Weichselian ice stream. Boreas 32, 167–177.

Rist, S., 1967. The thickness of the ice cover of Mýrdalsjökull, Southern Iceland. Jökull 17, 237–242.

Scharrer, K., Spieler, O., Mayer, Ch., & Münzer, U., 2008. Imprints of subglacial volcanic activity on a glacier surface – SAR study of Katla volcano, Iceland. Bulletin of Volcanology 70, 495–506.

Schomacker, A., Krüger, J., & Kjær, K.H., 2006. Ice-cored drumlins at the surge-type glacier Brúarjökull, Iceland: a transitional-state landform. Journal of Quaternary Science 21, 85–93.

Sigurðsson, F., 1990. Groundwater from glacial areas in Iceland. Jökull 40, 119–146.

Sigurðsson, O., 2005. Variations of termini of glaciers in Iceland in recent centuries and their connection with climate. In: Caseldine, C., Russell, A., Harðardóttir, J., Knudsen, Ó. (Eds), Iceland – modern processes and past environments. Developments in Quaternary Science 5, 241–255. Amsterdam, Elsevier.

Sigurðsson, O., Zóphóníasson, S., & Ísleifsson, E., 2000. Jökulhlaup úr Sólheimajökli 18, júlí 1999. Jökull 49, 75–80.

Sultan, L., 2002. Reconstruction of fan-shaped outwash in front of the Mýrdalsjökull ice cap, Iceland: architecture and style of sedimentation. M.Sc. Thesis 153, Department of Geology, Lund University, Sweden.

Thorsteinsson, Th., Waddington, E.D., Matsuoka, K., Howat, I., & Tulaczyk, S., 2005. Survey of flow, topography and ablation on NW-Mýrdalsjökull, S-Iceland. Jökull 55, 155–162.

Tómasson, H., 1996. The Jökulhlaup from Katla in 1918. Annals of Glaciology 22, 249–254.

van der Meer, J.J.M., Kjær, K.H., & Krüger, J., 1999. Subglacial water-escape structures and till structures, Sléttjökull, Iceland. Journal of Quaternary Science 14, 191–205.

van der Meer, J.J.M., Kjær, K.H., Krüger, J., Rabassa, J., & Kilfeather, A.A., 2009. Under pressure: clastic dykes in glacial settings. Quaternary Science Reviews 28, 708–720.

Wisniewski, E., Andrzejewski, L., & Olszewski, A., 1999. Relief of the Höfðabrekkujökull forefield, South Iceland, in light of geomorphological mapping. Jökull 47, 59–70.

9

Evidence for Subglacial Deformation and Deposition during a Complete Advance-Stagnation Cycle of Kötlujökull, Iceland – A Case Study

Knud Erik S. Klint[1,*], Niels Richardt[2] and Johannes Krüger[3]

[1]*The Geological Survey of Denmark and Greenland (GEUS), Øster Voldgade 10, DK-1350 Copenhagen K., Denmark*
[2]*Rambøll – Energy & Environment, Teknikerbyen 31, DK-2830 Virum, Denmark*
[3]*Department of Geography and Geology, University of Copenhagen, Øster Voldgade 10, DK-1350 Copenhagen K., Denmark*
**Correspondence and requests for materials should be addressed to K.E.S. Klint (e-mail: kesk@geus.dk)*

9.1. Introduction

During the past decades, research on subglacial conditions has been a rapidly developing field of study, which has resulted in the identification of the main processes acting in subglacial environments (Benn and Evans, 1998). This knowledge on subglacial conditions comes partly from observations of modern glacier beds by using data from boreholes, tunnels, natural cavities and remote sensing techniques, partly from study of subglacial landforms and sediments in modern glacier forefields, theoretical modelling, field and laboratory experiments and from the interpretation of former glaciated terrains (Boulton, 1975, 1979, 1996; Boulton and Dobbie, 1993; van der Meer, 1993; Krüger, 1994; Benn and Evans, 1998; Piotrowski and Tulaczyk, 1999; Fuller and Murray, 2002; Evans, 2003; Bennett *et al.*, 2003; Kjær *et al.*, 2003, 2006; Christoffersen *et al.*, 2005; Le Heron *et al.*, 2005; Nelson *et al.*, 2005; Evans *et al.*, 2006).

Much debate, however, has been focused on the problem to relate the sediment texture and architecture that appear in the geological records to specific subglacial processes (Boulton, 1978; Benn and Evans, 1996; Krüger and Kjær, 1999; van der Meer *et al.*, 1999; Fuller and Murray, 2002; Evans *et al.*, 2006). This discussion in turn has resulted in various genetic classifications of tills from the classical approach defined by Boulton (1968, 1971) and Dreimanis (1989) reflecting different depositional processes, such as lodgement till, flow till and melt out-till, into a more tectonic-related classification reflecting different degrees and types of deformation (Elson, 1989; Pedersen, 1989; Benn and Evans, 1998; van der Meer *et al.*, 2003; Evans *et al.*, 2006).

At the same time, the need for identification and classification of different types of tills with respect to their different hydraulic properties and their distribution in landscapes of different geomorphology has become more and more evident, since an increasing number of groundwater reservoirs covered by till have been contaminated by pollutants that have infiltrated the reservoirs through till (Jørgensen and Fredericia, 1992; Jørgensen *et al.*, 1998; Jakobsen and Klint, 1999). Among other things, the construction of regional hydraulic models requires data concerning spatial distribution, thickness and the bulk hydraulic conductivity of the different till types within the designated area (McKay and Fredericia, 1995; Brockman and Szabo, 2000). A differentiation of tills based on their hydraulic properties is, therefore, a key object for the assessment of especially pesticide and fertilizer migration into aquifers covered by till. Thus, at present, it is well known that the hydraulic properties of till is closely related to the occurrence of macro-pores and fractures within tills (Williams and Farvolden, 1967; Fredericia, 1990; Klint and Gravesen, 1999; McKay *et al.*, 1999; Klint *et al.*, 2001; Nilsson *et al.*, 2001; van der Meer *et al.*, 2009). The formation of fractures is closely related to specific tectonic processes during the deposition and deformation of the till (Haldorsen and Krüger, 1990; van der Meer *et al.*, 1999; Klint *et al.*, 2001). The fractures may thus be restricted to certain types of till, which are closely related to the tectonic regime in which they were formed. Proglacial thrust faulting and folding in front of an advancing glacier as well as subglacial loading and shearing produce various glaciotectonic fracture types (Klint and Pedersen, 1995; Klint, 2001; Le Heron and Etienne, 2005). But also de-icing of ice-cored moraines and dead-ice moraine development may produce fractures (Krüger, 1994; Kjær and Krüger, 2001). Thus, an improved understanding of the processes acting during till deposition and deformation is important to support the genetic interpretation of tills and their sediment architecture to perform a proper risk assessment for a given terrain in former glaciated areas.

The present-day retreat of many glaciers has exposed former glacier beds, which provide suitable conditions of studying sedimentological characteristics resulting from subglacial deposition and deformation. This chapter examines a 70-m-long geological section cut into hummocky moraine in front of Kötlujökull, the principal outlet glacier from the east flank of the Mýrdalsjökull ice cap (Fig. 9.1A). The high quality of the exposure allows detailed documentation of a complete sedimentary sequence representing a single glacier advance-stagnation cycle.

DEVELOPMENTS IN QUATERNARY SCIENCES
VOLUME 13 ISSN 1571-0866

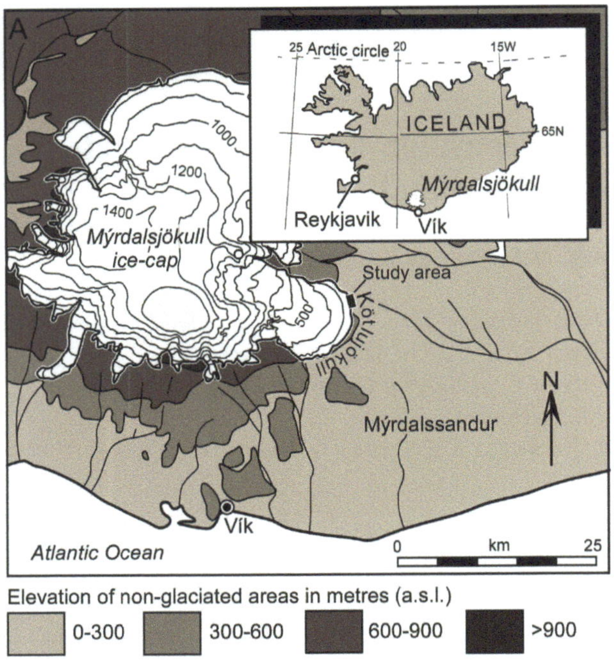

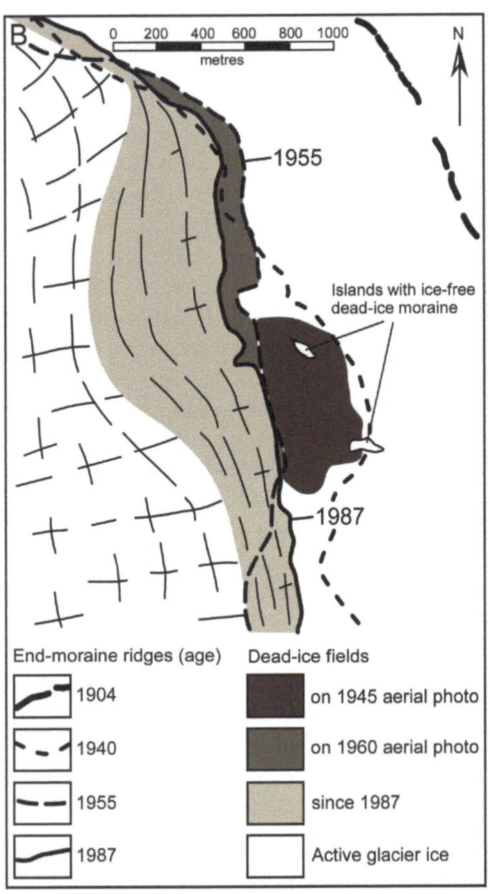

Fig. 9.1. (A) The location of the Mýrdalsjökull ice cap in Iceland. (B) Geomorphological map showing position and age of ice marginal ridges and location of research area.

9.2. Setting

Kötlujökull is a 15-km-long outlet glacier from the east flank of the Mýrdalsjökull ice cap. It descends from an elevation of 1200 to 220 m a.s.l. and drains ice from the eastern part of the 100 km^2 caldera where the subglacial volcano Katla is located. Below the 600-m level, the glacier spreads out to form an expanded piedmont lobe with a maximum width of 8 km terminating in low-relief terrain consisting of glacial and glaciofluvial sediments. The longitudinal compressive flow, generated when the glacier expands widely beyond its confining valley walls, is responsible for intense deformation with an upward-directed thrusting of the ice mass and production of many debris-loaded thrust-planes. Accordingly, the terminal 0.5–1 km of the glacier snout contains large volumes of supraglacial and englacial debris and appears as a fully dead-ice-cored moraine of irregular topography behind a 4–5 m high, well-defined frontal moraine ridge produced during a glacier advance in the 1980s (Krüger, 1994; Krüger and Aber, 1999; Krüger and Kjær, 2000; Kjær and Krüger, 2001).

Kötlujökull has advanced and retreated several times during the past 100 years, but its frontal fluctuations are often non-synchronous as some sections advance while others retreat during the same time period (Krüger, 1994). About 1.2 km beyond Kötlujökull, a system of end-moraine ridges produced by a glacier advance around 1900 separates the glacier forefield from the extensive Mýrdalssandur sloping gently towards the coast some 15–25 km to the south and southeast (Fig. 9.1). Behind the outermost moraine ridges, the glacier forefield consists of a complex pattern of ice-marginal ridges, fluted till, hummocky dead-ice moraine and debris-covered dead-ice fields dissected by proximal extensions of the outwash plain Mýrdalssandur (Krüger, 1994; Schomacker *et al.*, 2003).

The study site is a 200 m long and 70 m wide, NW-SE-directed, island-like remnant of low-relief, hummocky dead-ice moraine in a process of destruction by laterally migrating meltwater streams about 400 m in front of Kötlujökull (Fig. 9.1B). The dead-ice moraine, which represents a relict from ice stagnation around 1940, does not show any evidence of active ice melt and, therefore, demonstrates the nature of a fully ice-free dead-ice moraine terrain (Krüger, 1994; Kjær and Krüger, 2001). A well-exposed geological section, 170 m long and 3–4 m high, has been produced along the southwestern flank of the moraine-island due to lateral undercutting by a violent melt-water stream.

9.3. Methodology

A 70-m-long part of the geological section has been investigated (Fig. 9.2). The approach presented by Krüger and Kjær (1999) and Evans and Benn (2004) for detailed field description of glacial diamicts and associated sediments is adopted in this study. The field description of sorted deposits follows, in the main, the scheme proposed by Miall (1977). Clast fabric data were collected according to the criteria suggested by Kjær and Krüger (1998) and Krüger and Kjær (2005) and evaluated through a three-dimensional eigenvector analysis (Mark, 1973; Woodcook, 1977) using the program *Spheristat* for actual calculations. Clast fabric interpretation is validated and supported by contoured diagrams after Kamb (1959)

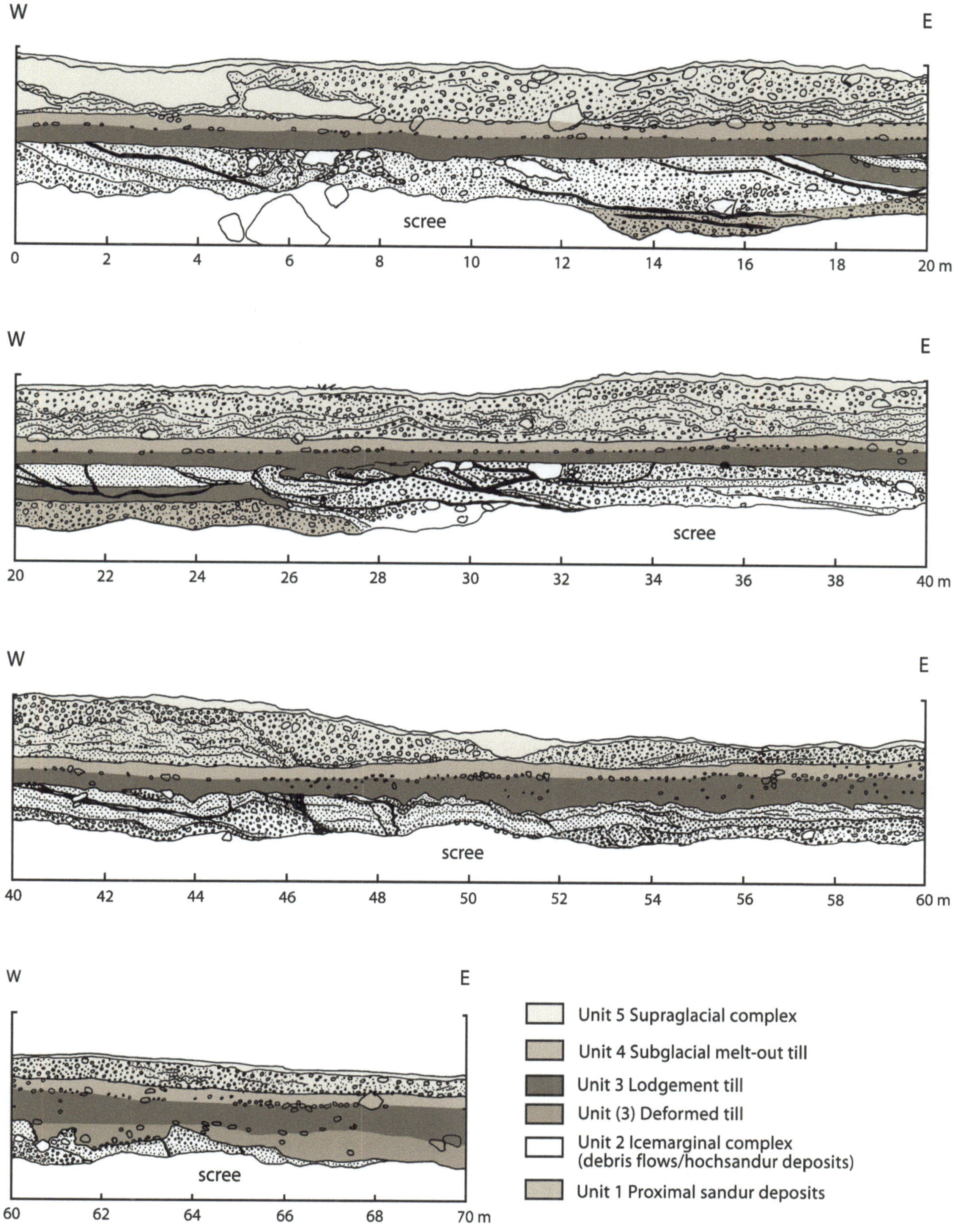

Fig. 9.2. Geological section in hummocky dead-ice moraine terrain in front of Kötlujökull.

using two times the standard deviation to identify non-random distribution. Clast-morphological properties such as shape and roundness are recorded for samples of minimum 50 clasts (Powers, 1953; Benn and Ballantyne, 1993, 1994).

9.4. The Geological Section

A sequence of five distinct sediment associations (units) is identified throughout the entire section (Fig. 9.3). Each unit represents a general facies association related to a

Fig. 9.3. Part of the geological section indicating the five stratigraphic units studied.

distinct depositional/deformational environment during a glacier advance-stagnation cycle. At the base of the section is (1) glaciofluvial outwash sediments more than 50 cm thick formed in a proglacial braided river system. Upwards, follows (2) an ice-marginal sediment association, 80-140 cm thick, of glaciofluvial sediments mixed with debris-flow deposits. This ice-marginal sediment association is sharply overlain by (3) basal till, 25–85 cm thick, capped by a clast-pavement. The till is classified as a type A-lodgement/deformation till according to classification by Benn and Evans (1996). The clast-paved till is overlain by a (4) basal melt-out till, 15–35 cm thick. A sediment association, 40–160 cm thick, of supraglacial origin (5) makes up the top of the section. All the units, except the lowermost one, are well exposed along the entire section and demonstrate the variations within the different facies associations.

9.5. Description and Interpretation of the Facies Associations

A data chart comprising a description of the sedimentary succession recognized in the geological section is given in Fig. 9.4. In the following, we describe the individual units in turn.

9.5.1. Unit 1: Glaciofluvial Outwash Sediments

The lowest unit, more than 50 cm thick, which was exposed irregularly at the base of the section, is poorly sorted, clast-supported medium- to coarse-grained stony gravel that shows large-scale planar cross-bedding indicating a palaeocurrent direction towards the E and ESE. Unit 1 is interpreted as being deposited in a proglacial environment; the unit comprises an outwash sequence (Boothroyd and Ashley, 1975; Miall, 1983), representing gravelly bar deposits from the upper reach of an outwash fan deposited by high-energy streams emerging from underneath the glacier (Krüger, 1994, 1997; Krüger and Kjær, 1999).

9.5.2. Unit 2: Ice-Marginal Sediment Association

This unit is highly variable along the section. It consists chiefly of poorly sorted gravel and horizontally laminated sand inter-fingering with lenses or beds of heterogeneous, sandy-gravelly, matrix- to clast-supported, loose diamict with irregular stone and boulder concentrates. In the western part of the profile (proximal to the ice), the deposits are dominated by down-glacier dipping (8–15°) strata that turn into sub-horizontal bedding in the distal part of the profile. Unit 2 also includes a bed of slightly clast-paved, firm, generally homogeneous, sandy-silty diamict, 25 cm thick (between 17 and 26 m in the profile; Fig. 9.2). In its up-glacier end, this diamict is discordantly overlain by Unit 3. Evidence of folding of the entire unit was found in the most distal part of the profile (between 45 and 65 m; Fig. 9.2), and series of drag folds were found along the contact towards the overlying basal till (between 25 and 28 m; Figs 9.2 and 9.5). The folds are clearly dragged into a thin sandy shear-plane that

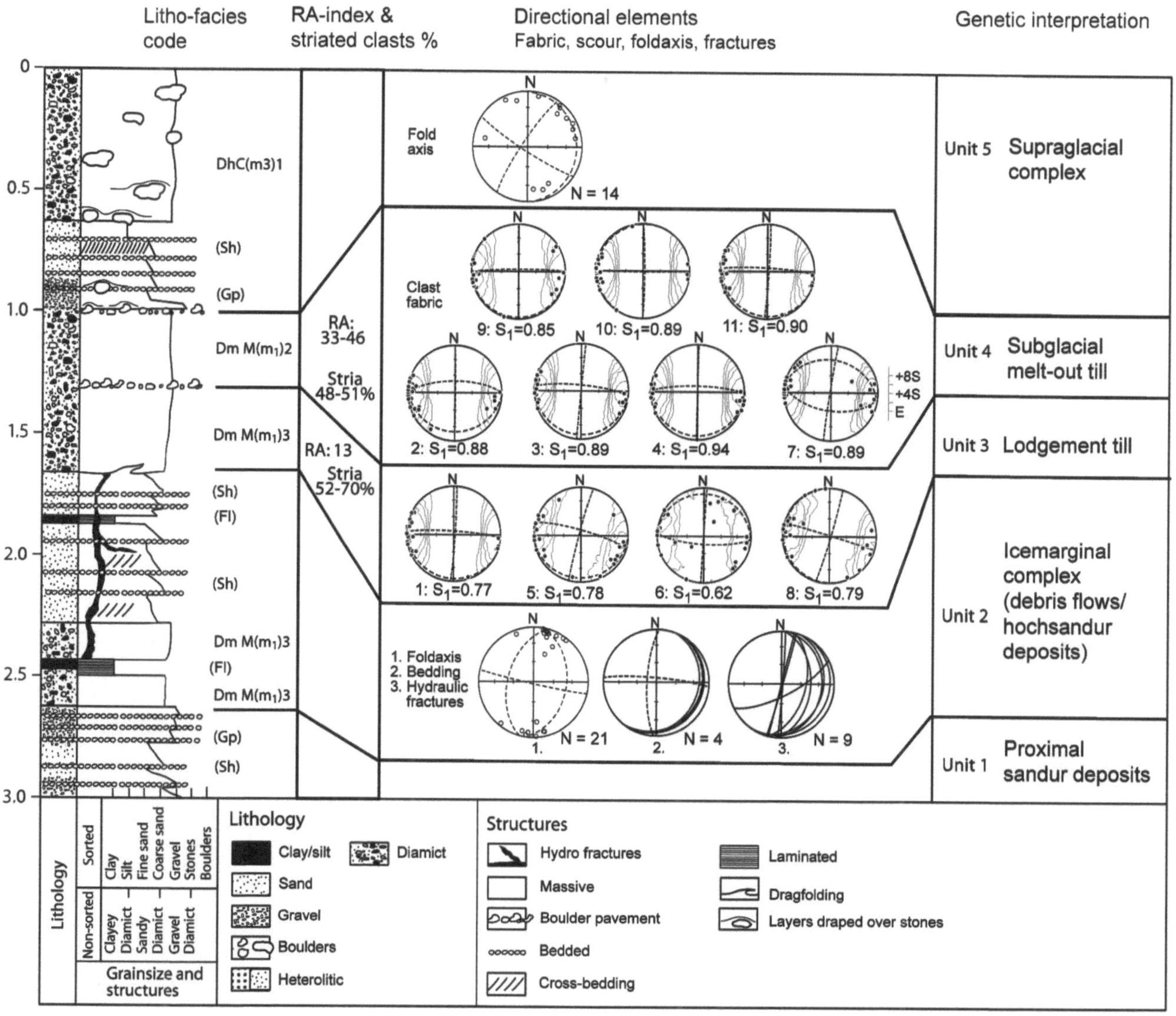

Fig. 9.4. Data chart comprising a description of the sedimentary succession recognized in the hummocky moraine terrain. Based on data chart developed by Krüger and Kjær (1999).

may be traced over a distance of 2–3 m where it caps the upper surface of two boulders situated at the contact between Units 2 and 3. A distinct fold-axis orientation perpendicular to the ice-movement direction indicates that Unit 2 has undergone subglacial deformation from a westerly direction.

Two types of hydrofractures are identified within Unit 2: downward intruded and upward intruded hydrofractures (Fig. 9.6) and in some places also clastic dykes occur (Fig. 9.7). The main part of the upward injected fractures was situated between 18 and 34 m on the profile, whereas another concentration of dominantly downward injected fractures were situated between 45 and 60 m on the profile. A total of 22 fractures are observed, generally 2–5 cm thick and 30–250 cm long. They are generally filled with finer material injected into coarser material in the surroundings. Some of the fill consisted of laminated silt/clay, whereas the fill in other fractures was dominated by more homogeneous sandy material. The fractures are generally dipping down-glacier with a preferred trend perpendicular to the ice flow direction (Fig. 9.4). One of the many upward injected fractures was also drag-folded at the base of the overlying diamict, indicating that these fractures were formed before the formation of the diamict. Two distinct clastic dykes (Fig. 9.7), approximately 25–35 cm thick and filled with diamict material, are injected from the overlying till into the deposits below at 46 and 64 m on the profile, indicating a younger origin than the upward injected hydrofractures (Fig. 9.2).

Unit 2 is interpreted as deposited in an ice-marginal environment; it comprises a mixture of frontal mass-flow sediments and hochsander fan deposits. In contrast to the large-scale fans fed by high-energy streams from underneath the glacier and therefore consisting of coarse-grained sediments of chiefly gravel and stone size in their proximal part (Boothroyd and Ashley, 1975), hochsander fans result from deposition of fine-grained material transported by low-energy streams from the supraglacial source area to the glacier forefield (Heim, 1983; Krüger, 1997; Kjær *et al.*, 2004). The bed of the slightly clast-paved till is interpreted as a remnant of basal till representing a former ice advance. This mixture of sediments was to some extent eroded and deformed by the overriding glacier. Most probably, folding took place during an initial stage after erosion of the topmost part of

Fig. 9.5. Dragfolds along the contact between Units 2 and 3.

Fig. 9.6. Upward injected hydrofracture in Unit 2 filled with laminated silt and clay.

Fig. 9.7. Clastic dyke in Unit 2.

Unit 2 and after deposition of the main part of the overlying till, but before deposition of the boulder pavement that drapes the till horizontally (Unit 3) throughout the profile. All fractures in Unit 2 are interpreted to have a glaciotectonic origin; their down-glacier dip indicates a systematic stress-field, suggesting that they originate from basal loading and shearing by an overriding glacier. Because they are truncated or deformed at the contact between Units 2 and 3, they must be older than the lowermost part of the deforming bed in the basal till.

9.5.3. Unit 3: Sediment Association Representing Deposition from Moving Ice

A clast-paved, very firm, generally homogeneous, grey, sandy-silty diamict, 25–85 cm thick, with a moderate content of clasts, rests on the ice-marginal sediment association, or on glaciofluvial outwash sediments. The lower boundary is erosive and subplanar (0–25 m and 30–42 m in profile; Fig. 9.2) or erosive and irregular with associated small-scale deformations in the substratum and the lower part of the diamict (25–30 m and 42–70 m in profile). Unit 3 shows a bipartition in the most distal part of the section (61–70 m in profile): a lower deformed and an upper non-deformed part separated by a weak boulder pavement. The matrix of the lower deformed zone is somewhat variable in texture. Two samples, an upper and a lower one, were collected in the deformed part of Unit 3 for grain-size distribution analysis. The upper analysis shows a grain-size distribution identical to that found in the upper, non-deformed part of the diamicts, whereas the lower distribution is dominated by very fine sand and very coarse silt. In contrast to this, the matrix in the upper non-deformed part of the diamict shows only little lateral or vertical variation. Five grain-size distribution analyses identified three maxima; one at very fine gravel to very coarse sand, one at very fine sand, and one at coarse silt to medium silt (Fig. 9.8). The non-deformed part of Unit 3 has abundant clasts with striated surfaces (52–70% in three samples), the orientation of striations being spatially consistent and parallel with ice-flow direction. The content of angular and very angular clasts is low; the RA value (Benn and Ballantyne, 1994) is 13%, and the C_{40} value, which identifies the percentage of clasts with c:a axis ratio ≤ 0.4 (Benn and Ballantyne, 1994) is 6–7%. Numerous bullet-nosed clasts with distinctly shaped stoss-and-lee morphology occur randomly scattered within the diamict, some of them with matrix infillings at their lee side. Four clast fabrics were sampled in the upper, non-deformed part of the diamict. They display strong fabric with low isotropy and high degree of clustering ($0.62 < S_1 < 0.79$) and with the principal vectors in a W to WNW direction. Two clast fabrics sampled in the lower, deformed part of the diamict show, respectively, a moderate fabric parallel with the direction of ice flow ($S_1 = 0.62$) and a weak fabric ($S_1 = 0.50$) approximately parallel with ice-flow direction. The clast pavement capping Unit 3 is distinct and dominated by hand-sized clasts with striated surfaces, the orientation of striation being spatially consistent and parallel with the direction of ice flow.

The diamict is interpreted as being deposited in subglacial environments. Particularly, diagnostic in substantiating this interpretation is the relatively uniform thickness and high degree of compactness of the diamicts, and the spatially consistent orientation of various sedimentary directional elements such as clast fabric, striation on clast surfaces, and stoss-lee morphologies, being parallel with ice-flow direction. There is also documentation to demonstrate that the glacier was moving during deposition of the diamicts. Particularly diagnostic is the presence of bullet-nosed clasts, an abundant content of clasts with striated surfaces, and elongated fabrics with a shallow upglacier plunge (Krüger, 1979, 1994; Dowdeswell and Sharp, 1986; Dreimanis, 1989; Schomacker *et al.*, 2006). Furthermore, the occurrence of stoss-lee morphologies in various sizes throughout the upper part of Unit 3 and the dominance of strong to very strong fabrics with a consistent orientation being parallel with the ice flow direction are features interpreted to result from deposition mainly by lodgement (Boulton, 1978; Dowdeswell and Sharp, 1986; Dreimanis, 1989; Krüger, 1994). Lodgement is here defined as a process by which the coarser components of debris, being transported at the ice/bed interface, are deposited particle-by-particle because of friction and local pressure melt-out, whereas the finer components are chiefly released by simple melt-out because of under-melting of ice (Krüger and Kjær, 1999). Tills produced by this process are classified as hybrid lodgement (coarse grains) and deformation (fine matrix) tills by Benn (1994), or a type A lodgement till according to classification by Benn and Evans (1996).

Generally, the contact between Units 3 and 2 is sharp and erosive in the up-glacier part of the section, whereas the contact is undulating and deformed in the down-glacier part of the section. Downward injected hydro-fractures are more frequent in the distal part of the section, whereas upward injected fractures are more frequent in the up-glacier part of the section. Absence of fractures in Unit 3 suggests a watersaturated ductile deformation environment.

9.5.4. Unit 4: Facies Association Representing Subglacial Deposition from Stagnant Ice

With a sharp conformable contact, the basal till deposited from dynamically active ice is overlain by a dark grey, crudely stratified, silty-sandy, matrix-supported, friable diamict, 15–35 cm thick, with a moderate content of clasts. The matrix shows only small, or none, lateral or vertical alterations of grain-size composition. Six grain-size analyses show a distribution with two maxima at very fine gravel to very coarse sand and fine sand to very fine sand. Unlike Unit 3, Unit 4 has no grain-size maximum at coarse silt to medium silt (Fig. 9.8).

The stratification consists of subhorizontal stringers of sand and clay/silt, which locally appear to bend symmetrically around clasts in between. There are zones with apparently homogeneous matrix. Transitions between zones with stratified and homogeneous matrix are gradual. Occasionally, more distinct down-glacier dipping strata occur. The diamict lacks distinct fissility, and stoss-and-lee morphologies are absent or rare. The clasts are commonly striated (48–51%) with most striae parallel to the direction of ice flow. The RA values are moderate, meaning that 33–46% of the clasts are angular or very angular. Seven clast fabrics sampled in Unit 4 display very strong fabrics ($0.85 < S_1 < 0.94$) with an extremely high degree of clustering. They show a spatially consistent orientation with the principal vectors in W directions, being parallel with ice-flow direction and all of them with an extremely shallow up-glacier plunge.

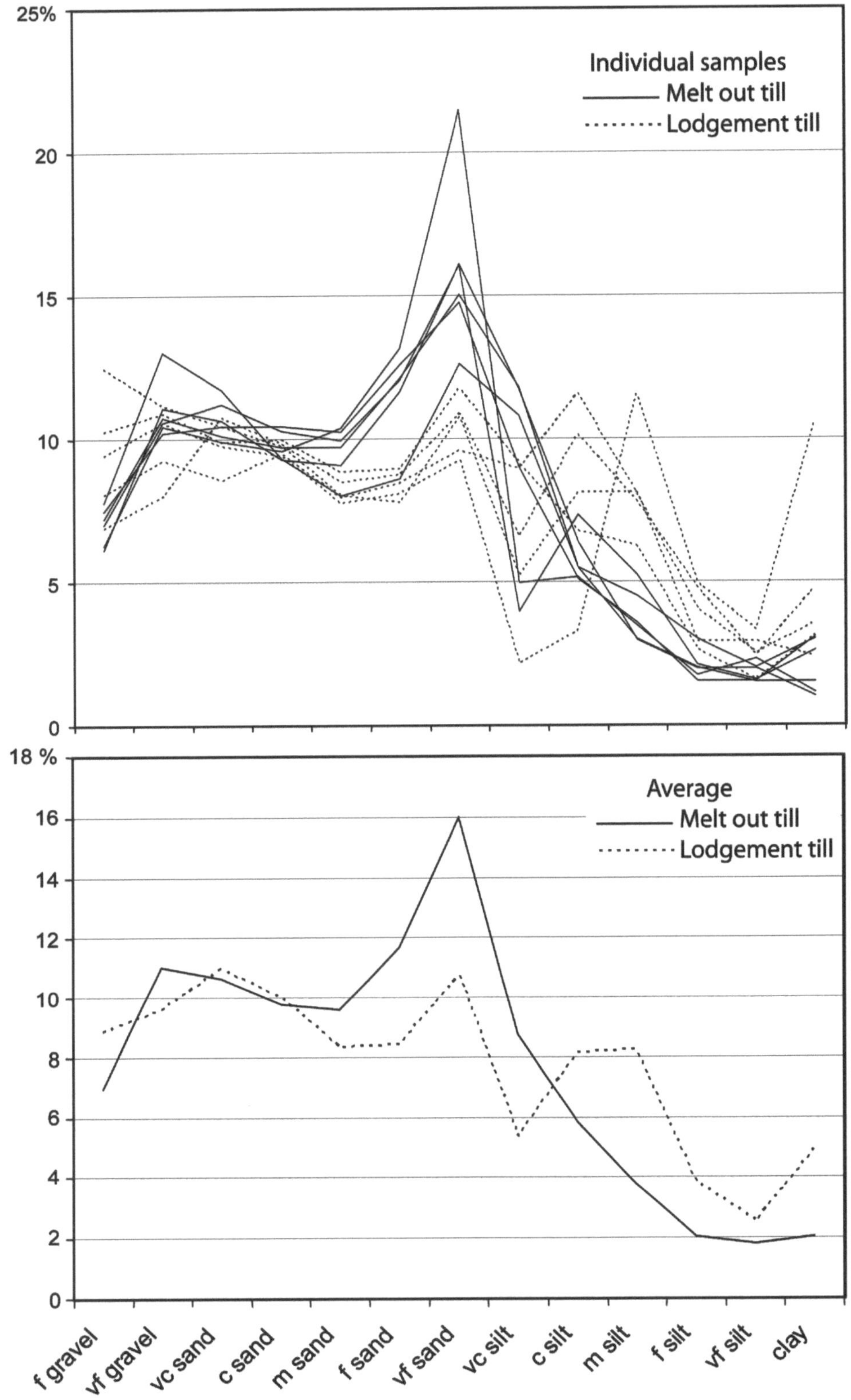

Fig. 9.8. Textural characteristics of lodgement till (Unit 3) and melt-out till (Unit 4).

The spatially consistent orientation of clast fabric and the draping of sand and silt strata over clasts suggest that Unit 4 was deposited in a subglacial environment by *in situ* melt-out of debris-loaded stagnant ice (cf. Krüger, 1994); the crude stratification is probably inherited from the compositional differences in basal ice-debris layers. The extremely high fabric elongation is similar to that reported by Lawson (1979) from fresh melt-out till from Alaska and remains higher than that generally associated with lodgement tills. Further properties associated with this diamict unit are also shared by melt-out tills from other localities; melt-out tills are moderately consolidated and are generally coarser in texture and contain more angular material than lodgement tills in the same area (Haldorsen, 1981, 1983), and stoss-and-lee morphologies are absent, because these morphologies develop during lodgement and ploughing beneath moving ice (Krüger, 1984; Clark and Hansel, 1989). Unit 4 is, therefore, classified as subglacial melt-out till, which is also supported by its stratigraphical position between a lodgement till beneath and an overlying supraglacial sediment association.

9.5.5. Unit 5: Facies Association Representing Supraglacial Re-Deposition in Dead-Ice Environment

The subglacial melt-out till is overlain by a facies association of variable thickness (50–120 cm). The unit shows a bipartition: a lower part of sorted sediments and an upper part of diamicts. Evidence of faulting, chiefly normal faults, and slumping found throughout Unit 5 illustrates that deposition of this unit was contemporary with the melting of underlying and adjacent ice. The sorted sediments consist of planar cross-bedded gravel overlain by horizontally laminated, or cross-bedded, silt and sand alternating with laminae of coarse sand, or fine gravel. The sorted sediments grade vertically into a loose diamict of variable thickness and exhibiting great lateral and vertical alterations of sediment composition. The wide range of textures includes clast-supported diamicts of sandy gravel and matrix-supported diamicts of gravely sand containing abundant sand lenses and numerous groups of bouldery materials. Interbeds of thinly laminated silt and sand, or granular materials, can be seen. At many places, they display deformation by minor folding, but with a random distribution of fold-axis orientation. Generally, diamicts in Unit 5 exhibit more large clasts, compared to the diamicts in Units 3 and 4.

The aforementioned description of the diamicts in Unit 5 is supplemented by data from a similar section in the nearby glacier forefield, where the facies association of supraglacial deposits (2.5–4.5 m in thickness) has been studied in details (Krüger and Kjær, 1999; Kjær and Krüger, 2001; Krüger *et al.*, this volume). Analysis of the loose diamicts displays a moderate RA value making up 32%. The C_{40} value is 12%. Less than half of the clasts are striated, the orientation of striation being highly dispersed. A number of clast fabrics sampled throughout the diamicts display very low to moderate eigenvector values. Although individual fabrics appear to show a preferred orientation, the azimuth of individual preferred orientations are widely dispersed and not related to ice-flow direction.

A supraglacial origin for the complex sediment succession overlying the basal melt-out till both in the present study field and the additional locality is indicated by the presence of slump structures and normal faults caused by ice melt-out collapse. Further arguments are the occurrence of loose chiefly heterogeneous diamicts with a high clast content and relatively high RA values intimately associated with sorted deposits. Particularly diagnostic is the high spatial variability of fabric maxima and fold-axis orientation (Lawson, 1979; Krüger, 1994).

The complex sequence of sorted sediments and diamicts in Unit 5 is interpreted as trough fillings caused by meltwater activity along troughs between ice-cored hummocks combined with the entry of diamict flows. The heterogeneous diamicts represents a series of sediment-flow units, but the final mass-movement processes – backslumping by rotational sliding and collapse – have given the sediment a somewhat chaotic appearance. This may also help to account for the highly dispersed clast orientations or weakly preferred orientations that are not related to ice flow (Lawson, 1979; Shaw, 1982; Krüger, 1994).

9.6. Discussion

The five units in the studied profile are interpreted to represent one single advance-stagnation cycle of Kötlujökull. The completeness of the cycle with deposits from the glacier forefield overlain by ice-marginal deposits, lodgement/deformation till, basal melt-out till and supraglacial deposits supports the genetic interpretation of the individual units. It also gives an opportunity to study details in the depositional and deforming processes relative to the stages in the advance-stagnation cycle. The following discussion focuses first on the cause of grain-size differences between lodgement till and melt-out till and second on the subglacial hydrological conditions during formation of the different units.

9.6.1. The Grain-Size Distribution in the Basal Tills

The lodgement till (Unit 3) and the basal melt-out till (Unit 4) differ in grain-size distribution with a higher content of medium sand to very coarse silt in the melt-out till (Unit 4) against a higher content of finer fractions in the lodgement till (Unit 3). The explanation could be differences in source area, in entrainment processes, or in the mode of transportation and deposition. The longitudinal compressive flow generated when the piedmont-like outlet glacier Kötlujökull expands widely beyond its confining valley walls is, however, responsible for intense deformation of the ice and thereby mixing of debris from different sources and different mode of entrainment and englacial transportation. The differences in grain-size composition between the two genetic types of till should therefore mainly reflect differences in mode of deposition.

Modifications of the grain-size distribution of the englacial debris could be caused by the removal of fine sediments in the melt-out till during the melt-out process or by abrasion and polishing of sand, gravel and clasts during lodgement/deformation of the basal till. Removal of fines during melt-out is less likely since the grain-size distribution in the upper and lower part of each unit is very uniform. Crushing clasts may produce gravel and sand, whereas abrasion and polishing may produce silt/clay in the lodgement till (Haldorsen, 1981). Sediment trapped in the ice and released by passive melt-out undergoes a smaller degree of erosion leading to a grain-size distribution much similar to that of englacial debris. Abrasion of sand during lodgement/deformation will lead to a decrease in sand coupled with an increase in silt/clay compared to the englacial debris. This relationship is, as mentioned earlier, observed with a decrease in medium sand to very coarse silt and an increase in coarse silt and finer fractions in the lodgement till (Unit 3) compared with the melt-out till (Unit 4), indicating that abrasion of sand during lodgement/deformation is the major control of grain-size differences between the till units.

9.6.2. Subglacial Hydraulic Conditions during Formation of the Different Units

Soft sediment intrusions or water escape structures such as clastic dykes, mud-diapirs or hydro-fractures are widely distributed in diamicts, or sorted sediments, in many former glaciated or periglacial areas (Berthelsen, 1978; Åmark, 1986; Dreimanis, 1992; Rijsdijk *et al.*, 1999; van der Meer *et al.*, 1999, 2009). Hydraulic fracturing takes place when high water pressure makes fractures propagate into different types of sediment. Such conditions may be initiated when the hydrostatic pressure increases. High water pressure may build up beneath warm-based glaciers, when ice melting injects water into permeable beds under maximum head equivalent to the total ice pressure (Boulton and Caban, 1995). In the present case, both upward injected and downward injected fractures are found in the ice marginal deposits (Unit 2). Silt- and clay-filled fractures apparently grew upwards from thin layers of laminated clay and silt embedded in flow-tills and melt-water deposits (Unit 2) below the basal till (Unit 3). The fractures penetrate bedded sandy and gravely melt-water deposits, but are truncated, and in some cases drag-folded, along the base of the basal till and none of them penetrate the base of the till. In general, the hydro-fractures have a systematic orientation striking perpendicular to the ice-flow direction with a steep to gentle down-glacier dip. This close relationship between strike of hydro-fractures and ice-movement direction suggests a glacio-tectonic origin of the fractures, but they must have formed during an early stage of ice transgression since they are cut and drag-folded at the base of the till. The presence of hydro-fractures below the till and lack of brittle deformation features such as fractures in the basal till, implies high hydrostatic pressure during both the transgression and the formation of the till.

The upward injected fractures indicate higher water pressure in the underlying sediments. This situation could very well be linked to a change in subglacial water recharge during a glacial meltwater outburst, with subsequent high water pressure in the high-permeable beds throughout Unit 1. The downward injected fractures observed in part of the geological section indicate high water pressure in the upper part of Unit 2 relative to its lower part. This may reflect a situation where the local water recharge chiefly originates from basal melting during periods of little recharge from the underlying glaciofluvial bed. It is assumed that recharge of melt-water to the sediments below the glacier must have been relatively high during deposition of the basal till, considering the overall ductile deformation style of the tills. This is also indicated by the relatively high melting rates of Kötlujökull (Krüger, 1994; Krüger and Kjær, 2000). The entrapment of water below the glacier could be favoured by the presence of permafrost in the sediment in front of and below the glacier as reported from Sléttjökull (van der Meer *et al.*, 1999). However, because of the low altitude and the general climatic conditions in the forefield of Kötlujökull, permafrost is absent (Krüger, 1994). It is suggested that subglacial conditions favourable for hydro-fracturing in general are restricted to areas close to subglacial drainage networks (Klint, 2001). Rapid changes in the water discharge from meltwater gaps over time indicate that the drainage pattern below the glacier is highly variable not only in time but also in position. In the Mýrdalsjökull region, subglacial volcanic activity creates a highly dynamic environment with sudden outbursts and rapid changes in subglacial hydrostatic pressure. Such a situation resulting in a three-step development of upward and downward injected water escape structures has been described by van der Meer *et al.* (1999) from the forefield of Sléttjökull, and they concluded that the glaciodynamic conditions strongly depend on the variability of the water-recharge and the drainage conditions in the deforming bed along the glacier sole.

Rijsdijk *et al.* (1999) describe various clastic dykes embedded in basal tills in Ireland. They concluded that the dykes were formed as a result of increasing, high pore-water pressure in a confined gravel layer between two fine-grained till units along with transgression of a fast moving glacier. The porewater was forced to drain by penetrating the overlying till through fractures, which were then filled with coarse sediment implying very high flow velocities (burst-out plumes). Rijsdijk *et al.* (1999) suggest that this type of "upward infilled" hydrofractures were related to submarginal processes, whereas "downward infilled" hydrofractures were related to subglacial environments located farther from the glacier margin. The nature of the older upward injected fractures and the younger downward injected fractures described in this chapter support this idea.

9.7. Event Stratigraphy

An idealized sequential model for the development of the present stratigraphy relative to glacier dynamic and landscape formation at Kötlujökull is shown in Fig. 9.9.

(A) Kötlujökull advances across proglacial outwash fans (Unit 1) and partially ice-cored terrain (Krüger *et al.*, 2002). Huge quantities of diamict sediments emerge on the glacier surface mainly from debris-loaded thrust-planes, because of intense upward-directed thrusting of the ice mass. From source areas on and above the frontal ice cliff material is delivered by dumping producing an ice-front scree with avalanche tracks and sediment flow tongues (Fig. 9.10). As the glacier advances, it simply pushes and overrides a mixture of frontal mass-flow sediments and hochsander fan sediments deposited by streams draining the debris-covered frontal slope of the advancing glacier (Unit 2). In the submarginal environment, high pore-water pressure and hydraulic fracturing under very rapidly changing water recharge conditions favour formation of both upward and downward injected

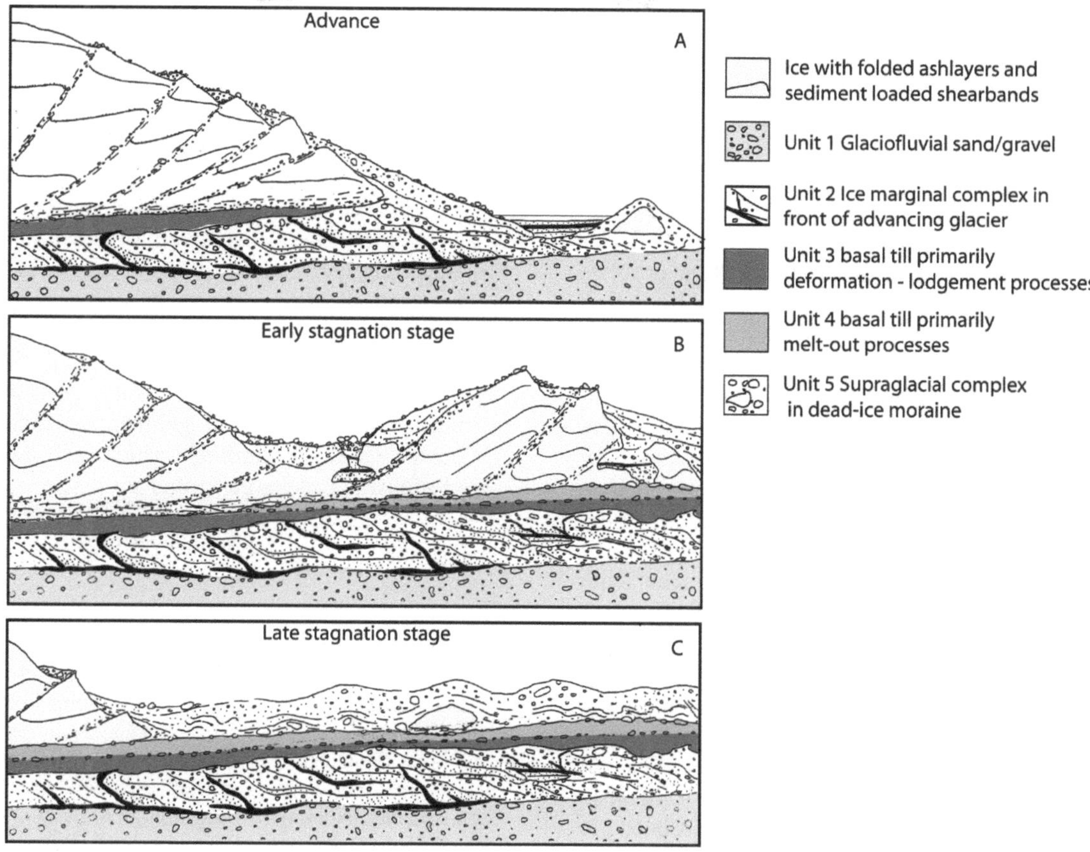

Fig. 9.9. Simplified sequential model for landscape formation and development of the stratigraphy at the margin of Kötlujökull.

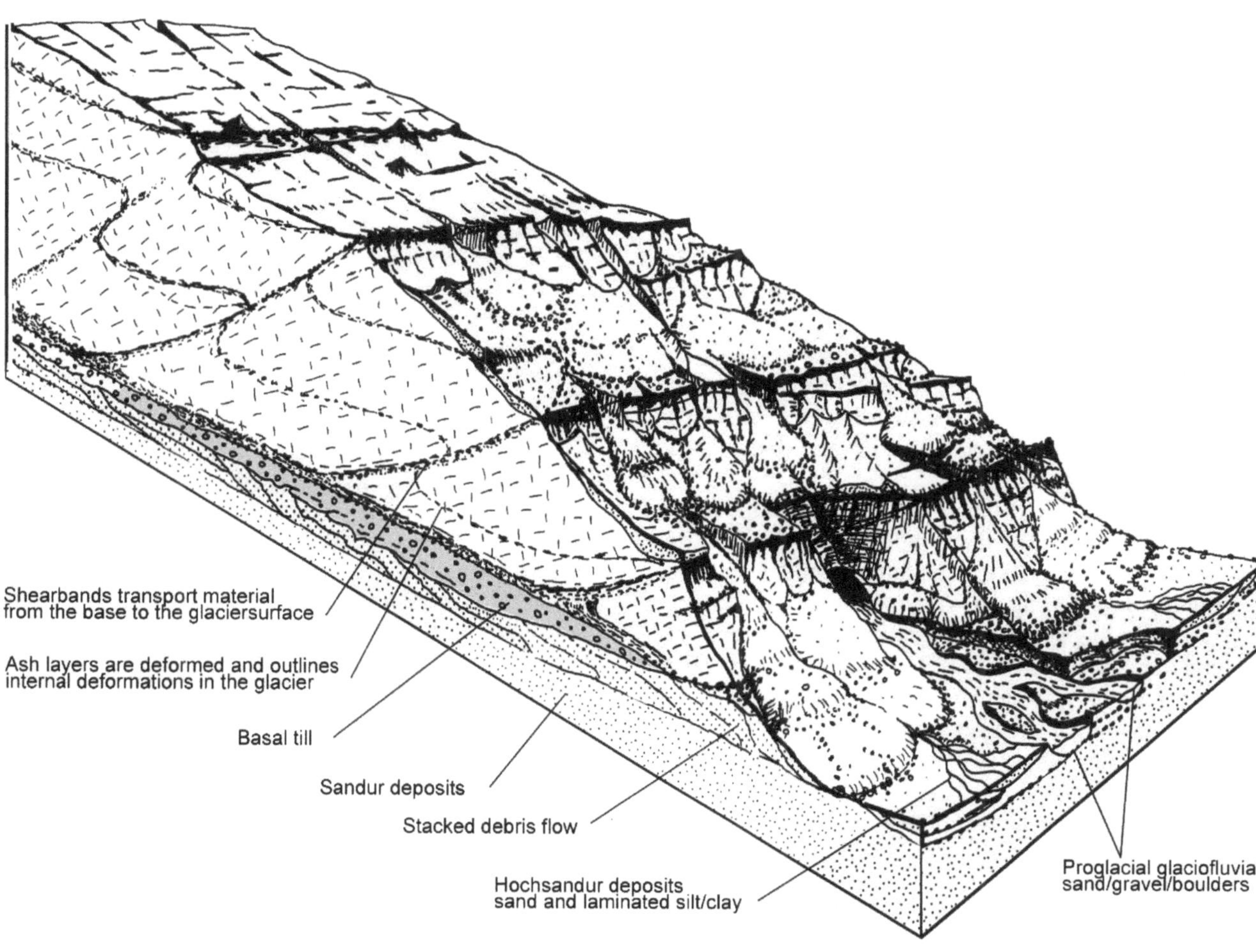

Fig. 9.10. Idealized characteristics of frontal processes acting in the terminal part of Kötlujökull.

fractures. The formation of fractures is facilitated by drag from moving ice. Subsequent erosion and deformation of the topmost part of Unit 2 take place along with deposition of deformation and lodgement till (Unit 3) during continued glacier advance.

(B) The glacier advance has stopped and the ice-front section terminates in a wide zone of debris-covered dead-ice. Meltwater escapes as groundwater and by subglacial flow and contributes to the subglacial melting process. From beneath stagnant ice and dead-ice, melting first releases all the material that was embedded in the glacier sole and that has escaped frictional deposition during the final deposition of lodgement till. This results in formation of the clast pavement on top of the lodgement till (Humlum, 1981; Krüger, 1994). Continuous melting of ice from beneath produces the basal melt-out till (Unit 4). In the supraglacial environment, the predominating processes are de-icing by backwasting of exposed ice walls and downwasting by top melt combined with cyclical re-working processes of the sediment cover (Krüger and Kjær, 2000).

(C) The glacier retreats. As ice-cores in the dead-ice moraine terrain gradually diminish, the effect of the latest re-sedimentation events overprint or destroy most sedimentary characteristics produced during earlier stages of the de-icing process (Kjær and Krüger, 2001). Thus, in the ice-free landscape, structures mainly related to slumping and collapse processes acting during the final phase of dead-ice moraine development remain imprinted on the sediment succession.

9.8. Conclusions

On the basis of the studies reported here from a 70-m-long geological section cut by meltwater in the hummocky dead-ice moraine terrain in the forefield of Kötlujökull, the following conclusions may be drawn:

1. The lithostratigraphy of the geological section is interpreted as a response to a glacier advance across the upper reach of a pre-existing outwash fan succeeded by stagnation and a following degradation of ice that finally melted away and left a mantle of sediments of supraglacial environments resting on beds of subglacial melt-out till and lodgement/deformation till.
2. The melt-out till displays moderate consolidation, moderate content of fines and angular clasts and a very strong clast fabric. The lodgement/deformation till displays strong to very strong consolidation, relatively high content of fines, low content of angular clasts, and a weak to strong clast fabric. The relatively high content of fines in the lodgement/deformation till compared with melt-out till probably reflects abrasion during the lodgement and deformation processes.
3. Subglacial deformations such as dragfolds and hydrofractures below the basal till strongly reflect ice-movement direction. The occurrence of water escape structures below the basal till and complete lack of ''brittle'' deformations in the lodgement/deformation till suggest more or less permanent water-saturated subglacial conditions during the advance of Kötlujökull.
4. Differences in the deformation pattern at the base of the lodgement/deformation till may reflect very heterogeneous hydraulic conditions below the glacier during the final stage of ice transgression. Accordingly, it is likely that the subglacial deformation pattern often changes between up and downward directed stress conditions due to changes in water recharge either due to local subglacial melting or because of general porewater pressure in the underlying glaciofluvial sediments, which depends on the subglacial drainage pattern and contribution of meltwater recharge further up-glacier.

References

Åmark, M., 1986. Clastic dykes formed beneath an active glacier. Geologiska Föreningens i Stockholm Förhandlinger 108, 13–20.

Benn, D.I., 1994. Fluted moraine formation and till genesis below a temperate valley glacier: Slettmarkbreen, Jotunheimen, southern Norway. Sedimentology 41, 279–292.

Benn, D.I. & Ballantyne, C.K., 1993. The description and presentation of particle shape. Earth Surface Processes and Landforms 18, 665–672.

Benn, D.I. & Ballantyne, C.K., 1994. Reconstructing the transport history of glacigenic sediments: a new approach based on the co-variance of clast shape indices. Sedimentary Geology 91, 215–227.

Benn, D.I. & Evans, D.J.A., 1996. The interpretation and classification of subglacially deformed materials. Quaternary Science Reviews 15, 23–52.

Benn, D.I. & Evans, D.J.A., 1998. Glaciers and glaciation. London, Arnold, 734 pp.

Bennett, M.R., Waller, R.I., Midgley, N.G., Huddart, D., Gonzalez, S., Cook, S.J., & Tomio, A., 2003. Subglacial deformation at sub-freezing temperatures? Evidence from Hagafellsjökull-Eystri, Iceland. Quaternary Science Reviews 22, 915–923.

Berthelsen, A., 1978. The methodology of kineto-stratigraphy as applied to glacial geology. Bulletin of the Geological Society of Denmark 27(Special Issue), 25–38.

Boothroyd, J.C. & Ashley, G.M., 1975. Process, bar morphology and sedimentary structures on braided outwash fans, northeastern Gulf of Alaska. In: Jopling, A.V., McDonald, B.C. (Eds), Glaciofluvial and Glaciolacustrine Sedimentation. Society of Economic Palaeontologists and Mineralogists, Special Publication 23, 193–222.

Boulton, G.S., 1968. Flow tills and related deposits on some Vestspitsbergen glaciers. Journal of Glaciology 7, 391–412.

Boulton, G.S., 1971. Till genesis and fabric in Svalbard, Spitbergen. In: Goldthwait, R.P. (Ed), Till, a symposium. Columbus, Ohio, Ohio State University Press, 41–72.

Boulton, G.S., 1975. Processes and patterns of subglacial sedimentation: a theoretical approach. In: Wright, A.E., Moseley, F. (Eds), Ice ages: ancient and modern. Liverpool, Seel House, 7–42.

Boulton, G.S., 1978. Boulder shapes and grain-size distribution of debris as indicators of transport paths through a glacier and till genesis. Sedimentology 25, 773–799.

Boulton, G.S., 1979. Processes of glacier erosion on different substrata. Journal of Glaciology 23, 15–38.

Boulton, G.S., 1996. The origin of till sequences by subglacial sediment deformation beneath mid-latitude ice sheets. Annals of Glaciology 22, 75–84.

Boulton, G.S. & Caban, P., 1995. Groundwater flow beneath ice-sheets part II – its impact on glacier tectonic structures and moraine formation. Quaternary Science Reviews 14, 563–587.

Boulton, G.S. & Dobbie, K.E., 1993. Consolidation of sediments by glaciers: relation between sediment geotechnics, soft-bed glacier dynamics and sub-glacial groundwater flow. Journal of Glaciology 39, 26–44.

Brockman, C.S. & Szabo, J.P., 2000. Fractures and their distribution in the tills of Ohio. The Ohio Journal of Science 100, 39–55.

Christoffersen, P., Piotrowski, J.A., & Larsen, N.K., 2005. Basal processes beneath an Arctic glacier and their geomorphic imprint after a surge, Elisebreen, Svalbard. Quaternary Research 64, 125–137.

Clark, P.U. & Hansel, A.K., 1989. Clast ploughing, lodgement and glacier sliding over a soft glacier bed. Boreas 18, 201–207.

Dowdeswell, J.A. & Sharp, M.J., 1986. Characterization of pebble fabrics in modern terrestrial glacigenic sediments. Sedimentology 33, 699–710.

Dreimanis, A., 1989. Tills: Their genetic terminology and classification. In: Goldthwait, R.P., Matsch, C.L. (Eds), Genetic classification of glacigenic deposits. Brookfield, A.A. Balkema, Rotterdam, 17–84.

Dreimanis, A., 1992. Downward injected till wedges and upward injected till dikes. Sveriges Geologiska Undersökning, Serie Ca 81, 91–96.

Elson, J.A., 1989. Comment on glacitectonite, deformation till, and comminution till. In: Goldthwait, R.P., Matsch, C.L. (Eds), Genetic classification of glacigenic deposits. Brookfield, A.A. Balkema, Rotterdam, 85–88.

Evans, D.J.A., 2003. Ice-marginal terrestrial landsystems: active temperate glacier margins. In: Evans, D.J.A. (Ed), Glacial Landsystems. London, Arnold, 12–43.

Evans, D.J.A. & Benn, D.I., 2004. A practical guide to the study of glacial sediments. London, Edward Arnold.

Evans, D.J.A., Phillips, E.R., Hiemstra, J.F., & Auton, C.A., 2006. Subglacial till: Formation, sedimentary characteristics and classification. Earth-Science Reviews 78, 115–176.

Fredericia, J., 1990. Saturated hydraulic conductivity of clayey till and the role of fractures. Nordic Hydrology 21, 119–132.

Fuller, S. & Murray, T., 2002. Sedimentological investigations in the forefield of an Icelandic surge-type glacier: implications for the surge mechanism. Quaternary Science Reviews 21, 1503–1520.

Haldorsen, S., 1981. Grain-size distribution of subglacil tills and its relation to subglacial crushing and abrasion. Boreas 10, 91–105.

Haldorsen, S., 1983. Mineralogy and geomchemistry of basal till and its relationship to till-forming processes. Norsk Geologisk Tidsskrift 63, 15–25.

Haldorsen, S. & Krüger, J., 1990. Till genesis and hydrogeological properties. Nordic Hydrology 21, 81–94.

Heim, D., 1983. Glaziare Entwasserung und Sanderbildung am Kötlujökull, Südisland. Polarforschung 53, 17–29.

Humlum, O., 1981. Observations on debris in the basal transport zone of Mýrdalsjökull, Iceland. Annals of Glaciology 2, 71–77.

Jakobsen, P.R. & Klint, K.E.S., 1999. Fracture distribution and migration of DNAPL in a clayey lodgement till. Nordic Hydrology 30, 285–300.

Jørgensen, P.R. & Fredericia, J., 1992. Migration of nutrients, pesticides and heavy metal fractured clayey till. DGU-reprint 79. Géotechnique 42, 67–77.

Jørgensen, P.R., McKay, L.D., & Spliid, N.H., 1998. Evaluation of chloride and pesticide transport in a fractured clayey till using large undisturbed columns and numerical modeling. Water Resources Research 34, 539–553.

Kamb, W.B., 1959. Ice petrographic observations from Blue Glacier, Washington, in relation to theory and experiment. Journal of Geophysical Research 64, 1891–1909.

Kjær, K.H. & Krüger, J., 1998. Does clast size influence fabric strength? Journal of Sedimentary Research 68, 746–749.

Kjær, K.H. & Krüger, J., 2001. The final phase of dead-ice moraine development: processes and sediment architecture, Kötlujökull, Iceland. Sedimentology 48, 935–952.

Kjær, K.H., Krüger, J., & van der Meer, J.J.M., 2003. What causes till thickness to change over distance? Answers from Mýrdalsjökull, Iceland. Quaternary Science Reviews 22, 1687–1700.

Kjær, K.H., Sultan, L., Krüger, J., & Schomacker, A., 2004. Architecture and sedimentation of outwash fans in front of the Mýrdalsjökull ice cap, Iceland. Sedimentary Geology 172, 139–163.

Kjær, K.H., Larsen, E., van der Meer, J.J.M., Ingófsson, Ó., Krüger, J., Benediktsson, Í.Ö., Knudsen, C.G., & Schomacker, A., 2006. Subglacial decoupling at the sediment/bedrock interface: an essential mechanism for rapid flowing ice. Quaternary Science Reviews 25, 2704–2712.

Klint, K.E.S., 2001. Fractures in glacigene diamict deposits; origin and distribution. PhD Thesis. Geological Survey of Denmark and Greenland, Special report no. 129.

Klint, K.E.S. & Gravesen, P., 1999. Fractures and biopores in weichselian clayey till aquitards at Flakkebjerg, Denmark. Nordic Hydrology 30, 267–284.

Klint, K.E.S. & Pedersen, S.A.S., 1995. The hanklit glaciotectonic thrust fault complex, Mors, Denmark. Geological Survey of Denmark Bulletin, Series A 34, 30 p.

Klint, K.E.S., Abildtrup, C.A., Gravesen, P., Jakobsen, P.R., & Vosgerau, H., 2001. Sprækkers oprindelse og udbredelse i moræneler i Danmark. Vand og Jord 8(3), 111–119.

Krüger, J., 1979. Structures and textures in till indicating subglacial deposition. Boreas 8, 323–340.

Krüger, J., 1984. Clasts with stoss-lee form in lodgement tills: a discussion. Journal of Glaciology 30, 241–243.

Krüger, J., 1994. Glacial processes, sediments, landforms, and stratigraphy in the terminus region of Mýrdalsjökull, Iceland. Folia Geographica Danica 21, 233 pp.

Krüger, J., 1997. Development of minor outwash fans at Kötlujökull. Quaternary Science Reviews 16, 649–659.

Krüger, J. & Aber, J.S., 1999. Formation of supraglacial sediment accumulations on Kötlujökull, Iceland. Journal of Glaciology 45, 400–402.

Krüger, J. & Kjær, K.H., 1999. A data chart for field description and genetic interpretation of glacial diamicts and associated sediments – with examples from Greenland, Iceland, and Denmark. Boreas 28, 386–402.

Krüger, J. & Kjær, K.H., 2000. De-icing progression of ice-cored moraines in a humid, subpolar climate, Kötlujökull, Iceland. The Holocene 10, 737–747.

Krüger, J. & Kjær, K.H., 2005. Fabric pattern in a basal till succession and its significance for reconstructing subglacial processes – discussion. Journal of Sedimentary Research 75, 323–326.

Krüger, J., Kjær, K.H., & van der Meer, J.J.M., 2002. From push moraine to single-crested dump moraine during a sustained glacier advance. Norwegian Journal of Geography 56, 87–95.

Lawson, D.E., 1979. Sedimentological analysis of the western terminus region of the Matanuska Glacier, Alaska. U.S. Army Cold Regions Research and Engineering Laboratory, Report 79-9, 112 pp.

Le Heron, D.P. & Etienne, J.L., 2005. A complex subglacial clastic dyke swarm, Sólheimajökull, southern Iceland. Sedimentary Geology 181, 25–37.

Le Heron, D.P., Sutcliffe, O.E., Whittington, R.J., & Craig, J., 2005. The origin of glacially related soft-sediment deformation structures in Upper Ordovician glaciogenic rocks: implication for ice-sheet dynamics. Palaeogeography, Palaeoclimatology, Palaeoecology 218, 75–103.

Mark, D.M., 1973. Analysis of axial orientation data, including till fabric. Bulletin Geological Society of America 84, 1369–1374.

McKay, L.D. & Fredericia, J., 1995. Distribution, origin and hydraulic influence of fractures in a clay-rich glacial deposit. Canadian Geotechnical Journal 32, 957–975.

McKay, L., Fredericia, J., Lenczewski, M., Morthorst, J., & Klint, K.E.S., 1999. Spatial variability of contaminant transport in a fractured till, Avedøre, Denmark. Nordic Hydrology 30, 333–360.

van der Meer, J.J.M., 1993. Microscopic evidence of subglacial deformation. Quaternary Science Reviews 12, 553–587.

van der Meer, J.J.M., Kjær, K.H., & Krüger, J., 1999. Subglacial water-escape structures and till structures, Sléttjökull, Iceland. Journal of Quaternary Sciences 14, 191–205.

van der Meer, J.J.M., Menzies, J., & Rose, J., 2003. Subglacial till: the deforming glacier bed. Quaternary Science Reviews 22, 1659–1685.

van der Meer, J.J.M., Kjær, K.H., Krüger, J., Rabassa, J., & Kilfeather, A.A., 2009. Under pressure: clastic dykes in glacial settings. Quaternary Science Reviews 28, 708–720.

Miall, A.D., 1977. A review of the braided river depositional environment. Earth-Science Reviews 13, 1–62.

Miall, A.D., 1983. Principles of Sedimentary Basin Analysis. Berlin, Springer-Verlag, 490 pp.

Nelson, A.E., Willis, I.C., & Ó Cofaigh, C., 2005. Till genesis and glacier motion inferred from sedimentological evidence associated with the surge-type glacier, Brúarjökull, Iceland. Annals of Glaciology 42, 14–22.

Nilsson, B., Sidle, R.C., Klint, K.E.S., Bøggild, C.E., & Broholm, K., 2001. Mass transport and scale-dependent hydraulic tests in a heterogeneous glacial till – sand lens – unconfined sandy aquifer system. Journal of Hydrology 243, 162–179.

Pedersen, S.A.S., 1989. Glacitectonite: brecciated sediments and cataclastic sedimentary rocks formed subglacially. In: Goldthwait, R.P., Matsch, C.L. (Eds), Genetic classification of glacigenic deposits. Brookfield, Rotterdam, 89–92.

Piotrowski, J.A. & Tulaczyk, S., 1999. Subglacial conditions under the last ice sheet in Northwest Germany: Ice-bed separation and enhanced basal sliding? Quaternary Science Reviews 18, 737–751.

Powers, M., 1953. A new roundness scale for sedimentary particles. Journal of Sedimentary Petrology 23, 117–119.

Rijsdijk, K.F., Warren, W.P., van der Meer, J.J.M., Owen, G., & McCarroll, D., 1999. Clastic dykes in consolidated tills; Evidence for sub-glacial hydro fracturing at Killiney Bay, Eastern Ireland. Sedimentary Geology 129, 111–126.

Schomacker, A., Krüger, J., & Larsen, G., 2003. An extensive late Holocene glacier advance of Kötlujökull, central south Iceland. Quaternary Science Reviews 22, 1427–1434.

Schomacker, A., Krüger, J., & Kjær, K.H., 2006. Ice-cored drumlins at the surge-type glacier Brúarjökull, Iceland: a transitional-state landform. Journal of Quaternary Science 21, 85–93.

Shaw, J., 1982. Melt-out till in the Edmonton area, Alberta, Canada. Canadian Journal of Earth Sciences 19, 1548–1569.

Williams, R.E. & Farvolden, R.N., 1967. The influence of joints on the movement of groundwater through glacial till. Journal of Hydrology 5, 163–170.

Woodcook, N.H., 1977. Specification of fabric shapes using an eigenvalue method. Bulletin Geological Society of America 80, 2342–2352.

10

Mýrdalsjökull's Forefields Under the Microscope. The Micromorphology of Meltout and Subglacial Tills

Jaap J.M. van der Meer[1,*], Simon J. Carr[1] and Kurt H. Kjær[2]

[1]*Centre for Micromorphology UoL, Department of Geography, Queen Mary, University of London, Mile End Road, London E1 4NS, UK*

[2]*Natural History Museum of Denmark, University of Copenhagen, Øster Voldgade 5-7, DK-1350 Copenhagen K., Denmark*

[*]*Correspondence and requests for materials should be addressed to Jaap J.M. van der Meer (e-mail: j.meer@qmul.ac.uk)*

10.1. Introduction

The well-exposed forefields of Icelandic glaciers have been the object of geomorphological and sedimentological studies for a long time. In general, one can recognise three types of studies: (i) those that describe the current situation or current processes (Boulton and Hindmarsh, 1987; Croot, 1988); (ii) those that study the development of landforms and sediment packages over time (Krüger, 1994; Evans, 2005) and (iii) those that try to understand current processes to explain Pleistocene landforms and sediment packages (Todtmann, 1960).

In this chapter, we will be looking at the micromorphology of sediments on the basis of observed processes and compare this to the micromorphology of subrecent sediments. The combination of these two (age and process) strengthens our interpretation of the micromorphology of Pleistocene or older sequences.

Theoretically, a large number of genetically different till types is recognized (Dreimanis, 1988; van der Meer *et al.*, 2003), and there is a wide range of literature on how to differentiate between these. However, the criteria for differentiating between these till types have been recently challenged, and the long-term survival potential of supraglacial meltout tills is in doubt (van der Meer *et al.*, 2003; Evans *et al.*, 2006). Some Icelandic glaciers have an extensive supraglacial debris cover, which offers the opportunity to study actively melting-out supraglacial sediments as well as the end result. Because in many forefields, meltwater streams are carving out channels, there are usually many exposures where sediments can be easily accessed for analysis. And because these exposures will also show underlying subglacial tills, it is easy to compare genetically different sediment types.

Thus, the aim of this chapter is to analyse modern and subrecent sediments from the forefields of outlet glaciers of the Mýrdalsjökull ice cap to define some key microscopic criteria for differentiating tills of different origins.

10.2. Micromorphology

Micromorphology is the microscopic examination of the composition and constituent structural elements and, in unlithified Earth materials, involves the impregnation of unlithified sediments. The technique was developed for soils in the 1930s but is now increasingly applied to the study of sediments (van der Meer and Menzies, submitted). The samples that are impregnated and used for micromorphology are preferably of the dimensions 8×14 cm, although samples of 6×10 cm samples are used regularly. Large thin sections have the important advantage that they provide a larger sample area and provide a better representation of macroscopically visible structures, including discontinuities such as inclusions, than small thin sections. For the purpose of studying sediments, standard petrographic thin sections are inadequate.

Samples were collected in the field in metal boxes of the dimensions given above and taped completely to preserve field moisture conditions. The latter enabled transport of the samples without collapse. The samples were taken to laboratories in Amsterdam and later to London where they were dried slowly, impregnated, cut and mounted following procedures described in Murphy (1986), van der Meer (1993), Lee and Kemp (1993), Carr and Lee (1998) and Menzies (2000).

The thin sections are analysed with a *Petroscope* (an amended microfiche reader in which the optics have been adapted to thin section analysis) and a low-magnification petrological microscope. Description and terminology follow the classifications of Brewer (1976), van der Meer (1993, 1997) and Menzies (2000).

Because of the application of a polarizing microscope, it is possible to use the optical properties of particles to derive information such as the stress history of the material.

The purpose of this chapter is to link micromorphological observations to observed processes and to

DEVELOPMENTS IN QUATERNARY SCIENCES
VOLUME 13 ISSN 1571-0866

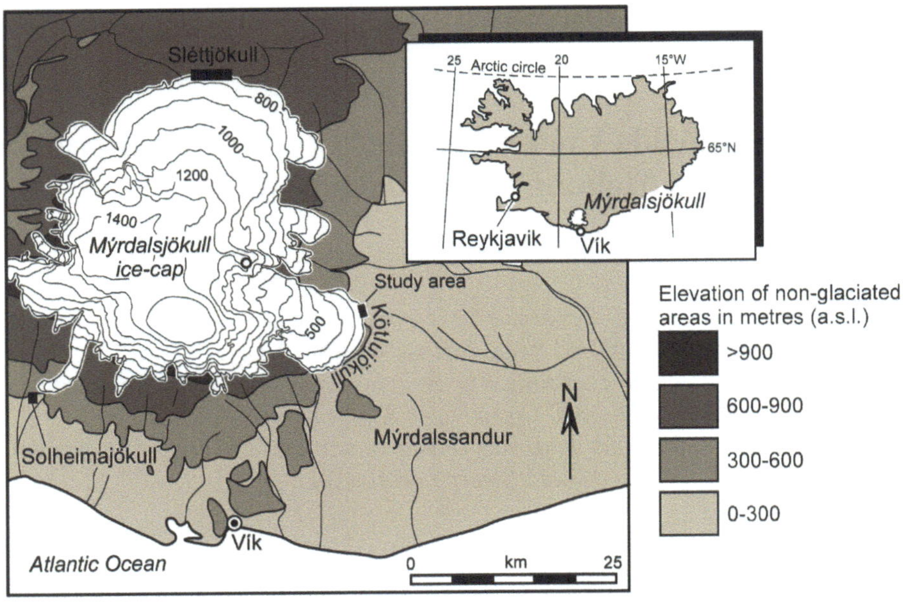

Fig. 10.1. Location of the Mýrdalsjökull ice cap in Iceland, and the Mýrdalsjökull ice cap with the glacier lobes Kötlujökull, Sléttjökull and Sólheimajökull indicated.

processes inferred on the basis of proximity to ice. This is most specifically the case for the establishment of meltout characteristics as glaciers in Iceland are the most accessible for collecting larger numbers of samples.

10.3. Sample Sites

The collection of thin sections from Iceland covers a number of sites from different settings across the majority of the main ice caps. In this chapter, we concentrate on outlet glaciers from Mýrdalsjökull (Fig. 10.1), which provide a comprehensive range of different glacial settings. Over the years, samples have been collected from Kötlujökull, a large lobate outlet in the east, from Sléttjökull, which forms the N sector of the ice cap, and from Sólheimajökull, a smaller topographically constrained outlet glacier in the south. All three are non-surging glaciers. The major Mýrdalsjökull outlet glaciers have been monitored by Krüger and co-workers between 1977 and 2009; see, for instance, Krüger (1994).

10.3.1. Kötlujökull

Since 1977, Kötlujökull has shown a retreat, an advance and renewed retreat. Details of this cycle and resulting landforms have been described, in for instance, Krüger (1995, this volume), Krüger *et al.* (2002), Krüger and Kjær (2000), Kjær and Krüger (2001) and Kjær *et al.* (2003, 2004). The margin of the glacier is extensively covered by supraglacial debris and is excellently suited for studies of meltout processes; supraglacial meltout tills (still in ice-contact position) are extant (Fig. 10.2), while in the forefield, 'islands' of meltout sediments (Fig. 10.3) are temporarily preserved in between meltwater streams. For the purpose of micromorphological investigations, Kötlujökull was visited and the forefield sampled in 1996. The forefield sample sites have since been extensively reworked and modified by meltwater activity, while the sample sites on the ice have changed unrecognisably due to continued downwasting. Lithostratigraphically, Kötlujökull exposures often show supraglacial meltout deposits with intercalated rhythmites overlying (occasionally a thin layer of basal meltout on) subglacial till, sometimes with more than one bed of the latter preserved. Rates of meltout are quite variable depending on position. In the partially ice-cored moraine, vertical down-melting by top melt is only 0.1 m/yr, while backwasting of ice-cored slopes is up to 2.4 m/yr (2.3 cm/day) (Kjær and Krüger, 2001). This melting releases substantial amounts of water, and it can come as no surprise that ice-cored slopes are highly unstable. Parts of slopes show rotational slumping, while other parts are continuously creeping.

10.3.2. Sléttjökull

Sléttjökull has retreated steadily since the early part of the twentieth century, leaving a record of annual moraines across the forefield. Although there are some outcropping ash layers on the glacier, there is no comparable supraglacial debris cover development as seen on Kötlujökull. There are well-developed debris cones on the ice, but the material is continuously redistributed as the cones melt down. The forefield of Sléttjökull consists of a fluted and drumlinized till plain (Krüger and Thomsen, 1984; Kjær, 1999; Kjær *et al.*, 2003), in which the till is well exposed along numerous incised meltwater streams. One of the peculiarities of Sléttjökull's forefield is the widespread existence of water escape structures (WES). These have been related to the Katla eruption of 1918, combined with permafrost under the glacier front (van der Meer *et al.*, 1999, 2009). The forefield of

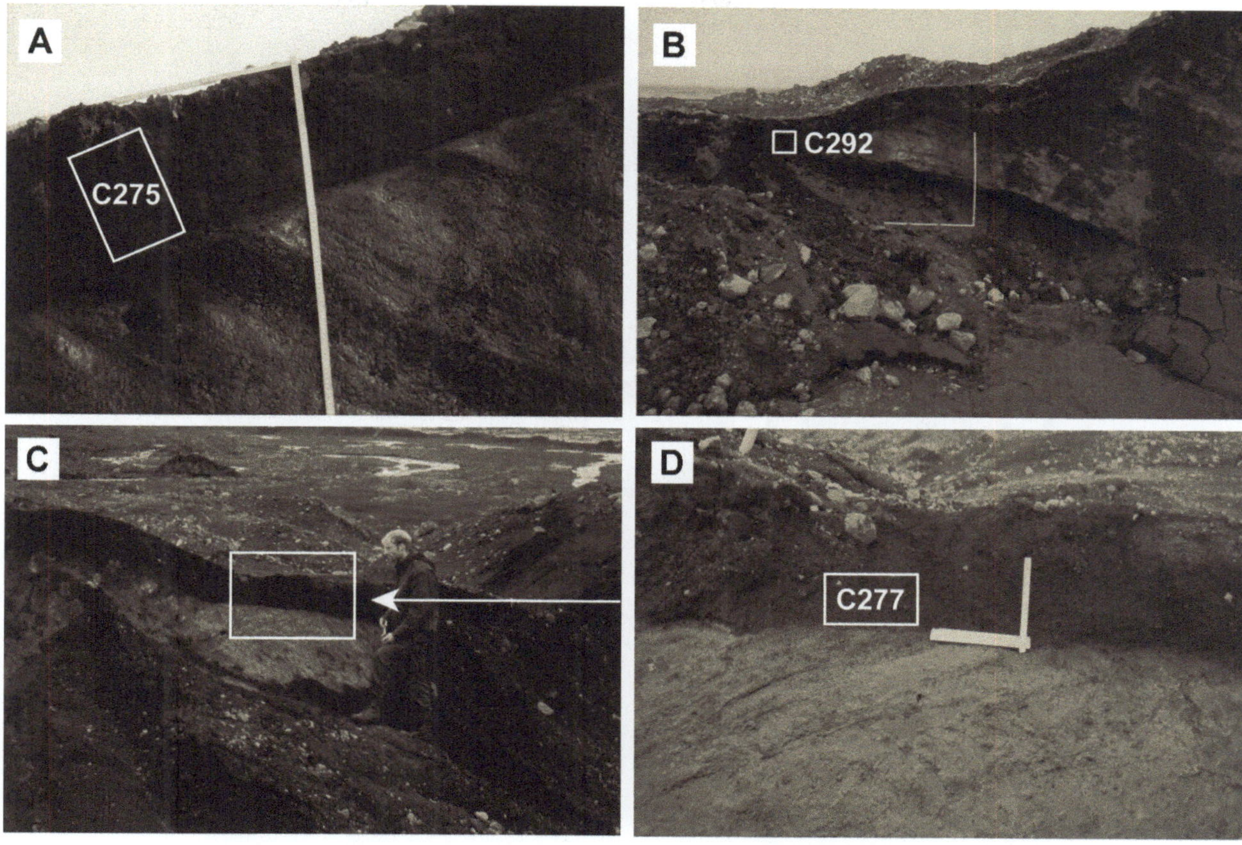

Fig. 10.2. Active meltout sample sites at Kötlujökull, sampled in 1996. (A) Site MO-1, note thickening of supraglacial debris layer with addition of debris melting out of englacial debris bands, preserving the lamination (B) Location of sample C.292 from a layer of subglacial till melting out in an englacial position (C) Site MO-3. The sample was taken in a horizontal position, but behind the face, the surface dips steeply away from the observer.

Sléttjökull was visited and sampled in 1996, 1997 and 2001. Here, the lithostratigraphy shows different sub-recent till beds, with WES emanating from the base, overlying fluvioglacial deposits with intercalated coarse tephras and rhythmites (Fig. 10.4).

10.3.3. Sólheimajökull

Sólheimajökull is a long and narrow outlet glacier, draining much of the southern sector of the ice cap, and has an incised meltwater channel running to the Atlantic, with Holocene tills exposed in the incisions. One site, consisting of different till beds overlying fluvioglacial deposits, was sampled in 1996 (Fig. 10.5). One bed boundary showed numerous fractured clasts. At the sample site, the till was almost 3 m thick and overlying outwash, which is quite different from the lithostratigraphy described by Le Heron and Etienne (2005), in a publication on WES, very similar to those at Sléttjökull. At their site, which cannot be far away from our sample site, they indicate some thin till beds on top of, intercalated within and at the base of thick outwash beds. The main difference between Sólheimajökull and the other two areas is that the Sólheimajökull tills are very hard, as if cemented, whereas all the other Mýrdalsjökull tills were soft, as if dilated.

10.4. Samples

A total of 15 samples were collected at Kötlujökull (Table 10.1). Of these, three samples were from meltout debris overlying and in immediate contact with the ice. Seven more samples were collected from slightly older sediments, supposed to be meltout till, but no longer in contact with the ice. One till sample was collected from a melting-out englacial band of basal debris, while three more subglacial till samples were collected in the forefield. Finally, one sample was collected from a subaerial, laminated, recent pond deposit overlying the ice. The rationale behind the latter sample is that meltout deposits in the forefield contain finely laminated units, which need to be compared to sediments of known origin.

At Sléttjökull, a total of 33 samples was collected. Of these, 18 samples are from tills and transitions between till beds. A further five samples are a combination of till and WES, while another five samples stem from wide WES only. To analyse the finely laminated WES, four samples of intercalated lacustrine units – two of which combined with WES themselves – were collected. Finally, two samples were collected from deposits intermittently fed by snow meltwater only.

At Sólheimajökull, three samples were collected from one profile.

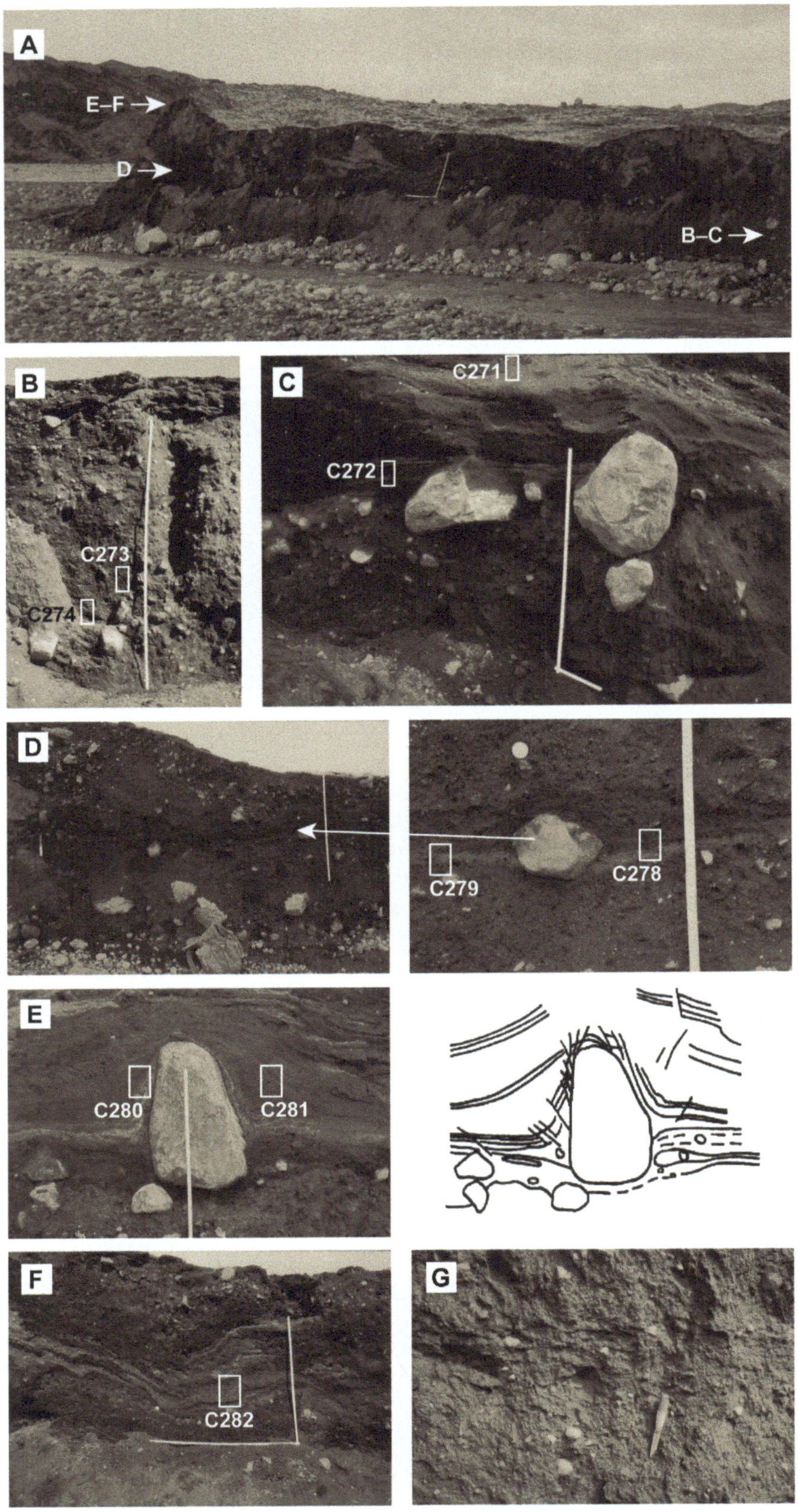

Fig. 10.3. Subrecent sample sites (position of thin sections indicated) at Kötlujökull, profile 3, see also Table 10.1. The site was swept away in 1997 (A) Overview of the site in 1996 (B) Locality 1 with supraglacial meltout deposits (brown) overlying subglacial till (grey) (C) Locality 2 with supraglacial meltout deposits. (D) Locality 3 with supraglacial meltout diamicts separated by a draped layer. (E) Locality 4 with laminated clays draped over a vertical boulder, overlying subglacial till. (F) Locality 5 with slumped fine-grained supraglacial meltout deposits. (G) Basal till with clear fissility underlying the meltout deposits at profile 3.

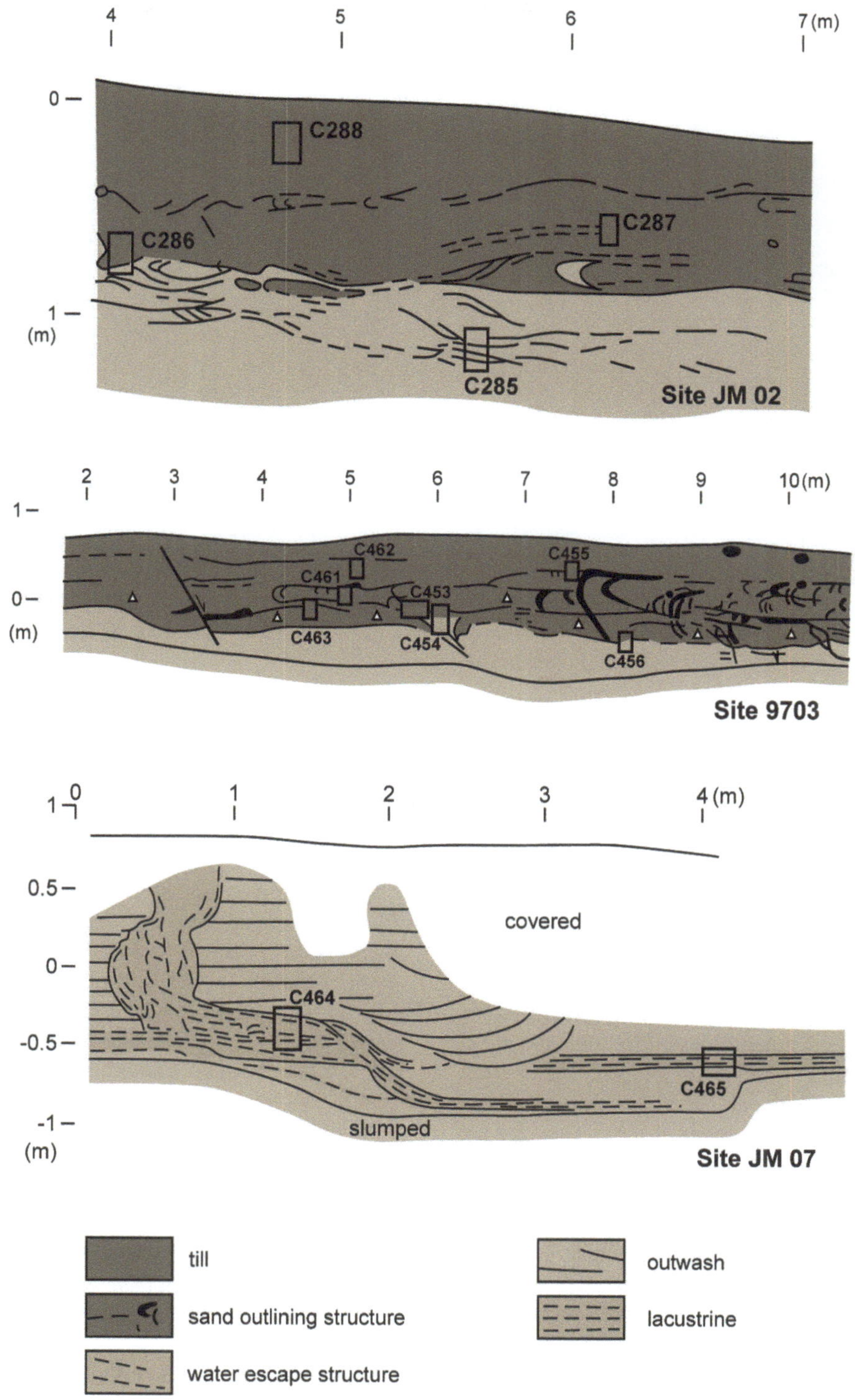

Fig. 10.4. Some of the sample sites at Sléttjökull studied in 1996 and 1997. Sites JM02 and 9703 are from tills overlying outwash. The structures in the till are mainly deformed water escape structure (WES) (see van der Meer et al., 1999; Kjær et al., 2003). Site JM07 is a large WES, possibly cutting through lacustrine deposits.

10.5. Description

It should be noted that many Icelandic sediments are compositionally particular in that they consist of basalt and tephra only, irrespective of grain size. Consequently, the fine fractions of sediments consist mainly of basalt, and thus, most particles are opaque when viewed under crossed polarizers. This makes it extremely difficult to observe birefringence. It is not necessarily absent, but difficult to detect. The other peculiarity is that because both bedrock and tephra have many vesicles, the glacial deposits are often very porous. Vesicles in rock fragments may not be connected to inter-particle pores, but in thin section, the sediment looks very porous (Figs. 10.6–10.8).

The observations in the thin sections are presented in Table 10.2, where they are arranged according to the glacier first, then to site and material. As a consequence, the samples are not in numerological order, but this way, it is easier to see the characteristics of different groups of samples. The first group consists of meltout sediments

Fig. 10.5. Sample site at Sólheimajökull in 1997. Note that at that time the glacier was almost overlying the sample site, but the vegetation shows that the till beds are not related to the most recent ice advance. A fourth sample was collected halfway between C.295 and C.297, but this failed impregnation.

from Kötlujökull, followed by a large group of tills and till structures from all three glaciers. Through a group of samples covering tills cut by WES, we come to a group of WES-only samples, to end with a group of samples from lacustrine (or pond) deposits and lacustrine deposits cut by WES.

The description starts with aspects of texture. The column 'mode in μm' indicates the most frequent grainsize, or if listed as '100–5,000' as a range either because there is no obvious mode or where giving a 'whole-sample' mode is meaningless. The most astonishing fact is the high variability of the grain size (Fig. 10.6A): from a mode at 50 μm for some meltout tills to a range of 500–2,500 μm in others. The tills are simpler in that for most it was possible to indicate a clear mode, although the degree of variation is high. Interestingly, the lacustrine beds and WES are equally variable in grain size and sometimes coarse grained. All the lacustrine samples are from well-bedded deposits, and some of these beds appear to be quite coarse. The coarseness observed in some of the WES is more astonishing as it is clear that cracks must be wide enough to accommodate coarse grains.

The next column deals with the shape of the skeleton particles, indicating it across the size range, from fine to coarse. Several trends are obvious: across the board, well-rounded particles are rare, whereas angular particles (Fig. 10.6A) form the majority of the particles in a number of samples. Furthermore, finer grains tend to be angular or more commonly subangular to subrounded, while larger grains tend to be subrounded. The peculiarity of all material being basaltic is that many of the grains are vesicular (Fig. 10.6B), and where these intersect the surface of the grain, there remains an angular aspect irrespective of the overall roundness of the grain.

There is no separate column for the composition of the samples as this is extremely uniform. The bedrock of the area is fully volcanic, mainly basaltic. Consequently, all particles are basalt – with brownish palagonite and (especially at Sólheimajökull) some crystalline rock types. In addition, volcanic eruptions have produced extensive ashfalls, and these particles are mainly translucent glassy tephra shards.

Because it is not possible to see individual particles that are smaller than the thickness of the thin section, there is no separate column for plasma. All samples contain plasma, albeit in different amounts. The meltout tills do not contain much plasma (Fig. 10.6C), except for the fine-grained (clayey, silty) laminae, while most of the till samples contain enough plasma to fill most of the spaces between skeleton grains (Fig. 10.6D). Plasma content of the WES is variable again, depending on individual laminae (Fig. 10.6E). The same applies to lacustrine deposits, although there are fewer laminae here from which plasma is excluded (Fig. 10.6F). Occasionally, there are light coloured, orange-tinted clay layers in which clay minerals appear to be the main composition, although they might be partly derived from the brown palagonite.

One of the most obvious structural elements is porosity, and given that vesicular basalts are widespread in these samples that is even more so under the microscope. This also makes it difficult to estimate total porosity (Fig. 10.6B). It should be noted that the smooth cracks developed during sample preparation are not taken into account. Pores come in different types, first of all packing voids (Fig. 10.6A), either simple (not in contact with other pores) or compound. Both types are present in almost all samples, with the exception of some of the fine-grained meltout and lacustrine samples. Partly this will be caused by the grain size, pores smaller than the thickness of the thin section will not be visible. Compound packing voids are much less frequent in meltout tills. In the first place, less frequent than simple packing voids, and secondly, less frequent than in the subglacial tills (Table 10.2). Packing voids are also absent

Table 10.1. Thin sections used in this study.

Number	Locality	Site	Material	Ice contact and age[1]
C.271	Kötlujökull	Profile 3-2	Meltout/debris flow	No, subrecent
C.272	Kötlujökull	Profile 3-2	Meltout?	No, subrecent
C.275	Kötlujökull	m-o 1	Meltout	Yes
C.276	Kötlujökull	m-o 2	Meltout	Yes
C.277	Kötlujökull	m-o 3	Meltout	Yes
C.278	Kötlujökull	Profile 3-3	Meltout	No, subrecent
C.279	Kötlujökull	Profile 3-3	Meltout	No, subrecent
C.280	Kötlujökull	Profile 3-4	Meltout	No, subrecent
C.281	Kötlujökull	Profile 3-4	Meltout	No, subrecent
C.282	Kötlujökull	Profile 3-4	Meltout	No, subrecent
C.273	Kötlujökull	Profile 3-1	Till	No, subrecent
C.274	Kötlujökull	Profile 3-1	Till	No, subrecent
C.292	Kötlujökull	m-o 1	Till	Yes
C.293	Kötlujökull	Profile 3-4	Till	No, subrecent
C.283	Sléttjökull	JM1	Till	No, < 100 yr
C.287	Sléttjökull	JM2	Till	No, < 100 yr
C.288	Sléttjökull	JM2	Till	No, < 100 yr
C.289	Sléttjökull	JM2	Till	No, < 100 yr
C.290	Sléttjökull	JM2	Till	No, < 100 yr
C.291	Sléttjökull	JM3	Till	No, < 100 yr
C.455	Sléttjökull	9703	Till+shear zone	No, < 100 yr
C.456	Sléttjökull	9703	Till transition	No, < 100 yr
C.457	Sléttjökull	JM04	Sheared till	No, < 100 yr
C.458	Sléttjökull	JM06	Lower till	No, subrecent
C.459	Sléttjökull	JM06	Till wedge	No, < 100 yr
C.461	Sléttjökull	9703	Middle till	No, < 100 yr
C.462	Sléttjökull	9703	Upper till	No, < 100 yr
C.463	Sléttjökull	JM06	Till transition	No, < 100 yr
C.465	Sléttjökull	9703	Trans. lower > middle till	No, < 100 yr
C.469	Sléttjökull	JM08	Lower till	No, subrecent
C.470	Sléttjökull	JM08	Till	No, < 100 yr
C.471	Sléttjökull	JM08	Till	No, < 100 yr
C.294	Sólheimajökull		Till	No, subrecent
C.295	Sólheimajökull		Till	No, subrecent
C.297	Sólheimajökull		Till	No, subrecent
C.284	Sléttjökull	JM1	Till+WES[2]	No, < 100 yr
C.286	Sléttjökull	JM2	Till+WES	No, < 100 yr
C.453	Sléttjökull	9703	Till+WES	No, < 100 yr
C.460	Sléttjökull	JM06	Till+WES	No, < 100 yr
C.475	Sléttjökull	735 m	Till+WES	No, < 100 yr
C.454	Sléttjökull	9703	WES	No, < 100 yr
C.468	Sléttjökull	JMX	WES	No, < 100 yr
C.476	Sléttjökull	865 m	WES	No, < 100 yr
C.477	Sléttjökull	865 m	WES	No, < 100 yr
C.270	Kötlujökull	pond	Lacustrine	n.a.
C.464	Sléttjökull	JM06	Lacustrine	n.a.
C.466	Sléttjökull	Snowlake	Lacustrine	n.a.
C.467	Sléttjökull	Snowlake	Lacustrine	n.a.
C.472	Sléttjökull	715 m	Lacustrine	n.a.
C.473	Sléttjökull	715 m	Lacustrine+WES	n.a.
C.474	Sléttjökull	715 m	Lacustrine+WES	n.a.

[1]yes, sediment on top of ice; no, ice contact not observed; subrecent, several hundred years in age; < 100 yr, usually several tens of years in age.
[2]WES, water escape structure, clastic dyke.

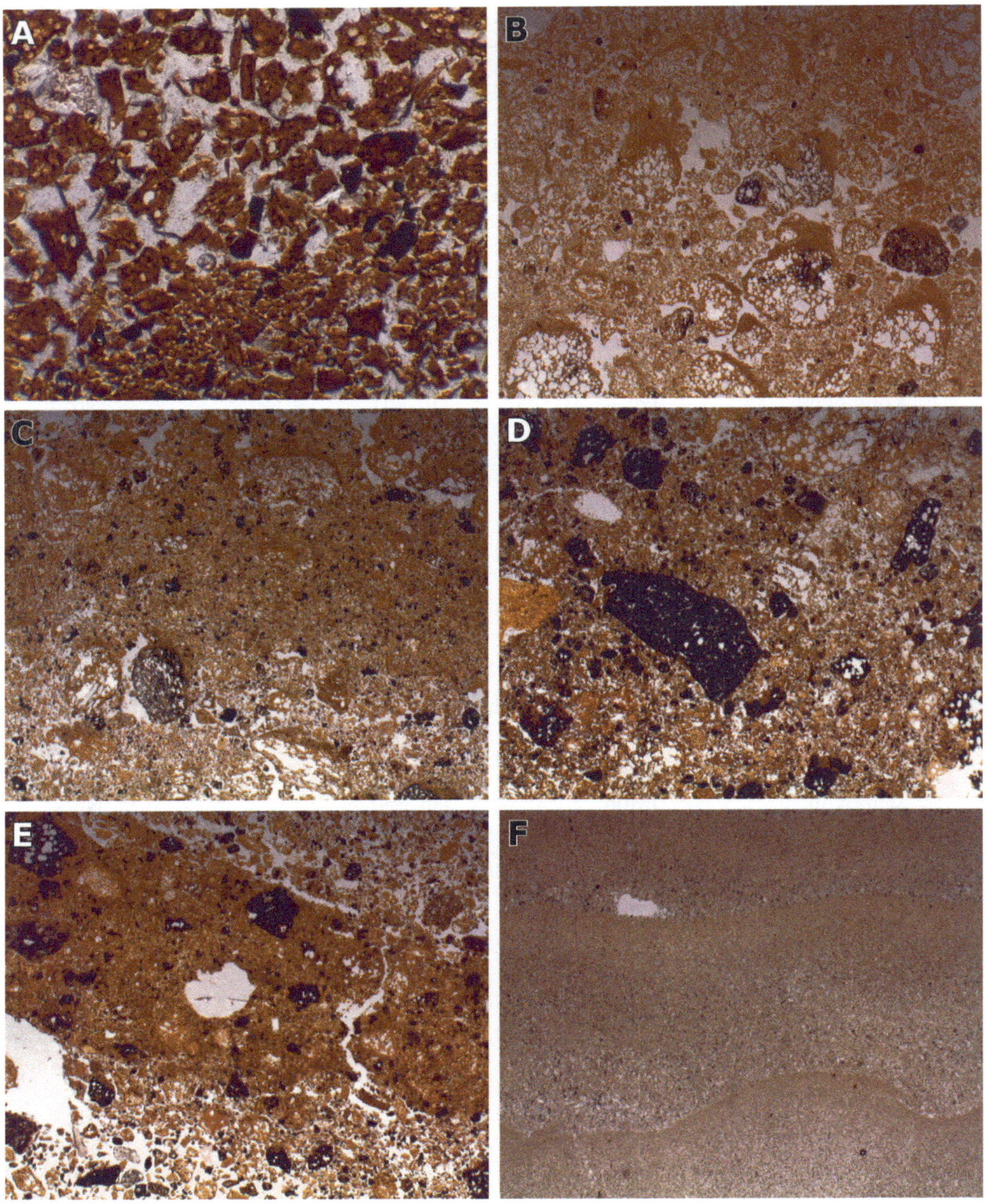

Fig. 10.6. Grainsize and shape. All images are in plane light, horizontal field of view (fov) is 18 mm and the top is up, unless stated otherwise. (A) C.272; fov 2.6 mm. Subrecent meltout. Note the range of grain sizes, organised in beds, as well as the lithological homogeneity and the angularity of the grains. (B) C.278. Subrecent meltout. This sample clearly demonstrates the vesicular nature of the rock fragments. (C) C.275. Ice-contact meltout. Differences in plasma content and thus porosity in one of the more plasma-rich zones. (D) C.283. Subglacial till. Uneven plasma content is evident by darker and lighter domains. (E) C.454. Lamination in water escape structure expressed by differences in plasma content. (F) C.270. Lacustrine/pond deposit in supraglacial position. Normal grading with upward increase in plasma content in each lamina.

in the Sólheimajökull samples (Fig. 10.7A) (except for compound packing voids in C.294). As the till at Sólheimajökull was of a much greater hardness than all other Mýrdalsjökull tills, it does not come as a surprise.

Equally common are channels, (sub-)horizontal primary fissures, which occur in almost all samples. In a number of Kötlujökull samples, these fissures are parallel to each other (Fig. 10.7B). In most cases, these are from sites that in the field had been described as being fissile or sheared. Vughs (irregular shaped pores of a rounded appearance; Fig. 10.7C are less common, but they still occur in about half the samples. They occur in almost all the meltout samples and in many of the WES and lacustrine samples but are rare in the subglacial till samples. Least frequent are vesicles (round to oval pores; Fig. 10.7D, which are typical of the meltout and lacustrine samples, but hardly occur in any of the tills. In sample C.463, which shows both vesicles and vughs, these are concentrated in narrow zones on either side of the transition between two till beds.

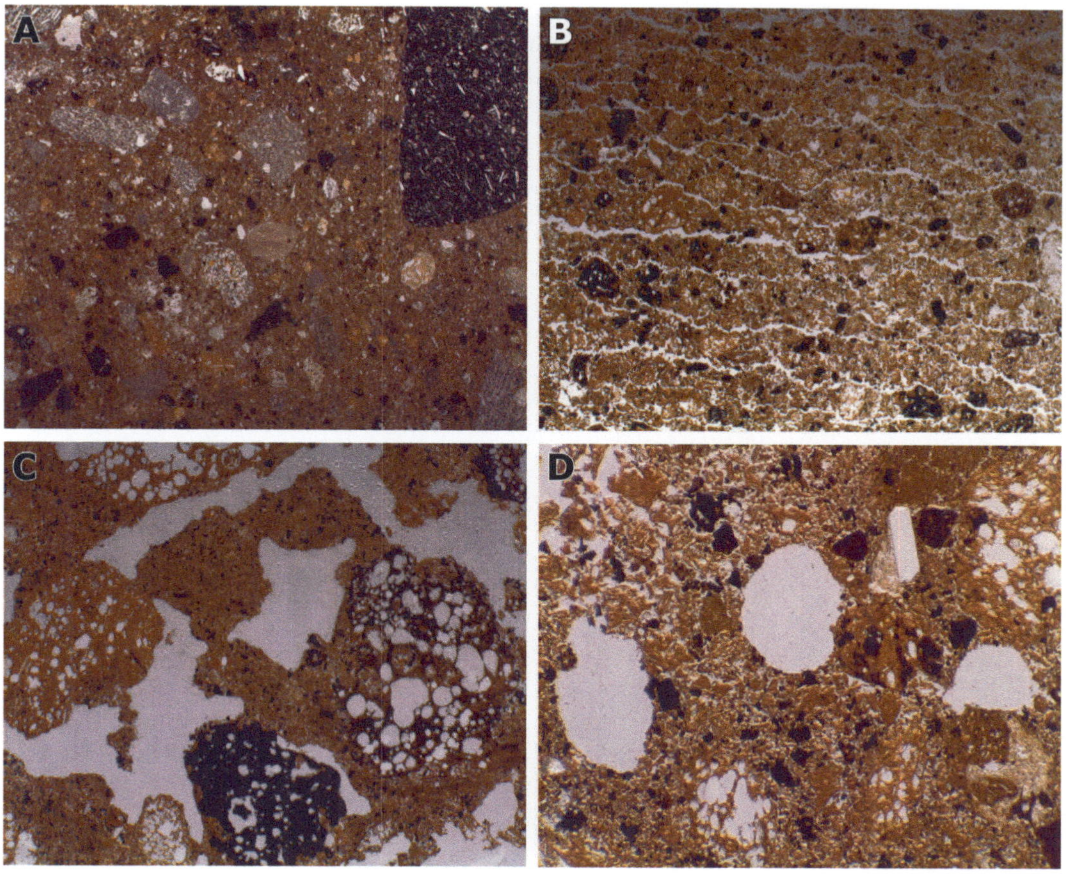

Fig. 10.7. Porosity. All images are in plane light, horizontal field of view (fov) is 18 mm and the top is up, unless stated otherwise. (A) C.295. Subglacial till from Sólheimajökull. Note the obvious absence of pores, which was found to be typical for the samples from this outlet glacier. (B) C.457. Subglacial till from Köttlujökull. High density of parallel channels. (C) C.276; fov 8.3 mm. Ice-contact meltout sediment demonstrate a high frequency of vughs. (D) C.277; fov 3.4 mm. Ice-contact meltout deposits with vesicles.

As is to be expected, bedding or lamination is most common in lacustrine (Fig. 10.8C) and WES (Fig. 10.8D) samples. But it is also common in almost all meltout samples (Fig. 10.8A). Where it is indicated for till samples, this is invariably because of the presence of textural banding (Fig. 10.8B). All the lacustrine samples and some of the meltout samples show normal grading of beds (Fig. 10.6F); reverse grading has never been observed. It implies that in bedded/laminated samples, it is not indicated, the beds are of uniform grain size though a different one for each bed.

Rotational structures like turbates or galaxy structures are rare; they occur only in some of the till samples (Fig. 10.8E) and in one of the meltout samples. In sample C.455, a marble bed structure is indicated (van der Meer, 1993). Although there are suggestions of this structure in some other samples, these were not strong enough to be listed, except for one of the Sléttjökull samples (Fig. 10.8F).

Aggregates are quite common in some of the Sléttjökull till samples as well as in the meltout samples investigated. In the latter, aggregates do not occur in the fine-grained samples and are only present in coarser grained units. In the subglacial tills, they have occasionally been observed in the Kötlujökull samples (Fig. 10.9A), while in the Sólheimajökull samples, they are only weakly represented in one of the samples, while prominent in another. In many cases, these aggregates are units of till mass that are separated by fissures and they can be of irregular or more rounded shape. In most cases, they are relatively small (< 1 cm), but larger ones do occasionally occur. Especially when the aggregates are rounded, they have a coating of fine-grained material (indicated with C or c in Table 10.2; Fig. 10.9), which can be a few millimetres thick. This coating material is either fine silt or clay (Fig. 10.9B), occasionally with some sand grains and can be finely laminated (Fig. 10.9C,D). Sometimes, the laminae are sharply bounded, sometimes boundaries are diffuse and this applies to each individual lamina. In the meltout samples, the coatings appear to be related to the bedding as they are more frequent in some, while also their size may change per bed (Fig. 10.11A). Sometimes, the coatings occur around the whole grain, more often they are only found on the upper surface, draping over the sides. Often, such partial coatings are irregular in thickness (Fig. 10.9B). When the coating is only partial, its presence on a side or the base demonstrates that the particle has moved since formation of the coating, especially when they occur on different sides on different grains (Fig. 10.9E). Movement is also suggested by erosion of the coatings, which appears to be most common in tills (Fig. 10.9F,G). Coatings in the latter are usually present as thin linings of adhering fine-grained

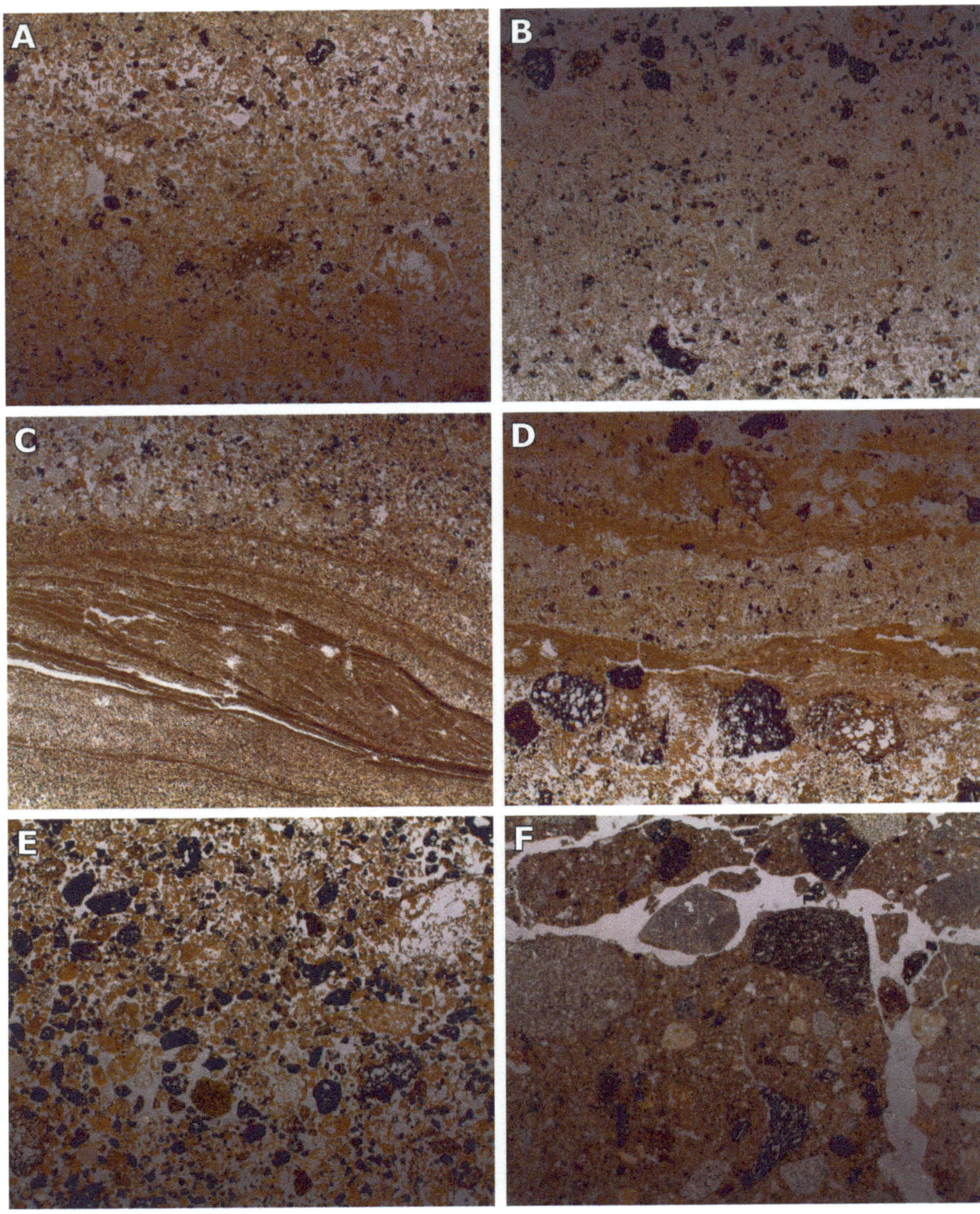

Fig. 10.8. Bedding and other structures. All images are in plane light, horizontal field of view (fov) is 18 mm and the top is up, unless stated otherwise. (A) C.275. Ice-contact meltout. Bedding is diffusely bounded and beds are characterised by uniform grain size. (B) C.470. Subglacial till demonstrating a fine-grained bed. Note the presence of coarser black particles above and below the bed and the presence of smaller black particles in the centre of the bed. (C) C.466. Lacustrine deposits fed by melting snow only. Multiple thin laminae in fine-grained bed. Breaking up of laminae most likely caused by freezing. (D) C.476. Bedding in water escape structure (WES). Rapid alternation of laminae in WES. In this case, the central, relatively uniform bed may have split a fine-grained bed, causing the latter to look like the lining of a sheet. (E) C.283. Turbate structures in subglacial till, outlined by black particles. (F) C.294 Channels outline a marble bed structure in one of the Sólheimajökull subglacial till samples; only known from deforming beds.

material preserving a bit of lamination. Although there are some aggregates in Sólheimajökull sample C.297, there are no coatings and neither are there any in the other samples from this glacier forefield.

Brittle deformation in the sense of fault structures has been observed only in some of the meltout samples (Fig. 10.10A,B), while they are common in the lacustrine samples (Fig. 10.10C). Brittle deformation has not been observed in any of the tills in this study. Most commonly, observed faults are only a few millimetres to a few centimetres in length, only affecting some laminae, although faults running through the whole sample do occur as well. In many cases, faults are associated with micro-WES (Fig. 10.10D–F), not indicated in Table 10.2. The micro-WES follow partly tortuous pathways, which can be up to a few centimetres long (Fig. 10.11B). Besides following the faults, they have also been observed between them.

Table 10.2. Micromorphology of thin sections from Mýrdalsjökull. Thin sections are ordered by (i) glacier, (ii) site and (iii) material.

Thin section number	Glacier lobe and site	Material	Texture		Pores					Structure					Plasmic fabric			
			Mode (μm)	Angularity fine>coarse	SPV	CPV	VE	CH	VU	bed	rota	agg	brit	plas	SK	MA	KINK	UNI
C.271	K, 3-2	Meltout	1,500	A > SA	X	X	X	X	X	X	X	X C	X					
C.272	K, 3-2	Meltout	100–400	SA > SR	X			X	X	X		X C	X					
C.275	K, MO-1	Meltout	100	A-SA > SR	X			X	X	X		C	?		X			
C.276	K, MO-2	Meltout	100–1000	SA > SR	X	x	X	X	X	X ↑		X C			x			
C.277	K, MO-3	Meltout	300	SA > SR	X	x	X	X	X	X		C			x			
C.278	K, 3-3	Meltout	1,000	SA > SR	X		X	X	X	X		X C						
C.279	K, 3-3	Meltout	500–2,500	SA > (S)R	X	X	x	X	X			X C						
C.280	K, 3-4	Meltout	50	A-SA > SR				X	X	X ↑			X					
C.281	K, 3-4	Meltout	50	A-SA > SA				X	X	X ↑			X					
C.282	K, 3-5	Meltout	50	A-SA > A/SA	X			X		X ↑			X			X		
C.273	K, 3-1	Upper till	100	SA-SR > SR			X		X			X C						
C.274	K, 3-1	Till	100	SA-SR > SR				X	X									
C.292	K, MO-1	Englacial till	100–5,000	SR > SR	X		X		X			C						
C.293	K, 3-4	Basal till	200	SA > SA-SR	X	X		X				C						
C.283	S, JM1	Till	100–500	SA-SR > SR	X	X	X	X	X		x	X c			?			
C.287	S, JM2	Till	1,000	SA > SR	X	X			X			x						
C.288	S, JM2	Till	50–500	SA > SR	X	X		X				C						
C.289	S, JM2	Till	40–120	SA	X	X		X										
C.290	S, JM2a	Till	1,000	SA-SR > SR	X	X		X				X C				X	X	
C.291	S, JM3	Till	50–800	SA-SR > SR	X	x		X				X			X			
C.455	S, 9703	Till+shear zone	500	SA > SR	X	X		X			MB	X			x			
C.456	S, 9703	Till transition	600	SA-SR > SR	X	X		X				X C			x			
C.457	S, JM04	Till	100–500	SR > SR	X	X		X o			X	X c						
C.458	S, JM06	Lower till	500	(S)R > (S)R	X	X		X			X	C			x			
C.459	S, JM06	Till wedge	100	(S)A > SR	X	X					X							
C.461	S, 9703	Middle till	400	SA-SR > SR	X	X		X				X			x			
C.462	S, 9703	Upper till	100–500	SR > SR	X	X		X o				X c						
C.463	S, 9703	Till transition	200	SA-SR > SR	X	X	X	X o	X		X	c						
C.469	S, JM08	Lower till	1,000	SA > SA(SR)	X	X		X		X		C			X		X	
C.470	S, JM08	Till	100–2,000	SA-SR > SR		X		X o		X		C						
C.471	S, JM08	Till	200	SA > SR	X	X		X o				C						
C.294	So	Till	1,500	SA > SR		X		X	X	x	MB	X			x			
C.295	So	Till	2,500	SA > SA-SR				X		x								
C.297	So	Till	3,200	SA > SA-SR				X		x		x			x	x		
C.284	S, JM1	Till+WES	800	SA-SR > SR		X	X					X		X		X		
C.286	S, JM2	Till+WES	1,200	SA-SR > SR	X	X			X	X		X	X			X		
C.453	S, 9703	Till+WES	500	SA > SA-SR	X	X		X	X				X		x		X	x

(*Continued*)

Table 10.2. (*Continued*)

Thin section number	Glacier lobe and site	Material	Texture		Pores					Structure					Plasmic fabric			
			Mode (μm)	Angularity fine > coarse	SPV	CPV	VE	CH	VU	bed	rota	agg	brit	plas	SK	MA	KINK	UNI
C.460	S, JM06	Till+WES	200	SA-SR > SR	X	X		X		X	X				x			X
C.475	S, 735 m	Till+WES	1,500	SA-SR > SR	X	X		X o	X	X		C			x			
C.454	S, 9703	WES	6,000	A-SA > SR	X	X		X		X		X			X		x	
C.465	S, JM07	WES ?	300	(S)A-SR > SR	X	X		X	X	X ↑			X	X	x			
C.468	S, JMX	WES	200	A > SR	X	X	X	X	X	X					x		x	
C.476	S, 865 m	WES	3,000	(S)A > SA-SR	X	X		X	X	X		c	X	X	x	X		
C.477	S, 865 m	WES	1,000	SA-SR > SR	X	X		X	X	X				X		X		
C.270	K, MO1	Lacustrine	50	A > SA	X		X	X	X	X ↑								
C.464	S, JM07	Lacustrine	3,000	SA-SR > R	X	X		X		X ↑			X	X	X	X		
C.466	S, snowlake	Lacustrine	1,500	SA-SR > SR	X	X	x	X	x	X ↑			X	X				x
C.467	S, snowlake	Lacustrine	200	SA-SR > SR	X	X		X	X	X ↑			X	X				
C.472	S, 715 m	Lacustrine	1,000	A > SR	X	X	X	X	X	X ↑		C	X	X		X		
C.473	S, 715 m	Lacustrine+WES	2,000	A > (S)R	X	X	X	X	X	X ↑					x	X		
C.474	S, 715 m	Lacustrine+WES	2,000	A > (S)R	X	X	X	X	X	X ↑		C	X	X		X		X

K = Kötlujökull
S = Sléttjökull
So = Sólheimajökull

A = angular
SA = subangular
SR = subrounded
R = rounded

SPV = simple packing voids
CPV = compound idem
VE = vesicles
CH = channels
o = oriented
VU = vughs
X = present
x = weak presence

Bed = bedding, lamination
↑ = normal grading
rota = rotational
agg = aggregates
C = as coatings
c = weak
MB = marble bed
brit = brittle deformation
plas = plastic
X = present
x = weak
? = sporadic

SK = skelsepic
MA = masepic
KINK = kinking
UNI = unistrial
X = present
x = weak

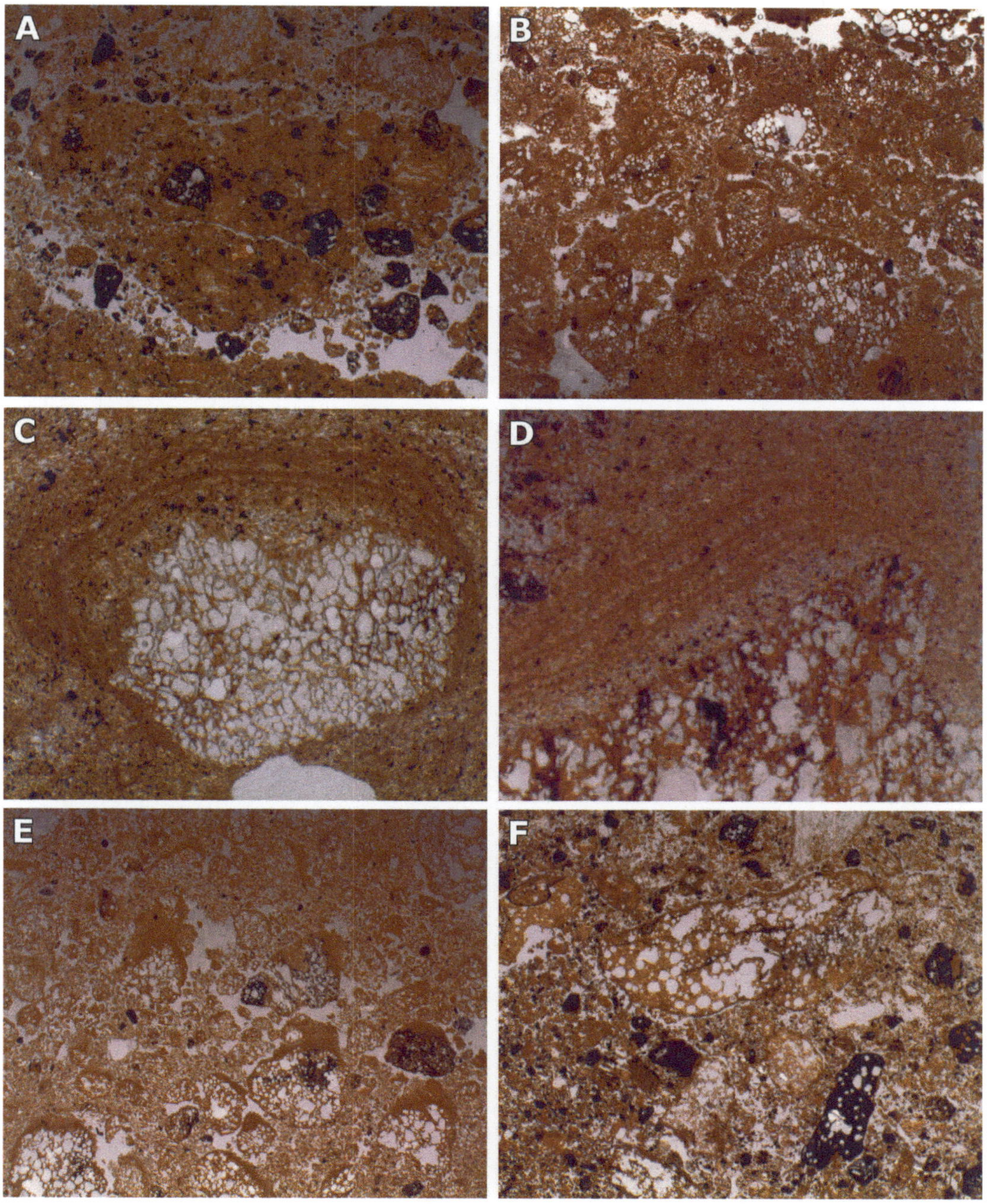

Fig. 10.9. Aggregates and coatings. All images are in plane light, horizontal field of view (fov) is 18 mm and the top is up, unless stated otherwise. (A) C.290; fov 10.4 mm. Fine-grained, dense aggregate in subglacial till. (B) C.271. Coatings in subrecent meltout deposits; the coatings are of a uniform nature. (C) C.276; fov 4.8 mm. Laminated coating in ice-contact meltout deposits. (D) C.276; detail of laminated coating. (E) C.278. Nonuniform coatings in subrecent meltout deposits. The coatings are very uneven in thickness and are on different sides of grains indicating movement of the aggregates. (F) C.283; fov 8.3 mm. Subglacial till. Thin coatings in till. The elongate grain just above the centre and the rounded grain below it show thin coatings. In the lower grain, it appears to be all around. (G) C.288; fov 4.8 mm. Subglacial till. Complex coating around rock fragment. (H) C.476, fov 7.0 mm. Remaining coating around a rock fragment in a water escape structure. (I) C.474. Remaining coating around a rock fragment in a lacustrine deposit. (J) C.277. Ice-contact meltout deposit. Coating filling embayments of a rock fragment. Such filled-in embayments are very difficult to remove. (K) C.272. Draped bedding in subrecent meltout deposits. (L) C.276. Laminated coatings in subrecent meltout deposits, suggesting that the coatings are derived from drapes as in (K).

Faulting is also present within some of the WES where laminae are seen to be displaced. It has not been observed in any of the tills, although there are most likely some examples in those tills that possess clay bands. Plastic deformation in the sense of folding of laminae has only been observed in some of the lacustrine and WES samples. In most cases, this concerns fine-grained laminae that are folded around larger hard particles.

Given the basaltic, opaque composition and nature of all the samples, it is surprising that any birefringence in

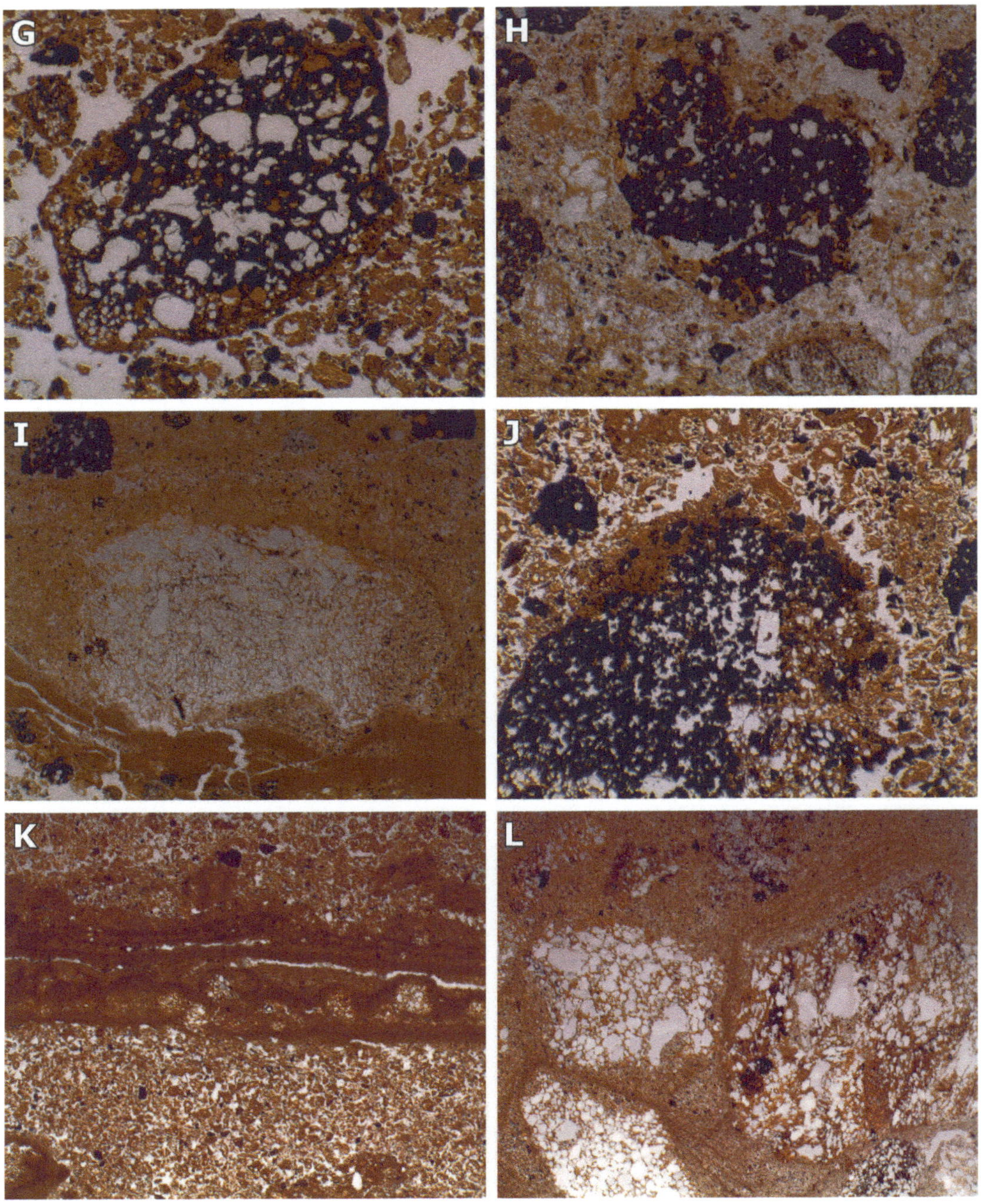

Fig. 10.9. (Continued)

the plasma has been observed and that different types of plasmic fabric can be discerned. Of the four different types observed, a skelsepic plasmic fabric is the most common. It has been observed in some of the meltout samples, where it is restricted to the coatings discussed above. In the subglacial tills, it has been observed a number of times, although in most samples, it is only weak. It furthermore occurs in some of the WES and even in some lacustrine samples. Whereas in subglacial tills it is related to rotational movement, in the lacustrine samples, it is probably related to compaction. In WES, it must be related to the movement of the particles in a slurry. A masepic plasmic fabric has mainly been observed in lacustrine and some WES samples. In this plasmic fabric type, oriented domains are straight and short, and in all the observed cases, these domains are oriented parallel to bedding, demonstrating that it is related to settlement and compaction. A kinking plasmic fabric (Bordonau and van der Meer, 1994) has been observed in some of the till samples and some WES. As this type of plasmic fabric is related to compression (van der Meer, 1993), this is not surprising. Finally, a unistrial plasmic fabric has been observed in a few WES and lacustrine samples. In this type, the oriented domains are long and thin, while sinuous in pattern, and it is related to shearing and faulting. In sample C.466, it is weakly present in one broken up clay layer, which explains why it occurs in a lacustrine deposit.

Because of the weak nature of the plasmic fabric and the inherent difficulty in picturing it properly, we have refrained from showing images of plasmic fabric in the Icelandic samples.

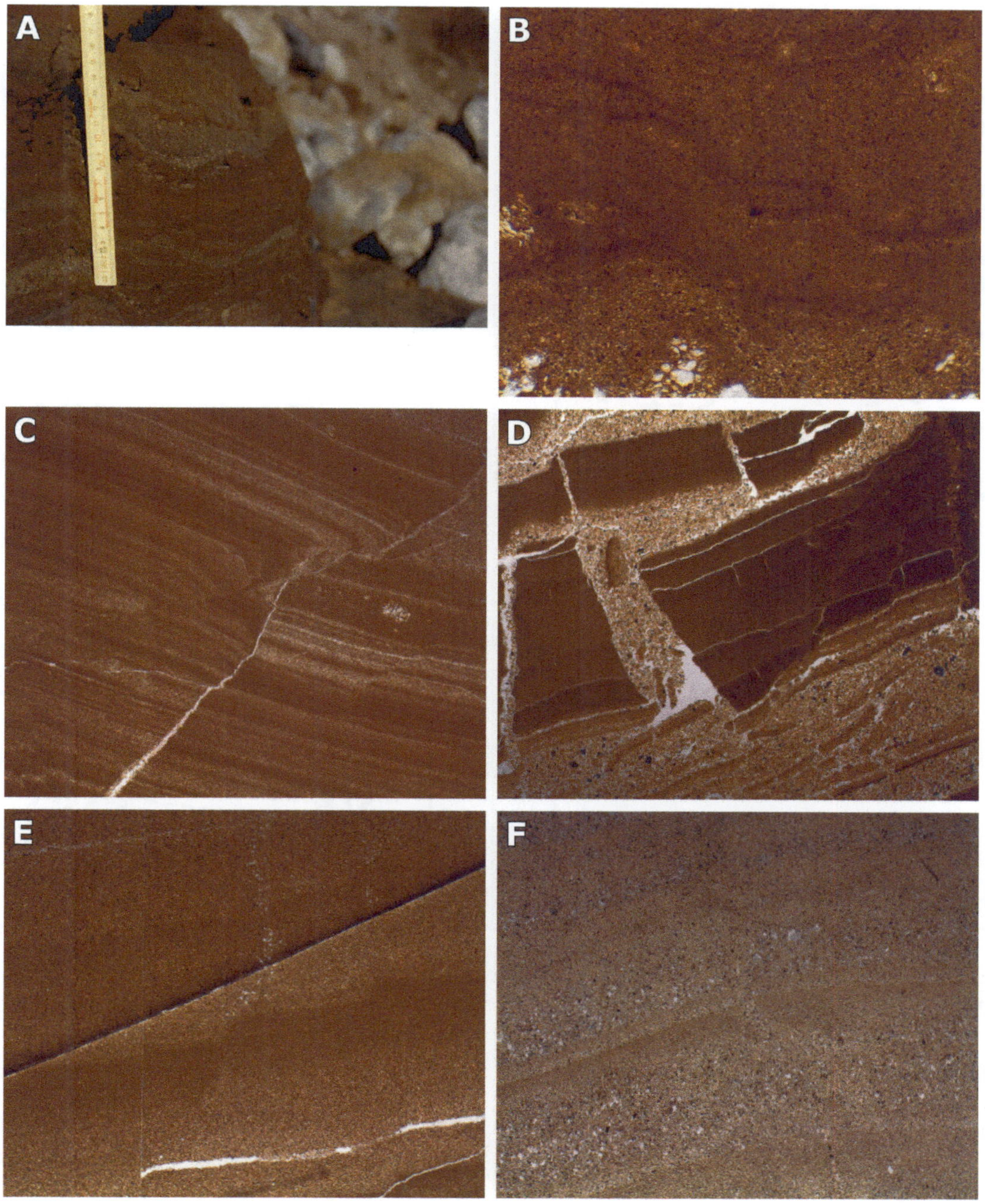

Fig. 10.10. Brittle deformations and water escape. All images are in plane light, horizontal field of view (fov) is 18 mm and the top is up, unless stated otherwise. (A) Pond deposits in the Kötlujökull meltout zone. Note multiple faults, also reflected in micrographs. (B) C.271, fov 2.6 mm. Micro-faulting in subrecent meltout deposits, faulting limited to laminae. (C) C.281. Faulting in subrecent meltout deposits, faulting cutting across multiple laminae. (D) C.282. Disruptive water escape in subrecent meltout tills. (E) C.282. Water escape in subrecent meltout deposits, associated with micro-faulting. In this case, the water escape seems to be highlighted by pores left behind by gas bubbles. (F) C.270. Water escape in lacustrine deposits.

10.6. Discussion

10.6.1. Grainsize and Shape

The grainsize distribution of the samples as observed in thin section is quite variable and that is true for the different groups of sediments. Although the data cannot be compared directly because of different methods, this variability of the tills from Sléttjökull is confirmed by Krüger (1994) and Kjær (1999) who found considerable variation in gravel, sand and silt content, with relatively little variation for the clay content. Because our samples were not collected following transects, it is not possible to see trends like those observed by Kjær (1999). Furthermore, thin section samples avoid gravel-rich deposits as these are almost impossible to sample. The grain size variability in the meltout tills is directly related to the grain size of the englacial debris bands (Eyles, 1979; Krüger and Kjær, 2000). As can be seen in Fig. 10.2A, the debris bands can be traced directly into the overlying

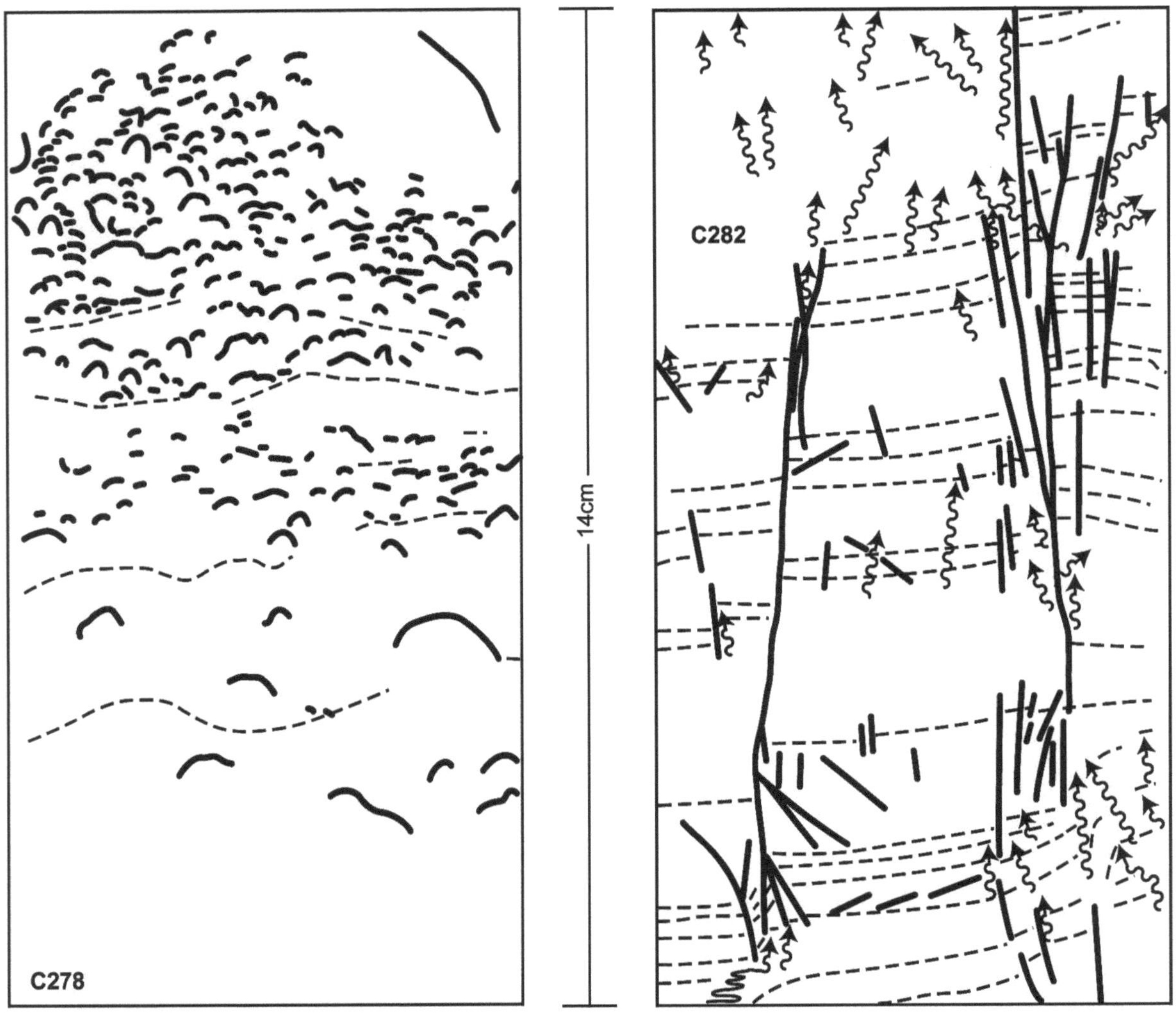

Fig. 10.11. (A) Distribution of coatings in sample C.278, ice-contact meltout deposits. (B) Micro water escape in sample C.282 in pond deposits intercalated in subrecent meltout deposits.

meltout till, and in association with this the incremental thickening of the supraglacial layer. Many of the englacial debris bands are airfall tephra and dust deposits, which have been incorporated in the ice by subsequent snows. There appear to be few basal sediments incorporated in the ice (Krüger and Kjær, 2000). Because bands can be traced from an englacial position into the overlying meltout deposits and down the slope, it is clear that very little mixing occurs within the sediment package. This is also obvious from the thin sections that show little or no mixing across bed boundaries. Consequently, if the package moves downslope on a lubricated, steep ice slope, it does so as a slab. However, as we will see later, this cannot be completely passive.

The finer grain-sized meltout samples come from pond deposits formed in sinkholes on the ice-cored moraines. Thus, it is obvious that they differ in many aspects from the other meltout samples. The subrecent meltout samples come from a more complex setting than the ones collected at the ice contact (C.275-C.276-C.277). Whereas the latter can be said to be single cycle – they have not been disturbed since melting out of the ice – the former ones have been recycled several times before final melting of the ice core. This is not only reflected in the overall structure (Fig. 10.3) but also in the nature of the individual beds as these comprise soliflucted, windblown, free fall, collapse and waterlain sediments.

Something similar is causing the variability of the grain size of the lacustrine deposits. There are no large, deep lakes in the forefield of either glacier, only small, often ephemeral lakes and ponds. These are all strongly influenced by surrounding slopes and the nature of the sediments on these. Given the strong influence of the wind, part of the lacustrine sediments must be airborne. Despite the higher rainfall in the area, visible windblown dust is not uncommon and silty accretion wards were regularly observed on moist surfaces. Whereas on exposed surfaces, such material can easily be removed again, which is not the case in lacustrine settings, unless they become fully drained.

The short transport routes are also highlighted by the angularity of the material. Apart from the short transport route, the angularity is also influenced by the composition of the material. In the finer grain sizes, we find a lot of glass tephra shards that are angular by nature. In the larger grain sizes, it is mainly vesicular basalt, the rock fragments of which can be become nicely rounded. However, upon breakage, they will not produce smooth surfaces but a very irregular surface with many sharp edges. As long as multigrain rock fragments are present, we will find a relatively high angularity. The observed increase in roundness with an increase in grain size is common; it has been observed in all our thin section studies. It must be related to the fact that it is easier to chip protuberances off larger particles compared to small particles. There is a direct consequence to the rough surfaces of most particles. Embayments on the surface tend to get filled (Fig. 10.9J) with fine-grained material (clay, silt), and this fill is very difficult to remove. Thus, it is relatively easy for this material to pick up a coating of fine-grained material forming a strong bond between grain sizes. Once such a coating has been formed, it is difficult to erode completely.

Our samples can be compared to observations reported by Fuller and Murray (2000) from Hagafellsjökull-Vestari and by Nelson *et al.* (2005) from Brúarjökull, both in Iceland. Both cases refer to tills produced by surging glaciers as opposed to the Mýrdalsjökull sediments reported on here. Both teams sampled tills in drumlins and flutes suggesting a strong coupling between glacier and bed. In that sense, we do not expect differences between these surging and non-surging glaciers as in this case they are both coupled to the bed. And although for both surging glaciers, the temporary presence of a waterfilm, decoupling the glacier, is discussed, the same can be said for Sléttjökull. The production of water may have been related to different processes, but the end result is the same.

10.6.2. Porosity

As said before, all samples, except those of the Sólheimajökull tills, have a very high porosity, more than 60% (by estimate) is no exception (Simons, 1998). This has also been measured by Kjær *et al.* (2003), who found that the Sléttjökull tills had porosities between 40 and 60%, while the lower, denser tills had porosities below 40%. This is still very high, as a study by Kilfeather and van der Meer (2008 and references therein) showed tills from Co Laois in Ireland to have a porosity between 1 and 19%, which compared well with porosities reported by Lind and Nyborg (1988) from Sweden. Some samples of immature tills from Co Laois had the highest porosity of the Irish group, which supports the notion that our Mýrdalsjökull tills are immature.

The simple and compound packing voids are –as the name implies – caused by the packing of particles. Their widespread occurrence in supraglacial meltout deposits suggests that after the disappearance of interstitial ice, the resulting space was only partially occupied by other particles, as we can expect that initially they were all compound. There must have been substantial compaction as can be seen in the difference in thickness between englacial debris layers and the continuation in the overlying deposits (Fig. 10.2). This compaction resulted in particles being forced into contact thereby reducing the share of compound packing voids. On the contrary, air must have been introduced into the system following de-icing and drainage of meltwater. This introduction of air resulted in the formation of vesicles (for a description of pore types, see Kilfeather and van der Meer, 2008), similar to vesicles in basalt. The combination of compaction and introduction of air must have led to the trapping of air into bubbles, creating the vesicles' rounded shape. Vesicles have also been observed in meltout tills from Antarctica (Lloyd Davies, 2004). As this is a stable shape, only the bigger ones have been squeezed into vughs by ongoing compaction. Note that all these pore types are also present in the one sample (C.292) that was taken from a basal till layer melting out supraglacially.

Curiously, basal tills in this study – with the exception of those from Sólheimajökull and Kötlujökull – also show a high percentage of simple and compound packing voids, with vesicles and vughs being rare in the Sólheimajökull samples. As mentioned above, the overall porosity is very high, it is as if the tills still occur in a highly dilated state. The question is where the packing voids come from as they are rare in basal tills from other areas (van der Meer *et al.*, 2006). The only reason why this dilated state and the packing voids are still present must be lack of compaction. And the lack of compaction can only be explained by high pore water pressure. As Sléttjökull is also the site of extensive WES (van der Meer *et al.*, 1999, 2009), a high pore water pressure during formation of these tills is highly likely. This high pore water pressure also explains other structural features. High percentages of porosity were found by Ronnert & Mickelson (1992) in basal meltout tills at Burroughs Glacier in Alaska. They did not find any evidence for deformation following slow meltout and concluded that a highly dilated state can be preserved for an unspecified length of time. Unfortunately, the material did not allow producing thin section, and thus, we can only compare numbers.

A high percentage of packing voids as well as the regular occurrence of vesicles and vughs in lacustrine deposits is not surprising. There has been gradual compaction and dewatering following continued sedimentation. As many of the lacustrine deposits sampled here come from ephemeral settings in which the sediments can be expected to have alternated between subaqueous and subaerial conditions, we also have to take the effect of snow into account. Snowfall and its subsequent melting are known to produce vesicles in surficial sediments (Fig. 10.12), although the process of air entrapment in this setting is not clear. But as a consequence, we can expect vesicles and vughs in all lacustrine sediments. The presence of vughs, but no vesicles in WES, must have a different cause. The absence of vesicles is clear as it is not obvious why air entrapment would occur in WES, which, by definition, form by subsurface water under pressure (van der Meer *et al*, 2009). Vughs are known to result not only from the collapse of air bubbles but also by collapse of pores formed by other processes. In case of WES, these initial pores most likely formed by enclosed water pockets.

The presence of channels in all samples is not problematic; these form by compaction and de watering (meltout, WES, lacustrine) as well as shear (WES, basal till). As all of these processes are known to occur in the studied sediments, there is no need for further discussion. In tills, parallel channels are an expression of fissility (Fig. 10.3G), which has also been described by Fuller and

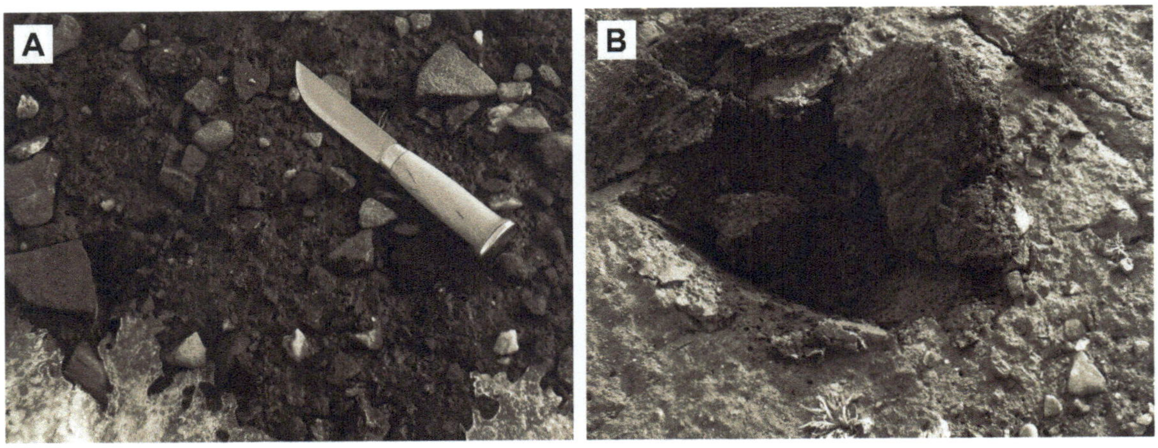

Fig. 10.12. (A) Surface re-emerging after recent snowfall demonstrating the high frequency of vesicles produced during such events. (B) Shallow cut showing that vesicles are not restricted to the immediate surface. Peary Land, northern Greenland.

Murray (2000) and Nelson *et al.* (2005). However, unlike Nelson *et al.*, Fuller and Murray describe fissility as formed by loading and unloading under normal stress and thus do not recognise this structure as formed by deformation, thereby reducing the thickness of their deforming bed.

The only exception is the marble bed structure observed in sample C.455. As discussed by Kilfeather and van der Meer (2008), a marble bed structure is thought to be the result of 'an initial subglacial compression and de-watering of the sediments followed by brittle failure along sub-horizontal planes, brecciation and rotation of aggregates caused by shearing'. This fits very well with the sequence inferred to have taken place in the Slettjökull forefield: a combination of drainage through WES, drying and stiffening of the tills followed by deformation. The only problem with this is that a marble bed structure has not been observed more often in this area. This can only be explained by the strongly dilated state of the tills that may have prevented complete dewatering. The latter may have been helped by interconnected vesicles in the basalt rock fragments.

10.6.3. Bedding

The bedding observed in the meltout deposits is partly related to intercalated waterlain beds, which show normal grading. The latter is also present in sample C.276, which was collected from an ice-contact position. In this case, the normal grading is undoubtedly derived from an airfall tephra layer melting out of the ice. All lacustrine layers show normal grading, as is to be expected. But it is only present in one WES sample, which indicates the difference between WES and lacustrine deposits. As indicated in van der Meer *et al.* (2009), grading in WES is caused by pressure gradients. The fact that it is shown here as normal grading is accidental as it refers to a sample collected over a horizontal bed, thought to be a WES, but could be an extension of a lacustrine bed (Fig. 10.4), the true relations could not be established in the field. Consequently, this sample is listed as WES in Table 10.2. Thin clay beds showing both brittle and plastic deformation have been observed in some of the till samples. Fuller and Murray (2000) explain the preservation of laminated beds in their samples as a lack of pervasive deformation, implying that the latter would destroy them. However, this does not take duration of deformation into account; if it is short, destruction will only be partial. It also does not take into account that as the clay beds are so different in, for instance, hydraulic conductivity, that they behave different from the host sediment, there is no need for destruction. In a study of tectonized sediments in the Netherlands, van der Meer *et al.* (1985) described a clay bed that had been rafted into a steeply inclined position, but in which only the top and basal 2–3 cm of the clay had been deformed. In between, the clays did not show any sign of deformation. Thus, clay beds can sustain pervasive deformation without undergoing intensive deformation themselves.

10.6.4. Deformation

Rotational structures have only been observed in some samples, including one meltout sample. Although turbates are common in many tills, they are not occurring in all of them (van der Meer, 1993; van der Meer *et al.*, 2003) nor are they exclusive to subglacial tills. All sediments that have undergone flow with different velocities – either random as in some debris flows or vertically arranged as in tills – demonstrate turbate structures. Usually, we can differentiate on the basis of frequency of turbate structures and the combination with other microstructures. In the Mýrdalsjökull tills, they are not only not occurring in all samples they are also infrequent when they occur. Van der Meer *et al.* (2003) argued that with increasing water content, we would see the disappearance of microstructures as particles more and more start to move independently. The highly dilated tills in this study support this argument. Rotational structures are also described by Fuller and Murray (2000) and Nelson *et al.* (2005), both indicating that such structures are rare in the upper parts of their sections. They ascribe this to a higher water content of these parts of the sequences. It should be noted that the coatings are also suggesting rotation, but when they occur all around particles that is more difficult to demonstrate as a change in orientation cannot be established. Consequently, this is not indicated in Table 10.2, although their widespread occurrence is indicative of rotation being more widespread than can be deduced on the basis of turbate structures alone.

10.6.5. Aggregates and Coatings

Aggregates in the sense of structural elements with the same composition as the host sediment, but separated by structural planes, occur in many meltout and till samples. Such aggregates can be derived from englacial debris bands because they had been produced before being incorporated, which is the most likely source in meltout deposits. In subglacial tills, they can be produced by incorporation as such, by breaking up cohesive layers like clay bands or by shearing of the till bed resulting in a marble bed structure as indicated above.

Coatings are another type of aggregate in the sense that they have literally been produced through aggradation of small particles onto the surface of large particles. They are not to be confused with the very thin layer of fine material that is always seen to separate skeleton particles in tills (van der Meer *et al.*, 2003). In the meltout samples (except for the waterlain intrabeds), the coatings sometimes surround the whole grain. The latter suggests that the grains have been rotating while accumulating the coatings, as each part of the surface must have been able to accumulate material (Kilfeather *et al.*, in press). Furthermore, this must have happened in a watery environment in which grains must have moved through or washed by pockets of different grain size. Where coatings are occurring on one side of the particle, only other processes come to mind. In the first place, it can be due to illuviation of fines, although the resulting cutans are usually found in pores, not on grain surfaces. The most likely process is the one described by, for instance, van Vliet-Lanoë *et al.* (1984), in which freeze/thaw processes combined with slope movement play a prominent role. We cannot exclude that during meltout, there have been regular cycles from thawing to freezing and vice versa. This would be very similar to nonglacial periglacial processes enhanced by sliding of the material over a wet, inclined ice surface. This makes these coatings unlike the ones described by Kilfeather *et al.* (in press) from marine debris flows on (glacial) trough-mouth fans. Alternatively, the coatings may form through the meltout of thin, fine-grained debris bands, which form a drape over the underlying material. If the latter consists of (widely) spaced grains, the drape will break up, adhering to the surface of the grains, while part of the drape may sink in between grains. Any movement of the grains will then produce coated grains. However, we do not claim to fully understand the processes that form these coatings.

The presence of coatings within a relatively large number of till samples is curious as they are not known from any non-Icelandic subglacial till. Similar coatings have been described from Iceland by Fuller and Murray (2000) and Nelson *et al.* (2005). In both publications, silt and/or mixed fine-grained coatings (partial and complete) as well as fines filling embayments on grains are described very similar to those in our samples. The presence of coatings in our till samples can be explained by different processes. In the first place, we know that both Kötlujökull and Sléttjökull have retreated and advanced in historic times (e.g. Kjær *et al.*, 2003). During retreat, supraglacial meltout deposits (containing aggregates) would be spread over the surface. During advance, such deposits would be reworked and partly incorporated in the newly formed till. Given the strong bond between skeleton particle and coating, the latter can at least partly be preserved.

The partial preservation of coatings in a deforming bed at Mýrdalsjökull is no surprise as the highly dilated nature of the bed would prevent forceful interparticle relations. Although van der Meer *et al.* (1999) reconstructed a period of drained and stiffened tills at Sléttjökull, this does not make the preservation of coatings impossible as this was also a period of non-pervasive deformation in the till bed.

Nelson *et al.* (2005) discuss the possibility of basal meltout in the formation of their tills but conclude that they must be a combination of lodgement and deformation. Similar clayey and dusty clayey coatings have been described from Taylor Glacier in Antarctica by Lloyd Davies (2004). The coatings he observed occur either partly or fully around skeleton grains and sometimes possess a skelsepic plasmic fabric. The microstructures were observed in a number of samples from what had previously been described as sublimation till, but what was found instead to be a meltout deposit effected by a complex array of displacement processes, very much like the processes observed at Kötlujökull.

The presence of coatings in WES and lacustrine deposits merely demonstrates that these coated particles must be widespread in this environment.

10.6.6. Plasmic Fabric

The weak presence of plasmic fabrics can be explained in two ways. The first one is the odd composition of these sediments in relation to other places. Whereas we are normally dealing with multiple lithologies in glacial sediments, in this case, they are almost mono-lithological. Apart from the basalt, there are some other volcanic crystalline rocks and there may be some clay minerals. As the basalt is opaque under crossed polarizers, plasmic fabrics simply cannot be observed unless there are clay minerals involved. The other reason for the absence of plasmic fabrics is a low strain level during deformation (Fuller and Murray, 2000). As discussed before, the effective stresses have not been very high because of the high pore water pressure, and this would hardly have affected the sediments. In the case of Fuller and Murray, they were dealing with a very low clay content – at only 2% much lower than in the Mýrdalsjökull samples – which is below the minimum requirements for the development of a plasmic fabric (van der Meer *et al.*, 2006).

The observed plasmic fabric is strongest in some of the coatings, which is where we expect them, especially the skelsepic plasmic fabric. Most other types of plasmic fabric have been observed in clay bands in till, and as these are regularly broken up, the masepic and sometimes kinking plasmic fabrics are the result of shearing. Other than that, plasmic fabric occurs a number of times in WES and lacustrine deposits. A masepic plasmic fabric in lacustrine deposits is usually the result of the reorientation of clay particles during compaction. As WES develop under pressure, the presence of plasmic fabrics in clay-rich layers is as expected.

10.6.7. Differentiation

When analysing Table 10.2 with the abovementioned criterion in mind, it is clear that we are dealing with three groups of sediments: (i) meltout deposits, (ii) subglacial tills and (iii) waterlain sediments. The meltout deposits can be said to be characterised micromorphologically by a high porosity (especially simple packing voids, channels and vughs with vesicles a normal presence), bedding, frequent aggregates and coatings and almost no plasmic fabrics. Within this group, we must realise that there are actually two subgroups: (i) the ice-contact samples and (ii) the subrecent completely de-iced samples. Where the first set is on its first cycle, just melted out and still overlying ice, the latter have a much more complex history. These sediments can be expected to have experienced several cycles of relocation while still overlying ice, interspersed by pond formation and aeolian accumulation. From Table 10.2 it is clear that some bands still have all or most of the characteristics of the first group, some others do not. In the Mýrdalsjökull settings, none of these groups has a high survival potential as most are swept away by meltwater; none of our sample sites exists any more.

Basal tills are characterised by high porosity (especially, packing voids and channels, occasionally, vughs), some rotational structures, common aggregates and/or packings and occasional skelsepic plasmic fabric. The three Sólheimajökull samples can be separated from the other till samples by showing least microstructure. Based on the microscopic observations, all tills have been deformed and thus have been part of a deforming bed.

The lacustrine and WES deposits appear to be comparable but can still be separated. Whereas lacustrine deposits exhibit all types of pores, including vesicles, the latter are lacking in the WES. Bedding is present in both, but the lacustrine deposits are characterised by normal grading. Lacustrine deposits also appear to be slightly more often affected by brittle deformation following compaction. On the contrary, the WES more often exhibit plasmic fabric, which also tend to be different from the masepic plasmic fabric of lacustrine deposits.

10.6.8. Deforming Bed

There is a continuous discussion about the thickness of the deforming bed (van der Meer *et al.*, 2003). Fuller and Murray (2000) come to the conclusion that they have evidence for a thickness of only 20 cm. However, as stated above, ascribing fissility to a non-deformational process automatically leads to such a shallow deforming layer. Nelson *et al.* (2005) come to the conclusion that the thickness of the deforming bed they are dealing with must have been between 50 and 90 cm, encompassing the full extent of their till beds. Our own studies at Brúarjökull (Kjær *et al.*, 2006) demonstrated that by detaching the sediment pile from the underlying bedrock, a deforming bed can be many metres thick. Truffer *et al.* (2000) also described deformation deep in the till bed, with no deformation recorded above this level. Both studies indicate that deformation can occur at different levels leaving parts or all of the overlying sediments uneffected. Our studies at Mýrdalsjökull reveal that tills have been deformed over their whole thickness, that is, it is more than 1 m. As the till profiles at Sléttjökull demonstrate that especially the higher part of the black WES has been intensely deformed, suggesting that this happened in a thinner deforming bed. However, there is enough evidence that the lower parts of the profile also have been deformed. This agrees with a decrease in deformation with increasing depth. It also agrees with the variability in space and time as discussed by van der Meer *et al.* (2003).

10.7. Conclusion

On the basis of the description and discussion above, we can come to the following conclusions:

- All glacial deposits from Mýrdalsjökull are highly porous.
- The porosity is partly determined by the local, vesicular basalt bedrock.
- In meltout deposits, the porosity consists of simple packing voids, channels and vughs with vesicles a normal presence.
- In subglacial tills, the porosity consists of packing voids, channels, occasionally, vughs.
- Meltout deposits are furthermore characterised by the widespread occurrence of coatings on skeleton grains.
- These coatings are the result of an ill-understood accretional process in sediments overlying ice in an unstable position.
- Coatings are difficult to destroy as evidenced by their presence in subglacial tills, WES and lacustrine deposits.
- Subglacial tills possess some rotational structures, common aggregates and/or packings.
- Because of the basalt lithology, plasmic fabrics are difficult to detect, if formed at all.
- Subglacial tills have all been deformed and have thus been part of a deforming bed.

Acknowledgements

JvdM would like to thank Johannes Krüger and Kurt Kjær to invite him to participate in the fieldwork at

Mýrdalsjökull and all the participants in the 1996, 1997 and 2001 field sessions for their good company. Furthermore, he would like to thank Jeroen Wijnen, Victor Simons and Aoibheann Kilfeather for helping with the sampling, preparation and description of the samples. Frans Backer is thanked for producing good quality thin sections, and Ed Oliver is thanked for his meticulous help with the figures.

References

Bordonau, J. & van der Meer, J.J.M., 1994. An example of a kinking microfabric in Upper Pleistocene glacio-lacustrine deposits from Llavorsi (Central Southern Pyrenees, Spain). Geologie en Mijnbouw 73, 23–30.

Boulton, G.S. & Hindmarsh, R.C.A., 1987. Sediment deformation beneath glaciers: rheology and geological consequences. Journal of Geophysical Research 92, 9059–9082.

Brewer, R., 1976. Fabric and mineral analysis of soils. Huntington, N.Y, R.E. Krieger Publ. Com.

Carr, S.J. & Lee, J.R., 1998. Thin section production of diamicts: problems, and solutions. Journal of Sedimentary research 68, 217–221.

Croot, D.G., 1988. Morphological, structural and mechanical analysis of Neoglacial ice-pushed ridges in Iceland. In: Croot, D.G. (Ed), Glaciotectonics. Forms and processes. Rotterdam, A.A. Balkema, 33–47.

Dreimanis, A., 1988. Tills: their genetic terminology and classification. In: Goldthwait, R.P., Matsch, C.L. (Eds), Genetic classification of glacigenic deposits. Rotterdam, Balkema, 17–83.

Evans, D.J.A., 2005. The glacier-marginal landsystems of Iceland. In: Caseldine, C., Russell, A., Harðardóttir, J., Knudsen, Ó. (Eds), Iceland – modern processes and past environments. Developments in Quaternary Science, 5, Amsterdam, Elsevier, 93–126.

Evans, D.J.A., Phillips, E.R., Hiemstra, J.F., & Auton, C.A., 2006. Subglacial till: formation, sedimentary characteristics and classification. Earth-Science Reviews 78, 115–176.

Eyles, N., 1979. Facies of supraglacial sedimentation on Icelandic and Alpine temperate glaciers. Canadian Journal of Earth Sciences 16, 1351–1361.

Fuller, S., Murray, T., 2000. Evidence against pervasive bed deformation during the surge of an Icelandic glacier. In: Maltman, A.J., Hubbard, B., Hambrey, M.J. (Eds.), Deformation of glacial materials. Geological Society Special Publication No. 176, London, 203–216.

Kilfeather, A.A., Ó Cofaigh, C., Dowdeswell, J.A., van der Meer, J.J.M., & Evans, D.J.A., in press. Micromorphological characteristics of glacimarine sediments: implications for distinguishing genetic processes of massive diamicts. Geo-Marine Letters.

Kilfeather, A.A. & van der Meer, J.J.M., 2008. Pore size, shape and connectivity in tills and their relationship to deformation processes. Quaternary Science Reviews 27, 250–266.

Kjær, K.H., 1999. Mode of subglacial transport deduced from till properties, Mýrdalsjökull, Iceland. Sedimentary Geology 128, 271–292.

Kjær, K.H. & Krüger, J., 2001. The final phase of dead-ice moraine development: proceses and sediment architecture, Kötlujökull, Iceland. Sedimentology 48, 935–952.

Kjær, K., Krüger, J., & van der Meer, J.J.M., 2003. What causes till thickness to change over distance? Answers from Mýrdalsjökull, Iceland. Quaternary Science Reviews 22, 1687–1700.

Kjær, K.H., Larsen, E., van der Meer, J.J.M., Ingólfsson, Ó., Krüger, J., Benediktsson, Í.Ö., Knudsen, C.G., & Schomacker, A., 2006. Subglacial decoupling at the sediment/bedrock interface – a new mechanism for rapid flowing ice. Quaternary Science Reviews 25, 2704–2712.

Kjær, K.H., Sultan, L., Krüger, J., & Schomacker, A., 2004. Architecture and sedimentation of outwash fans in front of Mýrdalsjökull ice cap, Iceland. Sedimentary Geology 172, 139–163.

Krüger, J., 1994. Glacial processes, sediments, landforms, and stratigraphy in the terminus region of Mýrdalsjökull, Iceland.. Folia Geographica Danica 21, 1–233.

Krüger, J., 1995. Origin, chronology and climatological significance of annual- moraine ridges at Mýrdalsjökull, Iceland.. The Holocene 5, 420–427.

Krüger, J. & Kjær, K.H., 2000. De-icing progression of ice-cored moraines in a humid, subpolar climate, Kötlujökull, Iceland.. The Holocene 10, 737–747.

Krüger, J., Kjær, K.H., & van der Meer, J.J.M., 2002. From push moraine to single-crested dump moraine during a sustained glacier advance. Norsk Geografisk Tidsskrift 56, 87–95.

Krüger, J. & Thomsen, H.H., 1984. Morphology, stratigraphy, and genesis of small drumlins in front of the glacier Mýrdalsjökull, south Iceland. Journal of Glaciology 30, 94–105.

Lee, J., Kemp, R., 1993. Thin sections of unconsolidated sediments and soils: a recipe. Centre for Environmental Analysis and Management (CEAM), Royal Holloway, University of London, CEAM Technical Report, 2, 1–32.

Le Heron, D.P. & Etienne, J.L., 2005. A complex subglacial clastic dyke swarm, Sólheimajökull, southern Iceland. Sedimentary Geology 181, 25–37.

Lind, B.B. & Nyborg, M.R., 1988. Sediment structures and the hydraulic conductivity in till. Urban Geohydrology Group 83, Chalmers Tekniska Högskola, Göteborg, 1–64.

Lloyd Davies, M.T., 2004. A polar paradise: the glaciation of South Victoria Land. Antarctica. PhD thesis, University of Amsterdam, 304 pp.

Menzies, J., 2000. Micromorphological analyses of microfabrics and microstructures indicative of deformation processes in glacial sediments. In: Maltman, A.J., Hubbard, B., Hambrey, M.J., (Eds), Deformation of glacial materials. Geological Society Special Publication No. 176, London, 245–257.

Murphy, C.P., 1986. Thin section preparation of soils and sediments. Berkhamsted, AB Academic Publishers, 149 pp.

Nelson, A.E., Willis, I.C., & Ó Cofaigh, C., 2005. Till genesis and glacier motion infered from sedimentological evidence associated with the surge-type glacier, Brúarjökull, Iceland. Annals of Glaciology 42, 14–22.

Ronnert, L. & Mickelson, D.M., 1992. High porosity of basal till at Burroughs Glacier, southeastern Alaska. Geology 20, 849–852.

Simons, V., 1998. Micromorphology of presentday glacial sediments from Mýrdalsjökull, Iceland. Unpublished MSc thesis, FGBL, University of Amsterdam.

Todtmann, E.M., 1960. Gletscherforschungen auf Island (Vatnajökull). Universität Hamburg, Abhandlungen aus dem Gebiet der Auslandskunde, Band 65, Reihe C (Naturwissenschaften), Band 19, 95 pp.

Truffer, M., Harrison, W.D., & Echelmeyer, K.A., 2000. Glacier motion dominated by processes deep in underlying till. Journal of Glaciology 46, 213–221.

van der Meer, J.J.M., 1993. Microscopic evidence of subglacial deformation. Quaternary Science Reviews 12, 553–587.

van der Meer, J.J.M., 1997. Particle and aggregate mobility in till: microscopic evidence of subglacial processes. Quaternary Science Reviews 16, 827–831.

van der Meer, J.J.M., Kjær, K., & Krüger, J., 1999. Subglacial water escape structures and till structure, Sléttjökull, Iceland. Journal of Quaternary Science 14, 191–205.

van der Meer, J.J.M., Kjær, K., Krüger, J., Rabassa, J., & Kilfeather, A.A., 2009. Under pressure: clastic dykes in glacial settings. Quaternary Science Reviews 28, 708–720.

van der Meer, J.J.M., Menzies, J., & Rose, J., 2003. Subglacial till: the deforming glacier bed. Quaternary Science Reviews 22, 1659–1685.

van der Meer, J.J.M. & Menzies, J., submitted. The micromorphology of unconsolidated sediments. Earth Science Reviews.

van der Meer, J.J.M., Rappol, M., & Semeyn, J.N., 1985. Sedimentology and genesis of glacial deposits in the Goudsberg, Central Netherlands. Mededelingen van de Rijks Geologische Dienst 39-2, 1–29.

van der Meer, J.J.M. , Rose, J., Menzies, J., & Domack, G., 2006. Micromorphology of glacial sediments, a workshop. Sixth International Workshop on the Micromorphology of Glacial Sediments, Department of Geosciences, Hamilton College, NY, USA.

van Vliet-Lanoë, B., Coutard, J.-P., & Pissart, A., 1984. Structures caused by repeated freezing and thawing in various loamy sediments: a comparison of active, fossil and experimental data. Earth Surface Processes and Landforms 9, 553–565.

11

Volcanogenic Jökulhlaups (Glacier Outburst Floods) from Mýrdalsjökull: Impacts on Proglacial Environments

Andrew J. Russell[1,*], Robert Duller[2] and Nigel P. Mountney[3]

[1]*School of Geography, Politics & Sociology, Newcastle University, Claremont Road, Newcastle Upon Tyne NE1 7RU, UK*
[2]*Department of Earth Sciences and Engineering, Imperial College, Royal School of Mines Building, London SW7 2AZ, UK*
[3]*Earth Sciences, School of Earth and Environment, University of Leeds, Leeds LS2 9JT, UK*
**Correspondence and requests for materials should be addressed to Andrew J. Russell (e-mail: andy.russell@ncl.ac.uk)*

11.1. Introduction

The Mýrdalsjökull ice-cap and its outlet glaciers have been the source of numerous glacier outburst floods or 'jökulhlaups' during modern, historic and prehistoric/geologic time (Fig. 11.1). Although some jökulhlaups from Mýrdalsjökull result from the drainage of ice-dammed lakes, most are generated by the Katla subglacial volcano, either by eruptions or by enhanced geothermal activity, and are termed 'Katlahlaups'. The modern proglacial landscape surrounding Mýrdalsjökull records the geomorphological and sedimentary legacy of Katlahlaups during Holocene (Larsen *et al.*, 2005) and historic times. Future Katlahlaups are a significant hazard to surrounding communities and infrastructure (Guðmundsson *et al.*, 2005; Elíasson *et al.*, 2006, 2007). Lower rates of sediment delivery to the coastline between major jökulhlaups will increase coastal erosion rates in the town of Vík, providing increased risk. Offshore, large-scale erosional and depositional impacts reflect cumulative evidence of jökulhlaup activity over Quaternary timescales (Lacasse *et al.*, 1995, 1996; Thordarson and Höskuldsson, 2002), highlighting the importance of Katlahlaups as a major source of sediment to the North Atlantic. The ability of subglacial volcanic activity to generate an almost instantaneous supply of both water and sediment is demonstrated by sediment-laden Katlahlaups which are released from the glacier margin almost immediately (<1 h) following Katla eruptions (Jónsson, 1982; Höskuldsson and Sparks, 1997; Guðmundsson and Högnadóttir, 2006). Katlahlaups have been used as a model for 'volcano-glacial' jökulhlaups (Thórarinsson, 1957, 1975; Rist, 1983) and 'volcano-glacial debris flows' (Jónsson, 1982). Subsequently, Judith Maizels used Katlahlaups and their deposits as the basis for a sedimentary model of 'volcano-glacial – Type III' sandur dominated by, relatively debris-rich, hyperconcentrated flows (e.g. Maizels 1989a, 1989b, 1991, 1993, 1997; Maizels and Russell, 1992). Recent work however has reinterpreted the same sedimentary evidence and provides a revised interpretation of much more dilute, fluidal flows (Duller, 2007; Duller *et al.*, 2008; Duller *et al.*, in press). It is clear that the proglacial area surrounding Mýrdalsjökull represents an important type-site for volcano-glacial jökulhlaups and holds the key to our understanding of jökulhlaup flow and depositional and geomorphic processes. This paper therefore provides a timely overview of current understanding of the characteristics and impacts of jökulhlaups from Mýrdalsjökull and its outlet glaciers.

11.2. Volcanic and Jökulhlaup History

The Katla Volcanic System (KVS) constitutes the most southerly mainland portion of the Eastern Volcanic Zone (Jakobsson, 1979; Larsen, 2000; Mattsson and Höskuldsson, 2003) and encompasses the Eldgjá and Katla fissures (Fig. 11.1). Holocene volcanism within the KVS has been focused along different parts of the main fissure system (Larsen *et al.*, 1998). By far the most frequent eruptive activities are highly explosive magma–water interactions, producing widespread tephra fall and giving rise to jökulhlaups (Larsen *et al.*, 1998, 2001). Rapid retreat of glacier ice following the Preboreal period, approximately cal. 9700 yrs BP, resulted in rapid isostatic rebound, estimated to have taken place in less than 1,200 years (Norðdahl, 1990; Guðmundsson, 1997; Thordarson and Höskuldsson, 2002; Norðdahl and Pétursson, 2005). Following recession of glaciers from their Preboreal limits, large areas of glacial outwash plain or 'sandur' aggraded and prograded rapidly into the North Atlantic, forming the constructional coastal landscape on Iceland's present southern coast.

The Katla volcano lies under the centre of the Mýrdalsjökull ice-cap and constitutes the most southerly portion of the KVS. The ice-filled Katla caldera is host to the Katla fissure and the Katla central volcano (1,450 m a.s.l.), which is 10–15 km in diameter and 600–700 m deep (Björnsson *et al.*, 1993, 2000). The Katla central

DEVELOPMENTS IN QUATERNARY SCIENCES
VOLUME 13 ISSN 1571-0866

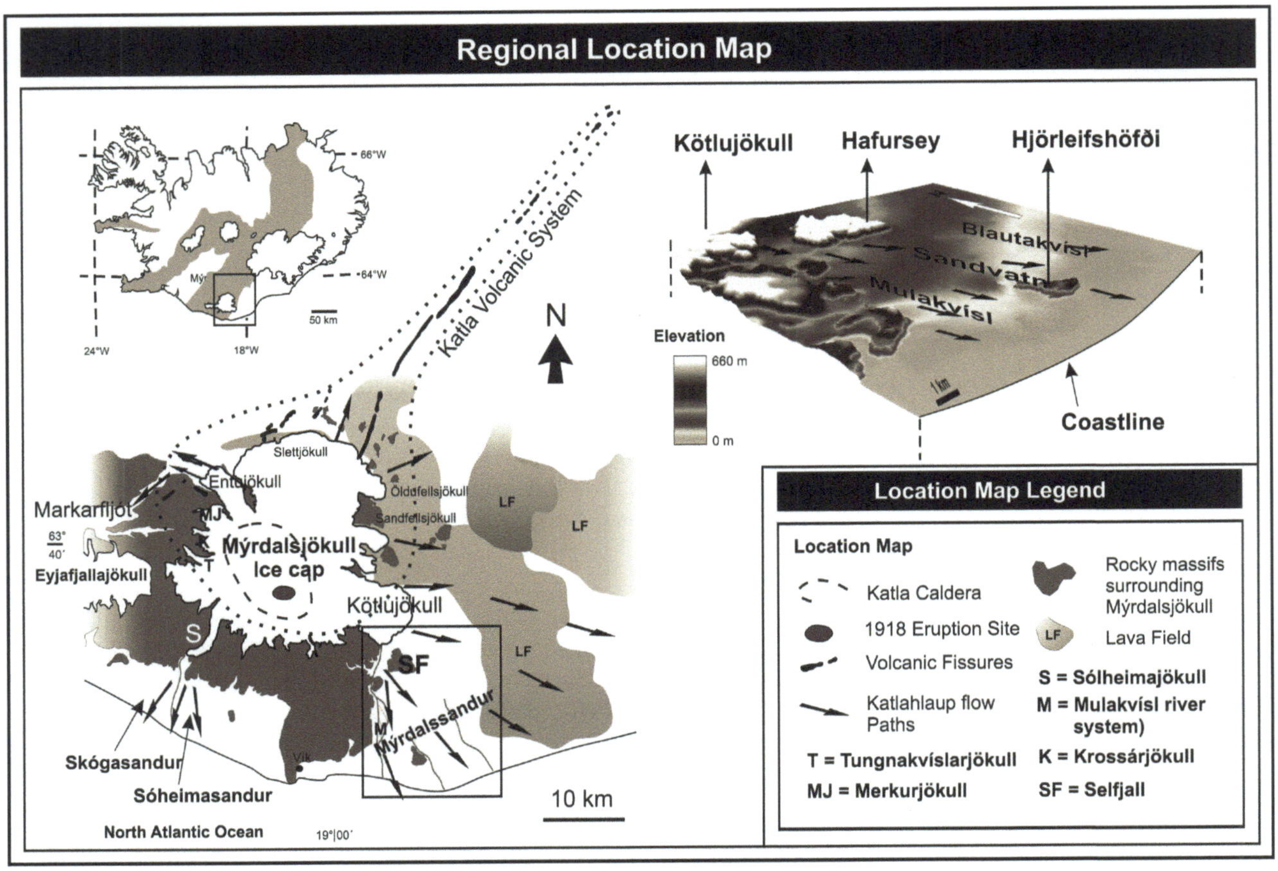

Fig. 11.1. Map of Mýrdalsjökull showing location within Iceland as well as main outlet glaciers and position of major volcanic features (modified after Duller et al., 2008).

volcano is the second most active volcano in Iceland, being the locus of predominantly basaltic volcanic activity on the KVS over the past 200,000 years (Larsen, 2000; Thordarson and Höskuldsson, 2002). The earliest recorded evidence of Katla volcanism dates back to AD 930, with the most reliable evidence, in the form of eyewitness accounts, dating from 1625 onwards (Thórarinsson, 1975; Jónsson, 1982). The duration of Katla eruptions documented since 1625 has varied from 2 weeks to over 5 months, and the last three Katla eruptions to have melted completely through the ice cover lasting between 20 and 28 days (Larsen, 2000). The recurrence interval for Katla eruptions varies from 35 to 80 years, with a longer return interval after the major tenth century Eldgjá fissure eruption, exceeding 200 years. However, eruptions during historic times (past 1,000 years) have occurred approximately twice a century and are commonly accompanied by sudden, high-magnitude jökulhlaups. A similar eruption frequency since ca. 7000 ^{14}C yrs BP is implied by the number of tephra layers in proximal soil sections (Larsen, 2000) and it is therefore no surprise that the calculated probability of a Katla eruption within 10 years is 20% (Elíasson *et al.*, 2006). Katla eruptions are the most voluminous and devastating in Iceland (Thórarinsson, 1957, 1975; Tómasson, 2002), producing very large volumes of meltwater as the magma and superheated water/steam melts through ~450 m of ice cover (Thordarson and Höskuldsson, 2002; Höskuldsson and Sparks, 1997; Larsen, 2000). Katla subglacial eruptions swiftly melt large volumes of ice, producing some of the largest floods in Iceland (Thórarinsson, 1957, 1975; Tómasson, 1996, 2002). Eruptions may be effusive in their initial stages, producing large volumes of water due to either the low volatile content of the magma (Höskuldsson and Sparks, 1997) or the confinement of volatiles by hydrostatic pressure imposed by the overlying water and ice column (Höskuldsson and Sparks, 1997; Smellie, 2000). At this stage a chasm full of water will result (Cas and Wright, 1987; Björnsson, 2002), which is able to overcome the hydrostatic pressure (Thórarinsson, 1959), causing rapid fracturing of the ice and rapid vesiculation and fragmentation of magma (Cashman *et al.*, 2000). Rapid heat transfer during Katla eruptions may be due to temporary storage of water at the start of volcanic activity, into which magma continues to erupt (Björnsson, 2002).

Katlahlaups have drained the Katla subglacial caldera via three low points occupied by the outlet glaciers Kötlujökull, Sólheimajökull and Entujökull (Björnsson *et al.*, 2000), impacting the sediments and landforms on Mýrdalssandur, Sólheimasandur/Skógasandur and Markarfljótsaurar, respectively (Haraldsson, 1981; Tómasson, 2002; Guðmundsson *et al.*, 2005; Russell *et al.*, 2005) (Fig. 11.2). The Markarfljótsaurar contains well-defined erosional and depositional evidence for Katlahlaups (Haraldsson, 1981; Björnsson *et al.*, 2000; Tómasson, 2002; Smith and Haraldsson, 2005; Smith and Dugmore, 2006). van der Meer *et al.* (1999) provided sedimentary evidence of subglacial jökulhlaup discharge to the northern margins of Mýrdalsjökull. The Eldgjá

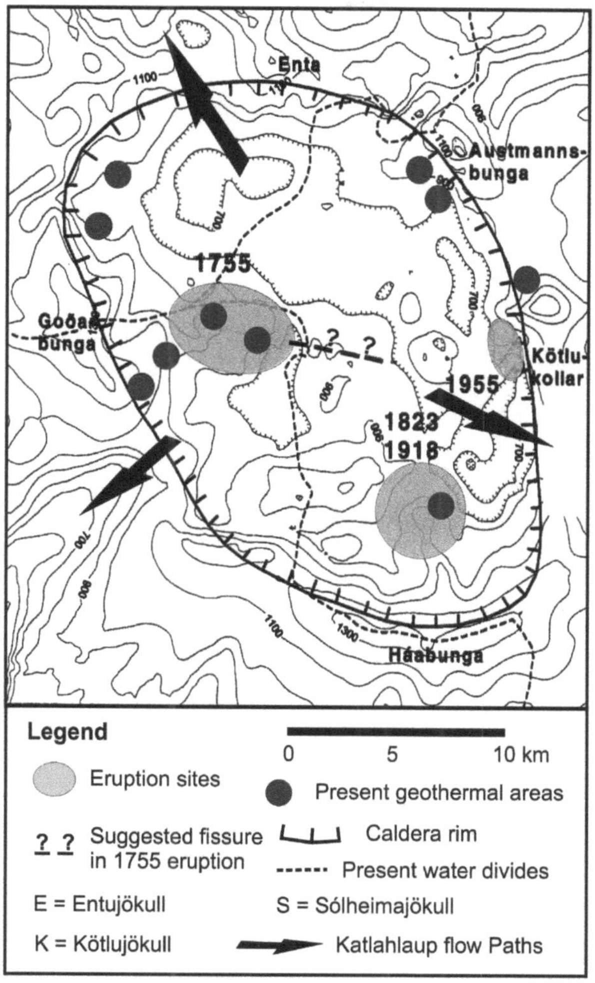

Fig. 11.2. Locations of the 1755, 1823, 1918 and 1955 Katla eruptions beneath Mýrdalsjökull (after Björnsson et al., 2000).

fissure eruption in AD 934 extended under the glacier-covered outer rim of the Katla caldera, generating jökulhlaups from Sandfellsjökull and Öldufellsjökull that drained to the northeast and east into the river Holmsá (Björnsson *et al.*, 2000; Larsen, 2000; Guðmundsson and Högnadóttir, 2006). A change in Katlahlaup routing may be a net result of the interplay between several variables, including changes in the location geometry of the eruptive centre, in the glacier configuration, and in the absolute magnitude of the eruption and ensuing jökulhlaup (Björnsson *et al.*, 2000; Larsen, 2000; Elíasson *et al.*, 2006).

11.3. Jökulhlaups from Kötlujökull Draining across Mýrdalssandur

11.3.1. Introduction

Jökulhlaups draining across Mýrdalssandur usually emerge from the snout of Kötlujökull from two main loci, one in the south and one in the east (Figs. 11.1–11.3). The flows of 1660, 1721 and 1755 are believed to be the most voluminous and violent in history (Magnússon, 1626; Larsen, 2000). Although volumetrically less significant, records indicate that sequence of events during the 1918 eruption was similar to those of the 1721, 1755 and 1823 eruptions (Jónsson, 1982).

11.3.2. The 1918 Katlahlaup

The 1918 eruption of Katla volcano generated a jökulhlaup that inundated ca. 400 km^2 of Mýrdalssandur (Fig. 11.1). The rheology of the 1918 jökulhlaup is somewhat ambiguous, and has been categorised as both a water flow (Björnsson, 1993; Tómasson, 1996) and a hyperconcentrated flow (Maizels 1989a, 1989b, 1991, 1993, 1997). Jónsson (1982) presented a model for Katla jökulhlaups as volcanogenic debris flows (Fig. 11.4), suggesting their similarity to debris flows. Eyewitness accounts of the jökulhlaup, however, point to a more turbulent fluidal flow than laminar debris flow (Tómasson, 1996).

Jökulhlaups inundating Mýrdalssandur typically rise to peak discharge within hours of the onset of an eruption (Jónsson, 1982; Guðmundsson and Högnadóttir, 2006). Although most of the volume of Katlahlaups is discharged subglacially from Kötlujökull, the onset of some jökulhlaups has been witnessed breaking through large fractures in the outlet glacier and travelling supraglacially for distances of up to 10 km to the snout of Kötlujökull (Jónsson, 1982; Tómasson, 1996; Roberts, 2005; Guðmundsson and Högnadóttir, 2006). This sequence of events was observed during the 1918 jökulhlaup, which initially broke out supraglacially shortly before subglacial discharge commenced from the snout of Kötlujökull (Tómasson, 1996; Roberts, 2005). The passage of the subglacial flood waves to the snout of Kötlujökull typically results in hydrofracturing of glacier ice leading to the removal of large parts of the glacier snout and liberating large volumes of ice blocks including individual blocks up to 60 m in diameter (Fig. 11.5).

The 20 h long 1918 Katla eruption released a very large jökulhlaup onto Mýrdalssandur that reached a peak discharge estimated as 1×10^5 m^3/s to 4×10^5 m^3/s (Hannesson, 1934; Thórarinsson, 1957, 1974; Tómasson, 1996, 2002) and 1.6×10^6 m^3/s (Jónsson, 1982; Maizels, 1992, 1993). The most detailed eyewitness accounts of the 1918 Katla eruption and jökulhlaup are given by Jóhannsson (1919) and Sveinsson (1919). These accounts have been translated and discussed by several authors (Jónsson, 1982; Tómasson 1996; Larsen, 2000; Thordarson and Höskuldsson, 2002), as follows: 'At 13:00, October 12th, 1918, a sudden sharp earthquake was felt in the village of Vík, which signalled the onset of subglacial eruptive activity. This initial earthquake was followed by tremors for about 30 minutes (Sveinsson, 1919). Subaerial eruptive activity commenced at around 1500 h as observed from Vík and Hjörleifshöfði, producing an eruption cloud that reached an elevation of 14 km a.s.l. as measured from Reykjavík (Eggertsson, 1919)' (Larsen, 2000). Jóhannsson (1919) recalls: 'at the same time as the eruption cloud was noticed here, or a little later, the glacier flood could be seen racing along the course of the Múlakvísl all the way to the sea'. From Hjörleifshöfði, shortly after he saw the eruption cloud, Markússon recalls: 'a terrible flood had then broken away between

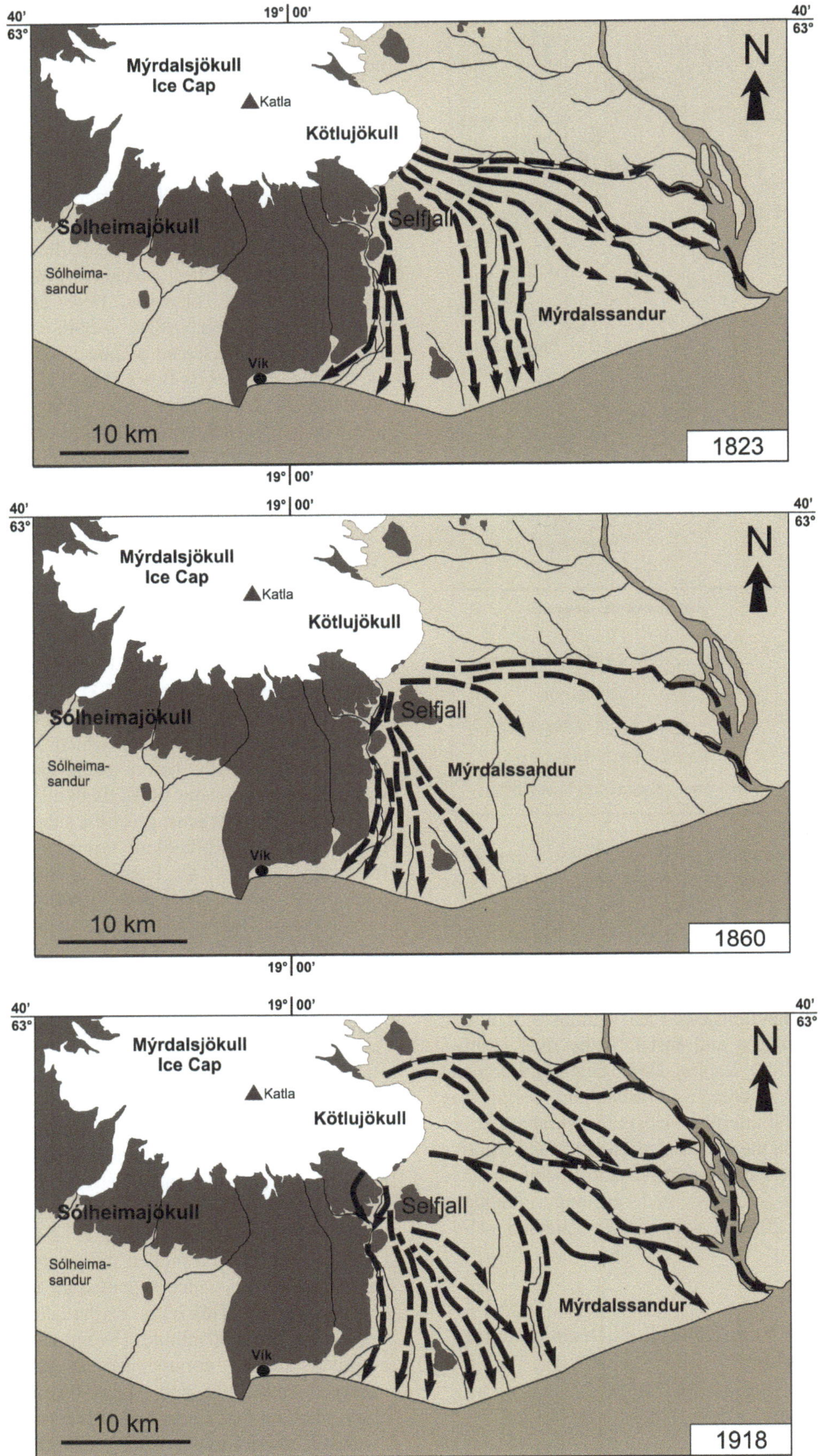

Fig. 11.3. Position of flood outlet and subaerial routeways taken by the 1823, 1860 and 1918 jökulhlaups on Mýrdalssandur. All jökulhlaups emerging from Kötlujökull since AD 1600 have done so through the southwest corner, inundating western Mýrdalssandur (after Larsen, 2000).

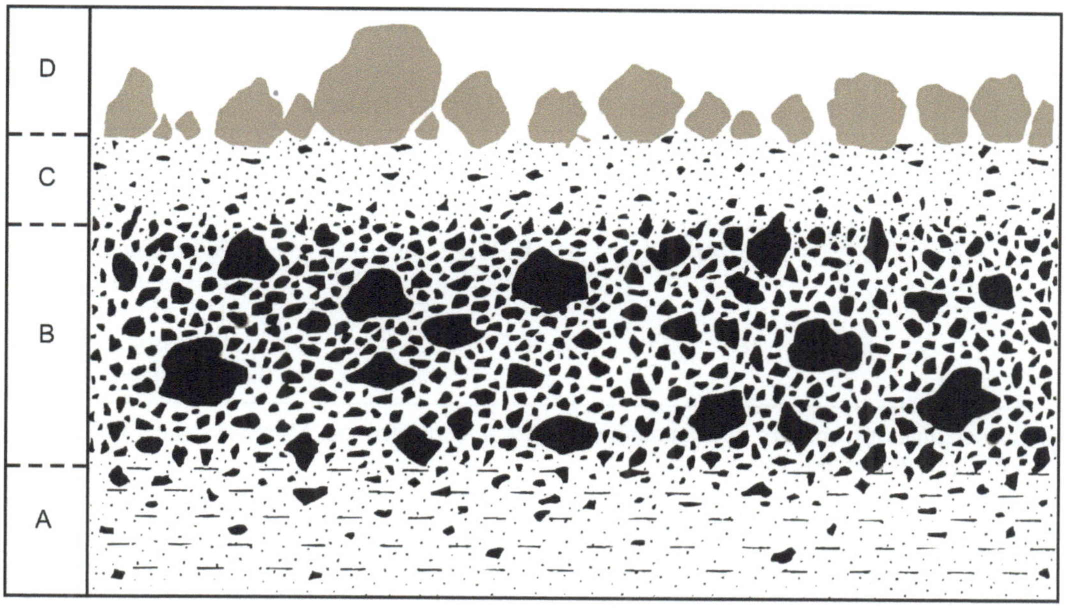

Fig. 11.4. A hypothetical cross section of a Katla flow. (A) Fine-grained base of the flow, mainly pumice. (B) Coarse-grained central section consisting of clasts in all dimensions, and pumice. (C) Top section containing a mixture of fresh volcanic material, clasts, soil fragments picked up by the flow, fragmented ice and ice blocks. (D) Large ice blocks. Figure and caption after Jónsson (1982).

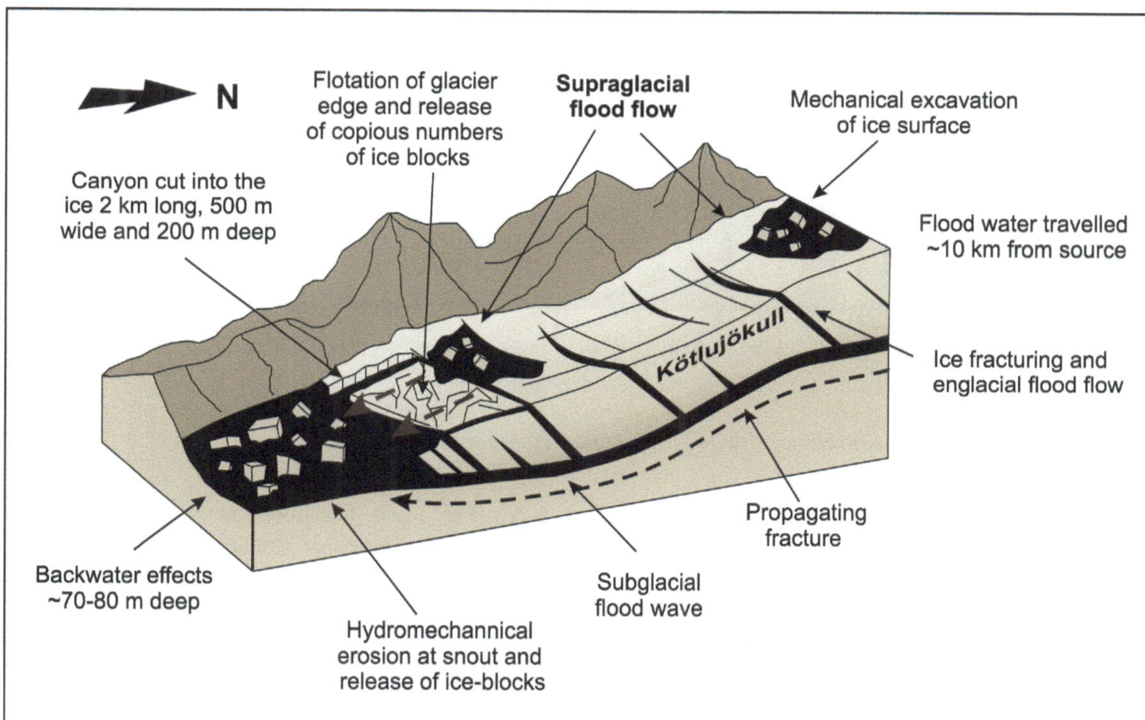

Fig. 11.5. An illustration of the likely scenario of supraglacial floodwater routing and mechanical removal of glacier ice during the 1918 jökulhlaup, Mýrdalssandur.

Hafursey and Selfjall and pushed forward with incredible speed over all the course of the Sandvatn'.

Following its onset, the flood magnitude was steady for up to 2 h, after which it suddenly increased. From Vík, Jóhannsson (1919) described this second flood surge: 'There is no halt in the torrential jökulhlaup at Múlakvísl until 1730 h. But as soon as it seemed to be easing, an even greater jökulhlaup seems to rush forward to the west of Hjörleifshöfði'. This more voluminous second flood surge removed and transported vast numbers of ice blocks (50–60 m in height) from the snout of Kötlujökull, and these became stranded in the main course of the Sandvatn (Jónsson, 1982; Tómasson, 1996). Jóhannsson (1919) wrote: 'we saw clearly over the sand plain nearest to us (Háfell). It was like one continuous outlet glacier all the way from Mýrdalsjökull to the sea. Large deep channels

Table 11.1. Estimations of hydraulic and physical parameters associated with the 1918 jökulhlaup.

Discharge (m^3/s)	Velocity (m/s)	Solid concentration (%)	Flow type	Volume of material *	Source
275×10^3	7–10	0–20	WF	0.95	Tómasson (1996)
1.5×10^6	10–15	20–47	HF		Maizels (1992, 1993)
$\geq 1.5 \times 10^6$		60–80	DF		Jónsson (1982)
		15–20	WF/HF		Björnsson (1993)
				0.25	Larsen and Ásbjörnsson (1995)
$>100 \times 10^3$					Thórarinsson (1957)
		38	HF		Einarsson (1979)
350×10^3					Hannesson (1934)

Note: Volume estimations concern the amount of material deposited on Mýrdalssandur. WF, water flow; HF, hyperconcentrated flow; DF, debris flow.

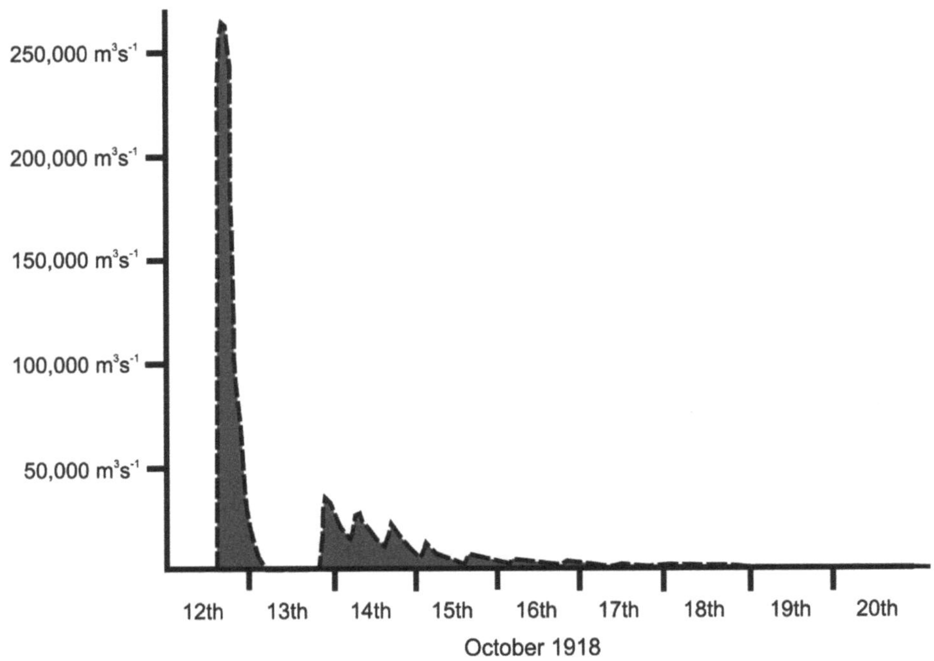

Fig. 11.6. Reconstructed 1918 Katlahlaup hydrograph (after Tómasson, 1996).

could be seen in places where the water had cut into the sand and swept away all the greater icebergs. But on both sides of these channels were high sand benches covered by gigantic icebergs'.

Using the above eyewitness accounts of the 1918 Katlahlaup, Tómasson (1996) estimated the initial flood wave velocity as ~10 m/s (Table 11.1). Following the eruption, peak jökulhlaup discharge was reached within 4–5 h (Jónsson, 1982; Tómasson, 1996, 2002), with total volumes of meltwater of the order of 8 km^3 draining over a period of 3–5 days (Thórarinsson, 1957, 1975; Tómasson, 1996; Höskuldsson and Sparks, 1997; Larsen, 2000). The rate of discharge increase for the 1918 event is an order of magnitude greater than that for any known 'storage-release' related jökulhlaup (Björnsson, 2002; Roberts, 2005). Maizels (1993) and Tómasson (1996) used slope–area techniques in their palaeohydraulic reconstructions for the 1918 event. Tómasson (1996) reconstructed a hydrograph for the 1918 Katlahlaup, showing a main flood peak of $2-3 \times 10^5$ m^3/s, lasting for 5–6 h (Fig. 11.6). The floodwaters then abruptly declined, after which a series of smaller jökulhlaups (~10^4 m^3/s) occurred over the next 2 weeks. Tómasson (1996) attributed the sudden halt in the flood to subglacial ice tunnel collapse and temporary blockage, which occurred repeatedly resulting in the observed discharge fluctuations (Tómasson, 1996) (Fig. 11.6).

11.3.3. Impact of the 1918 Katlahlaup

The main 1918 Katlahlaup carved out a large ice-walled channel 'Rjúpnagil' (Jóhannsson, 1919; Sveinsson, 1919; Tómasson, 1996) (Fig. 11.5). This channel was 1,460–1,830 m in length and 366–550 m in width, with walls up to 145 m in height (Jóhannsson, 1919; Sveinsson, 1919; Tómasson, 1996; Roberts, 2005). Photographs taken during the waning stage of the 1918 jökulhlaup show vigorous flows filling Rjúpnagil completely, with only a few bar surfaces emergent (Roberts, 2005). It is highly probable that Rjúpnagil represents a former subglacial or englacial conduit whose roof was subject to rapid failure and evacuation by the flow. Jóhannsson (1919) described the release of a huge mass of ice blocks from the Rjúpnagil area after 2 h of the 1918 jökulhlaup. Little is known about the sedimentology of the Rjúpnagil channel-fill and whether subsequent waning stage flows or glacier advance have removed or over-ridden these features. The main phase of the jökulhlaup resulted in large-scale hydrofracturing of Kötlujökull and the entrainment of large ice blocks (Tómasson, 1996). The flood carried so much ice that one observer commented 'it looked like snow-covered hills were rushing forward' (Sveinsson, 1919). These ice blocks were transported in the flow to the ocean, where they were seen to rapidly sink (Jónsson, 1982). However, a large proportion of these ice blocks were stranded in the main course of the Sandvatn and accumulated against the northern side of Sellfjall, enhancing backwater or temporary ponding effects upstream of the Hafursey–Selfjall flow constriction. These ice blocks were 50–60 m high, according to photographs taken after the event. Downstream, stranded ice blocks also contributed to the formation of longitudinal bars (Tómasson, 1996), which have been found to show surface meltout features (Maizels, 1992; Olszewski and Weckwerth, 1999; Duller, 2007). Along with ice blocks, the 1918 jökulhlaup transported huge boulders, one of which, 'Kötluklettur', is estimated to weigh 1,000 tons and is believed to have been transported over a distance of 14 km (Jónsson, 1982) (Fig. 11.7).

Attempts have been made to estimate the total flux of sediment liberated during the eruption and deposited on Mýrdalssandur. Larsen and Ásbjörnsson (1995) estimated that 0.64 km^3 of sediment was transported onto Mýrdalssandur, of which 40% was deposited into the ocean. Similar estimates (0.7 km^3) were reached by Eggertsson (1919). Tómasson (1996) used historical maps (1904 and 1946) and eyewitness descriptions of the 1918 Katlahlaup to calculate net sediment deposition and probable sediment concentration. The volume of sediment deposited above sea level, as calculated from the difference between the contours on the two maps, is 0.95 km^3 corroborated by a more recent analysis by Duller (2007) (Table 11.1). Total sediment and ash volume for the 1918 Katlahlaup is estimated by Tómasson (1996) at 3.6 km^3 with a total meltwater volume of 8 km^3. Sveinsson (1919) and Jóhannsson (1919) estimated that the coastline was extended seawards by up to 3 km in places (Fig. 11.8), accounting for a sediment volume of 0.3–0.4 km^3 (Tómasson, 1996). The delivery of huge volumes of material to the coastal areas of south Mýrdalssandur by regular Katlahlaups has influenced the rate of shoreline progradation in historic times and the coastline now lies 2.2–2.5 km further south than it did in 1660 (Nummedal *et al.*, 1987; see Fig. 11.8).

Maizels (1989a, 1989b, 1991, 1992, 1993) related vertical lithofacies sequences to flow rheology, the jökulhlaup hydrograph and sediment supply. She undertook sedimentary analysis of 1918 Katlahlaup deposits, focusing on the textural interpretation of localised, one-dimensional sedimentary profiles (Maizels, 1989a, 1989b, 1991, 1992, 1993). In 1989, she identified a massive black scoria granule lithofacies (unit 2, Fig. 11.9), interpreted as the product of 'hyperconcentrated flow' deposition which she regarded as being characteristic of volcano-glacial drainage (Maizels, 1989a, 1989b). She also identified a 'widespread' four-fold sedimentary facies sequence at Mýrdalssandur (Fig. 11.9) linked to the passage of the 1918 flood hydrograph (Maizels, 1991, 1992) and the lateral variation in lithofacies successions (Types B1–B4) on Mýrdalssandur (Figs. 11.10 and 11.11), associated with the 1918 jökulhlaup, which are thought to record differential flow paths and boundary

Fig. 11.7. Kötluklettur (Katla boulder). This 1,000 ton clast was transported for over 10 km and deposited on Mýrdalssandur during the 1918 jökulhlaup (Jónsson, 1982).

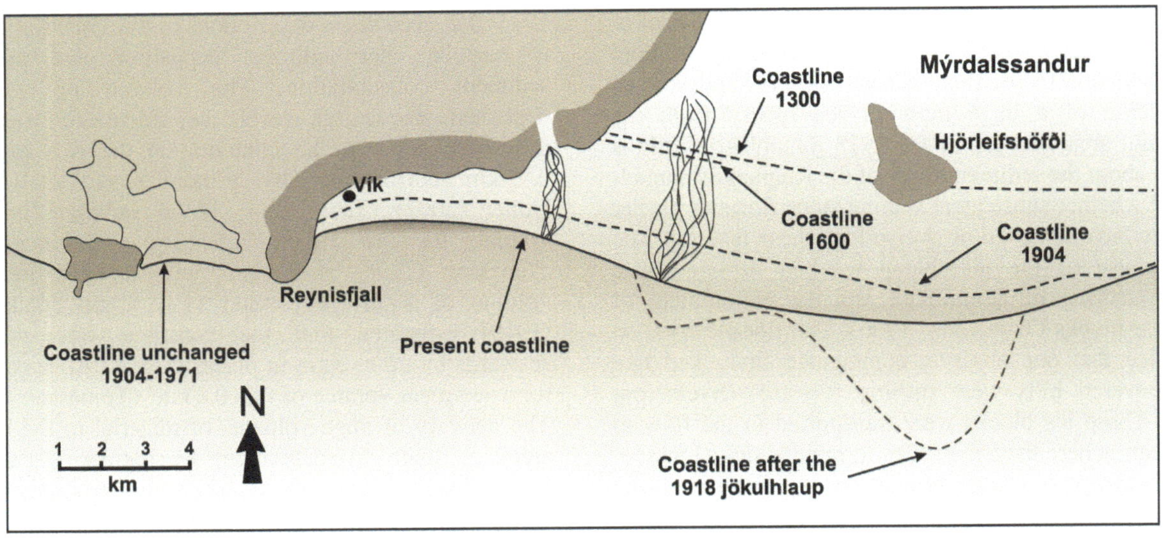

Fig. 11.8. Observed and inferred progradation of the coastline of southwestern Mýrdalssandur since AD 1300. Note the extension of the coastline inferred by eyewitnesses, indicated by the red dashed line, following the 1918 jökulhlaup. Although the headland of Reynisfjall may contribute to rapid shoreline progradation in this area, due to the inhibition of longshore drift, there is no doubt that the sediment flux that occurs during jökulhlaups is a major control of shoreline progradation (reproduced from Nummedal et al., 1987).

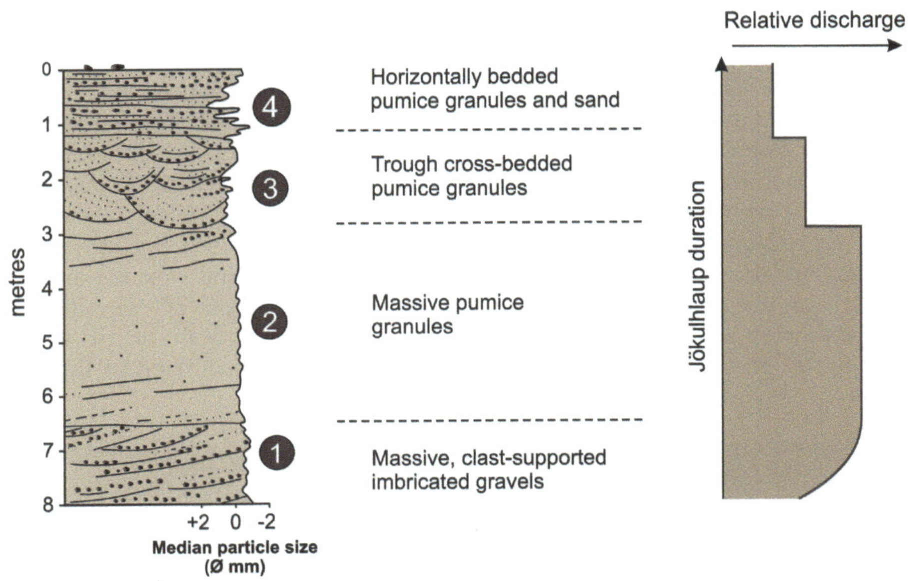

Fig. 11.9. The four-fold vertical sedimentary sequence associated with the 1918 jökulhlaup deposits as defined by Maizels (1992). Pre-surge gravels (1) are overlain by main-surge (2) and post-surge (3 and 4) pumice pebbles. Schematic diagram to the right shows the relative discharge of the jökulhlaup during the deposition of each unit (modified from Maizels, 1992).

conditions (Maizels, 1993). Maizels (1993) linked each lithofacies succession to the inferred jökulhlaup behaviour, at positions on the sandur relative to the main flow path and surrounding topography (Figs. 11.10 and 11.11).

More recently, Duller (2007) and Duller *et al.* (2008) used the analysis of large-scale sedimentary architecture to reinterpret unit 2 of Maizels' lithofacies Type B3 exposed in the Selfjall river cliff section (Fig. 11.12). Well-defined large-scale sedimentary structures in scoria deposits, hitherto classified as massive, structureless and the product of hyperconcentrated flow deposition, have been reinterpreted as the product of large-scale antidune migration and the development of stationary chute and pool structures associated with localised hydraulic jumps (Duller, 2007, Duller *et al.*, 2008) (Figs. 11.13a and 11.13b). As such it is clear that all of the 1918 Katlahlaup deposits described by Jónsson (1982) and Maizels (1991, 1992, 1993) are the product of deposition from a highly concentrated fluidal turbulent flow and also that previous textural facies and hydrodynamic schemes are unsatisfactory.

Although large-scale 1918 Katlahlaup flow paths and interaction with topography are described by Maizels (1993), relatively little is known about the internal

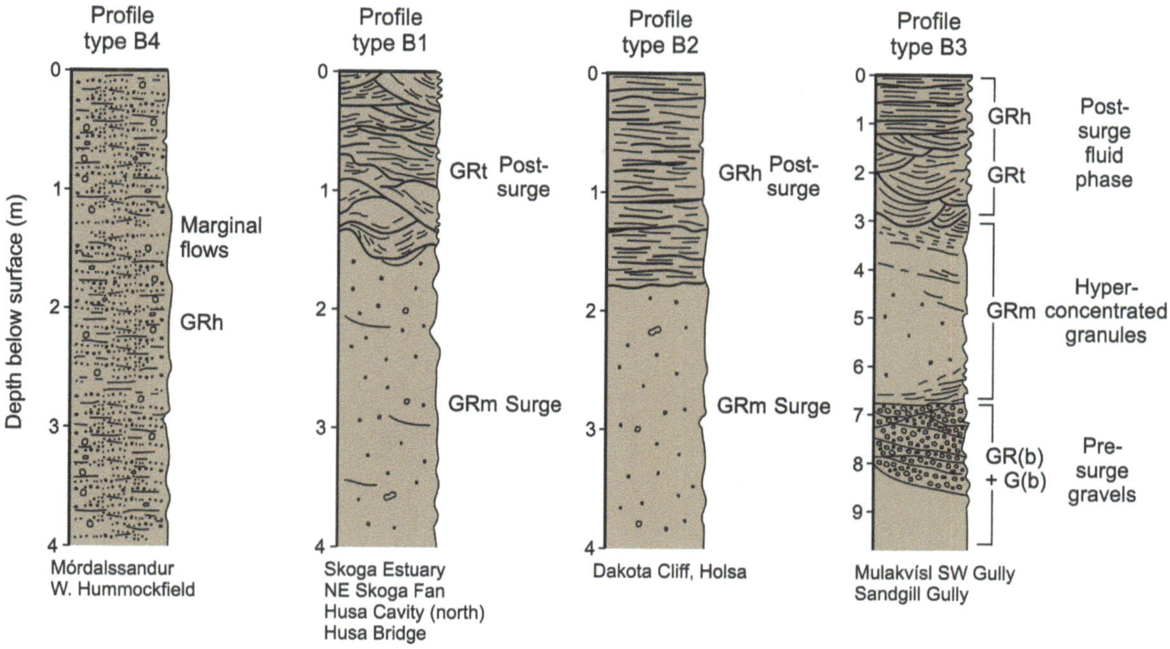

Fig. 11.10. Type B vertical lithofacies profiles within the jökulhlaup sandurs of southern Iceland (after Maizels, 1993). Refer to Fig. 11.11 for key to lithofacies codes.

composition and formation of large-scale bar surfaces and deposits. Maizels' (1992) map of 1918 residual bar surfaces indicates that ice proximal flows between Hafursey and Selfjall may be controlled by local backwater conditions but that flow expansion is dominant in the mid and distal reaches of Mýrdalssandur (Fig. 11.11). A large bar, upflow of Selfjall may reflect large-scale eddy circulation during the jökulhlaup within relatively slackwater conditions. Duller *et al.* (2008) presented a depositional model demonstrating how the sequence of depositional process and bedform configuration during a jökulhlaup is unique to any one location on the sandur and will therefore account for the spatial variability of vertical lithofacies associations and large-scale sedimentary architecture (Fig. 11.14).

Elevated 1918 Katlahlaup bar surfaces display evidence of ice block stranding, grounding and burial (Maizels, 1992; Olszewski and Weckwerth, 1999; Duller, 2007; Duller *et al.*, 2008). Maizels (1992) reported up to 150 distinctive boulder ring structures found on the surfaces and stoss-side slopes of residual jökulhlaup bars (Fig. 11.15). Boulder rings with rims of up to 4 m in height and 40 m in diameter are lined with diamicton and contain laminated sediments. The ring rims and diamicton fill are attributed to the meltout of coarse sediment from debris-rich ice blocks, whereas the laminated sediments reflect deposition within a kettle hole melt pond within the centre of the ring structure (Maizels, 1992; Olszewski and Weckwerth, 1999). Maizels (1992) identified four distinct ice-block-related structures in order of increasing ice block sediment concentration: 'normal kettles' (Type 1), 'rimmed kettles' with deep hollows and narrow rims (Type 2), 'crater kettles' with broad hollows and high rims (Type 3) and 'till-fill kettles' displaying a central mound morphology (Type 4). Olszewski and Weckwerth (1999) used detailed sedimentary evidence from 16 meltout kettle holes in the ice-proximal area to provide a detailed model of the development of a 'rimmed-normal' kettle equating to Types 1 and 2 of Maizels' model. Local disturbance and faulting exposed within a river bank exposure provide evidence of the meltout of an ice block incorporated within 1918 Katlahlaup deposits (Duller *et al.*, 2008; Fig. 11.5d). The deposition of relatively small ice blocks, which would be expected to float within a water flow, suggests strong downward sediment flux and high sedimentation rates during the Katlahlaup overcoming ice bock buoyancy.

11.3.4. Pre-1918 Katlahlaup Evidence on Mýrdalssandur and Surrounding Landscape

The flanks of Katla are riven by numerous deeply incised gorges such as Þakgil and Remundargil, which are currently only occupied by misfit streams feeding into the Múlakvísl. Despite the spectacular nature of these gorges and their probable association with Katlahlaups, little is known about their origin and age. A distinctive high terrace (Terrace I) on the northern, eastern and southern flanks of Hafursey lies 12–15 m above the highest 1918 Katlahlaup surface and shows signs of prolonged aeolian activity suggesting a substantial age (Maizels, 1992; Wisniewski *et al.*, 1999) (Fig. 11.16).

Some of the material left by the 1721 and 1755 jökulhlaups is still retained in Höfðabrekkujökull ice block meltout complex (Jónsson, 1982; Tómasson, 1996), where relatively small, discrete blocks of ice derived from the 1755 jökulhlaup remain (Everest and Bradwell, 2003). Jónsson (1982) reported large rip-ups or 'intraclasts' composed of soil near Vík airstrip at Höfðabrekkujökull (Fig. 11.17). Downstream of Selfjall, a section rich in scoria intraclasts, many of which themselves contain smaller intraclasts of diamicton and soil, most probably relates to a pre-1918 Katlahlaup (Fig. 11.18). The curious

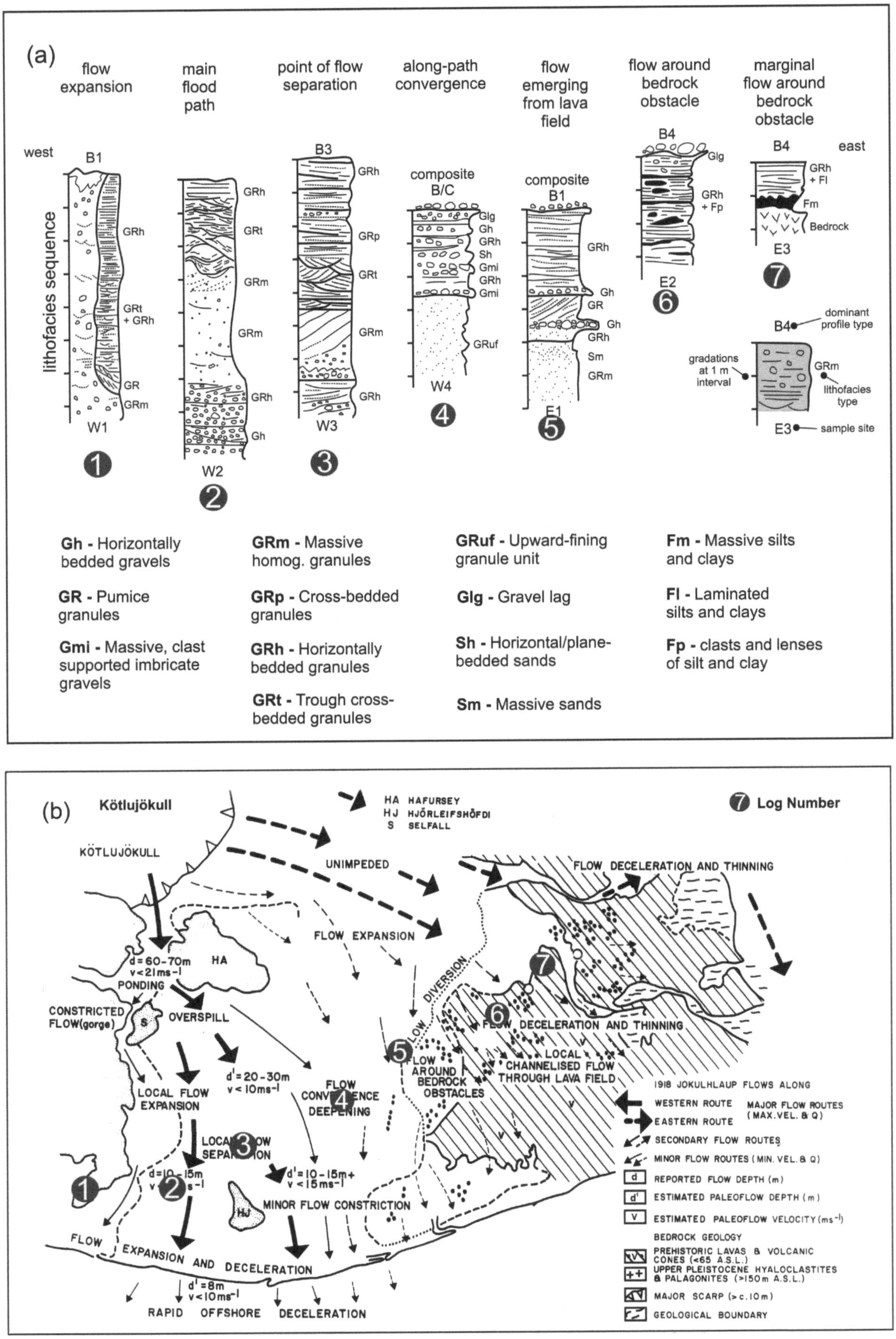

Fig. 11.11. (a) Sedimentary sections of the 1918 jökulhlaup deposits (see (b) for location), showing the lateral variations in facies assemblage in relation to the main flow path and local- and regional-scale topographic influences (after Maizels, 1993). (b) Log locations (a) and reconstructed flood pathways and palaeoflow conditions during the 1918 jökulhlaup peak on Mýrdalssandur (after Maizels, 1993).

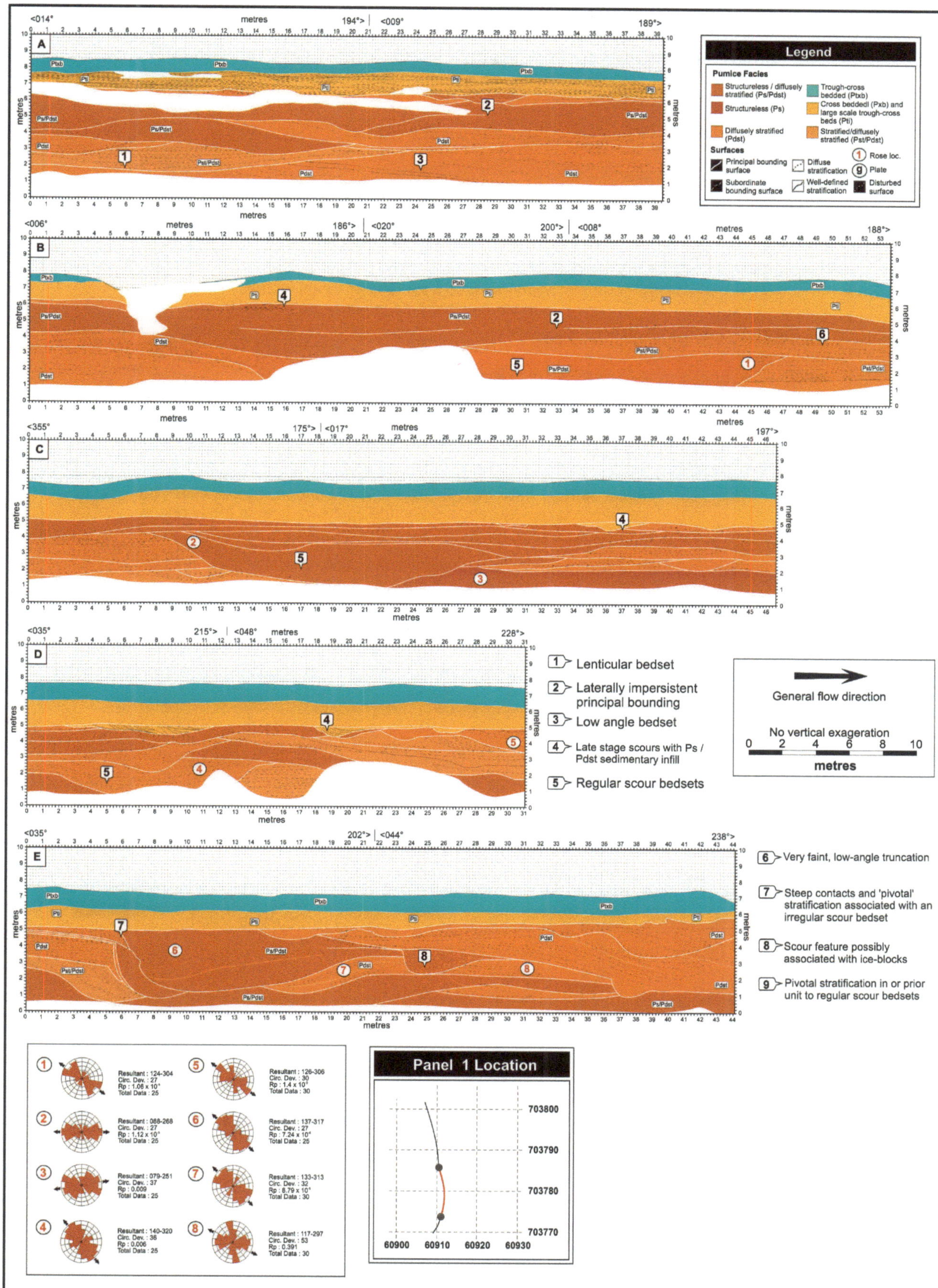

Fig. 11.12. Architectural panels depicting the sedimentary architecture of the 1918 jökulhlaup deposits exposed downstream of Selfjall. All panels are adjoined spatially. CFA = clast fabric analysis of a-axes *measurements* (n = 30). *Sub-panels A–E are adjoined spatially (modified after Duller, 2007).*

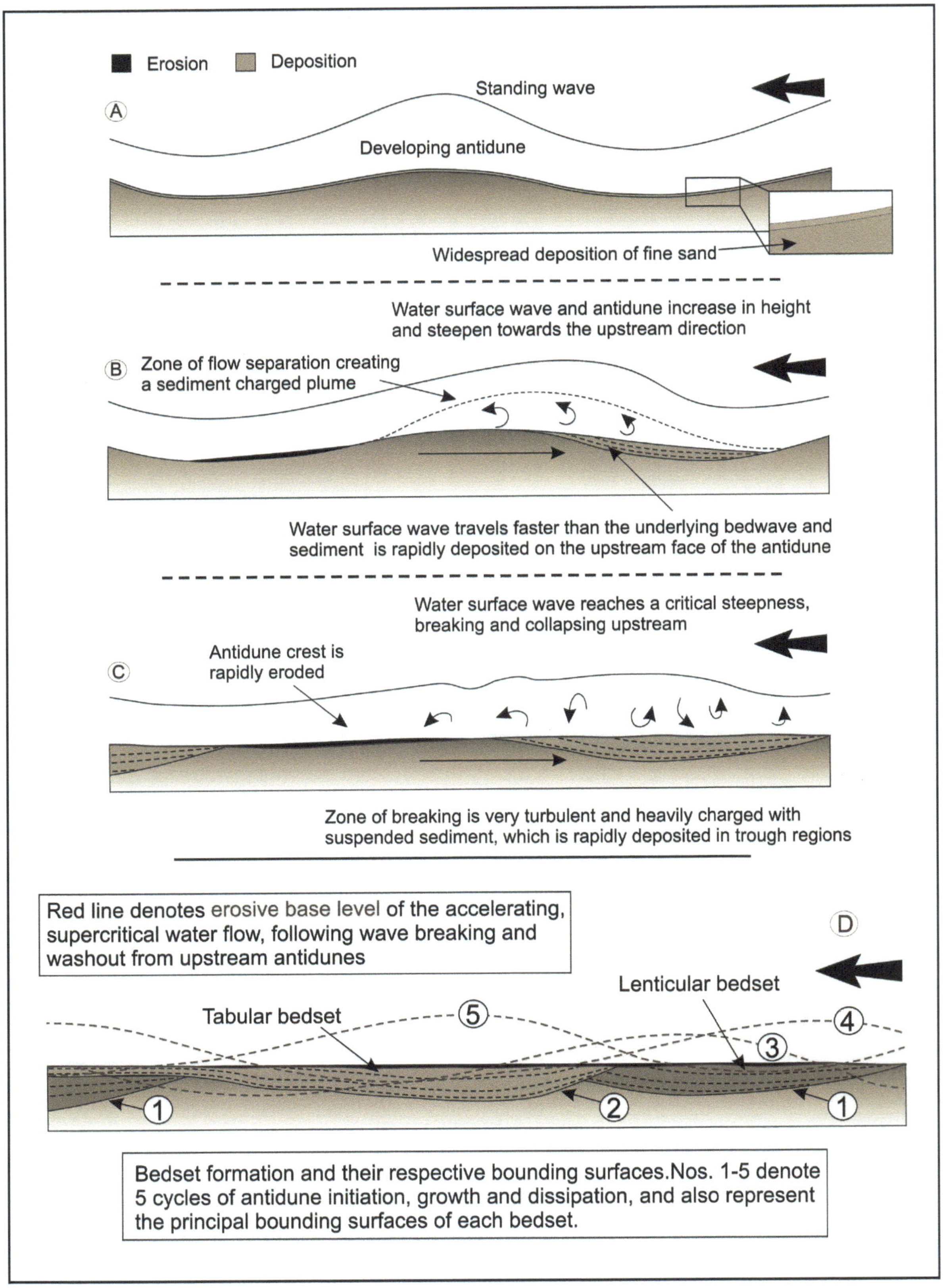

Fig. 11.13a. (A–C) Diagram illustrating the process involved in the formation and destruction of an antidune during a single cycle. (D) Schematical illustration of the formation of bounding surfaces and bedsets by the superimposition of several antidune cycles (after Duller, 2007, and Duller et al., 2008).

melange in this section points to a complex origin possibly related to several jökulhlaups. Sedimentary evidence upstream of Selfjall suggests the passage of up to four large jökulhlaups (Duller, 2007; Duller *et al.*, in press).

11.3.5. Recent Katlahlaups (1955–Present)

Numerous historic eruptions of Katla subglacial volcano have generated high-magnitude jökulhlaups (10^5–10^6 m^3/s) capable of inundating the vast sandar surrounding

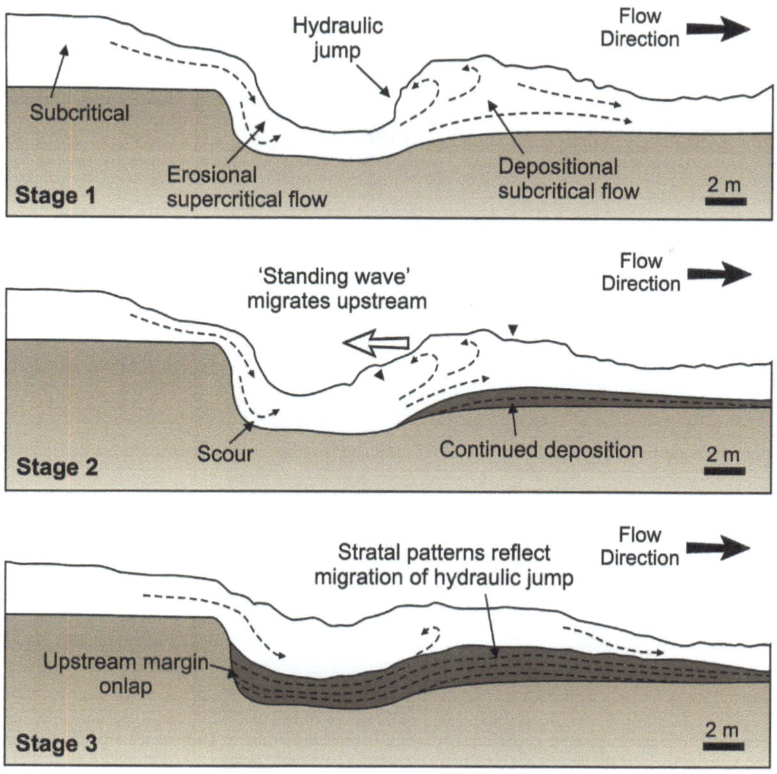

Fig. 11.13b. Initial scour and formation of the basal surface is associated with a mature hydraulic jump, during an increase in flow discharge (stage 1). Initial sedimentation takes place immediately down-current of the hydraulic jump, beneath the 'stationary' wave (stage 2). Deterioration of the hydraulic jump, caused by a decrease in flow discharge or by the effect of topographic modification during continued deposition, enables sedimentation to migrate up-current (stages 2 and 3) (adapted from Duller, 2007, and Duller et al., 2008).

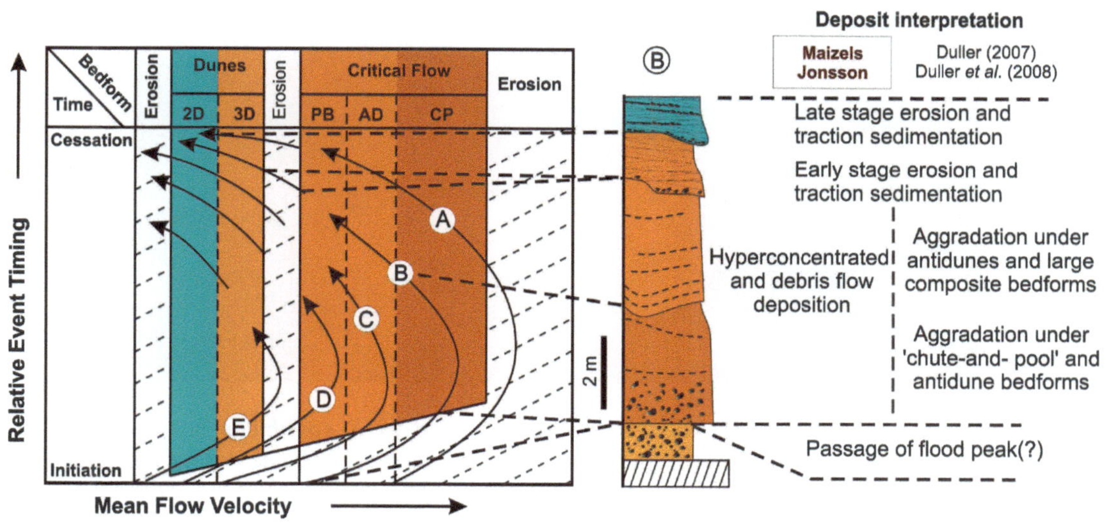

Fig. 11.14. Depositional model illustrating the temporal and spatial effects of bedform configuration on the character of vertical sedimentary sequences following a depletive (attenuating) waning flow (top left). Arrows A–E represent five separate hydrographs of differing magnitude. Each hydrograph can be interpreted to represent discrete events, or the spatial attenuation down-current of a single event, as illustrated below. Figure at top right represents a schematic sedimentary succession, not dissimilar to that observed on Mýrdalssandur, of the temporal flow path 'B' in top left figure. Previous interpretations that involved one-dimensional sedimentary analysis are shown (adapted from Duller, 2007, and Duller et al., 2008).

Mýrdalsjökull (Jónsson, 1982; Maizels, 1991, 1993; Tómasson, 1996, 2002; Roberts, 2005; Duller *et al.*, 2008) (Fig. 11.1). Following the last large jökulhlaup in 1918, intermediate magnitude (10^3 m^3/s) jökulhlaups have drained from Kötlujökull in 1955 and 1956 (Thórarinsson, 1957; Rist, 1967). Heim (1983) noted that 11 jökulhlaups emanated from Kötlujökull between 1945 and 1975 including the well-documented 1966 event and smaller jökulhlaups were reported in 1975 and 1976 in the Múlakvísl (Björnsson, 1976; Björnsson and Einarsson, 1976). The subglacial outlet for these lower magnitude (10^2 m^3/s) jökulhlaups is thought to have become progressively more concentrated on the western margin of Kötlujökull near the present subglacial outlet at Rjúpnagil (Heim, 1983).

Jökulhlaups in June 2002, July 2003, July 2005 and July 2006 resulted from the drainage of accumulated subglacial meltwater beneath ice surface cauldrons

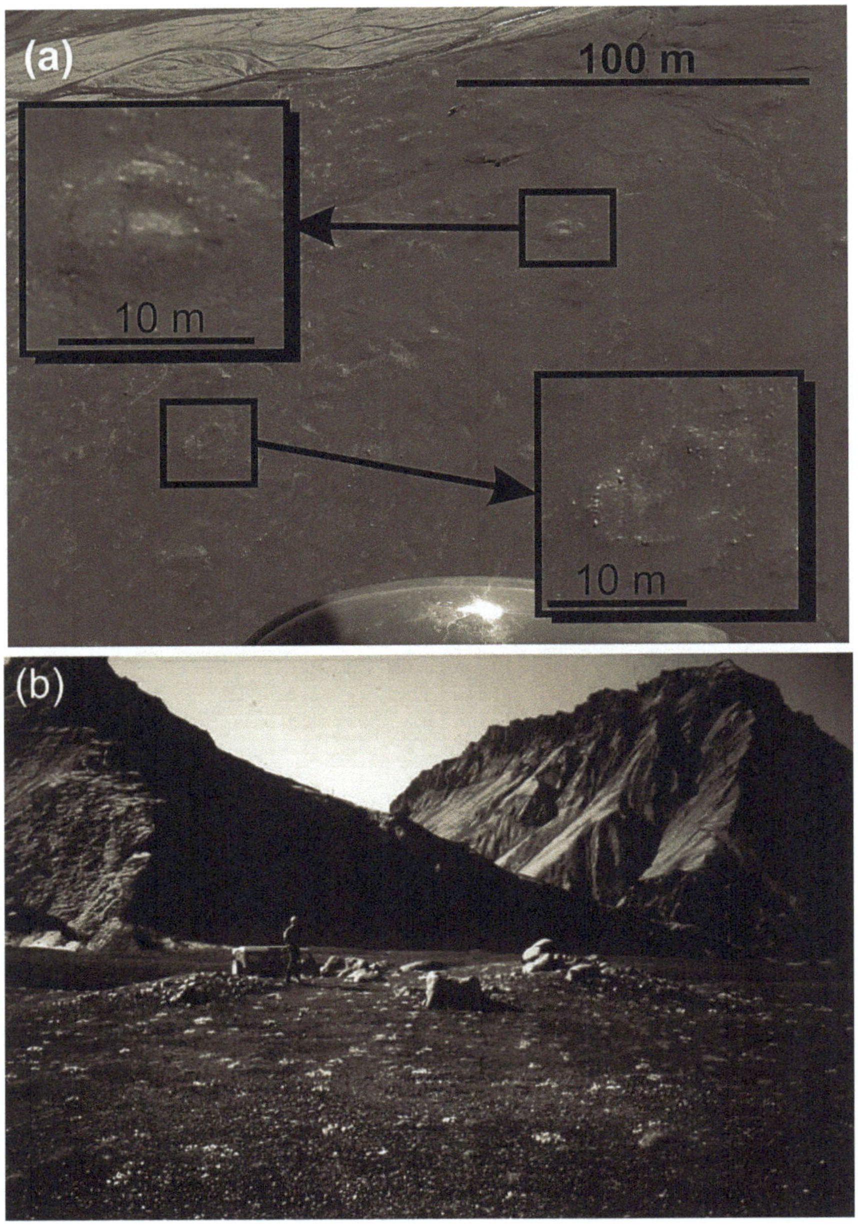

Fig. 11.15. (a) Aerial view of well-defined boulder ring structures on 1918 Katlahlaup bar surface. Note also the generally hummocky appearance of the bar surface which indicates the presence of numerous ice blocks deposited on the waning flow stage. (b) View of typical boulder ring structure within the proximal reach of the Sandvatn (Hafursey in the background).

(Roberts, 2008; Russell *et al.*, 2008) (Fig. 11.19). Jökulhlaups with peak discharge of $10^2\,m^3/s$ have been relatively frequent during the current period of volcanic unrest associated with enhanced geothermal activity after July 1999 (Soosalu *et al.*, 2006; Sturkell *et al.*, 2008).

The 2003 and 2005 jökulhlaups eroded their channel margins and deposited ice-proximal expansion and longitudinal bars. Both jökulhlaups were characterised by rapid ice-proximal deposition from high-energy turbulent fluidal flow, with the 2005 event showing evidence of more rapid deposition from a hyperconcentrated flow. Deposition during the 2005 jökulhlaup buried larger numbers of ice blocks resulting in post-flood kettle hole development (Fig. 11.19). Both floods deposited distinctive debris-rich ice blocks, resulting in the formation of boulder ring structures and till-fill kettles (Fig. 11.19) (Maizels, 1992; Olszewski and Weckwerth, 1999). Flood impacts decreased markedly downstream, suggesting rapid downstream attenuation of ice-proximal peak discharges of $500 \pm 150\,m^3/s$ calculated for the 2005 jökulhlaup. No increase in stage was recorded at the Múlakvísl bridge. The 2003 and 2005 jökulhlaups had a noticeable ice-proximal geomorphic and sedimentary impact which diminished rapidly during their downstream passage.

It is clear that recent jökulhlaups from Kötlujökull have played a major role in adjusting the proglacial fluvial system in the wake of massive sandur aggradation during the 1918 jökulhlaup and recent glacier margin fluctuations (Krigström, 1962; Heim, 1983; Russell *et al.*, 2008). Detailed geomorphological mapping work

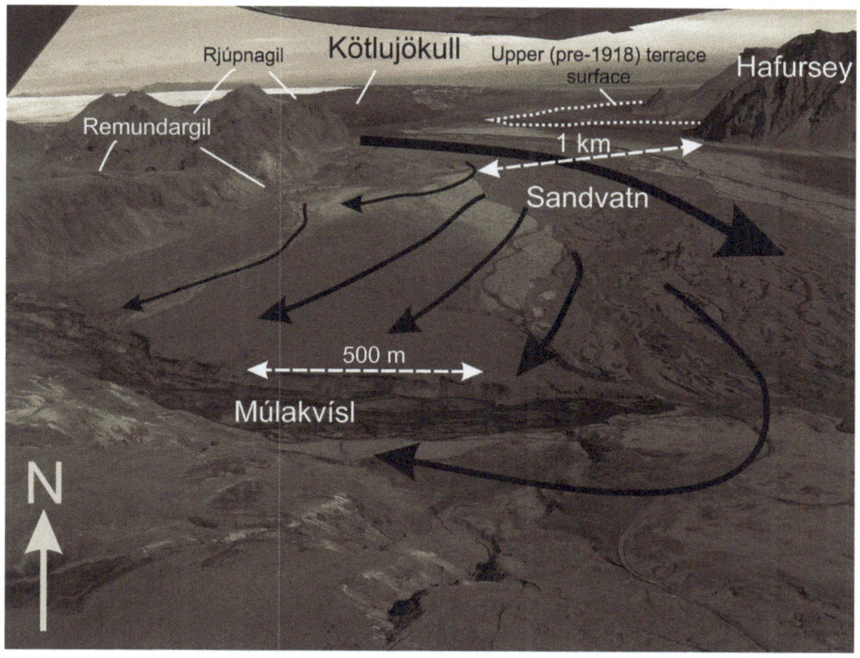

Fig. 11.16. Oblique aerial photograph showing the main 1918 jökulhlaup route to the west of Hafursey. Note the flow expansion into the area at the confluence of the Sandvatn and Múlakvísl, upstream of Selfjall, which was subject to backwater conditions during the 1918 jökulhlaup. Remundargil bedrock gorge and Rjúpnagil ice-walled canyon are indicated. The pre-1918 Katlahlaup terrace is indicated upstream and flanking Hafursey.

Fig. 11.17. Picture of soil rip-up at Höfðabrekkujökull, near Vík airfield. The ranging pole is 2 m in length. The intraclast, reported by Jónsson (1982), is cigar in shape. Numerous smaller intraclasts seen within the surrounding flood transported scoria deposits are thought to have broken off from the main intraclast during flood transport.

indicates eight, post-1918 jökulhlaup, terrace surfaces in the ice-proximal proglacial area, indicating late-waning stage and post-flood incision (Wisniewski *et al.*, 1999). Proglacial fluvial incision occurs predominantly in areas fed by subglacial meltwater outlets during the late jökulhlaup. By contrast, adjacent areas of ice proximal outwash on Mýrdalssandur are dominated by the development of supraglacially fed minor outwash, so-called 'hochsander' fans, which are deposited during periods of glacier advance and standstill (Heim, 1983, 1992; Krüger, 1994, 1997; Kjær *et al.*, 2004). In the event of another high-magnitude Katlahlaup, hochsander fans provide a readily available supply of sediment to supraglacial and subglacial jökulhlaup flows.

11.4. Sólheimasandur and Skógasandur

11.4.1. Introduction

Sólheimasandur and Skógasandur constitute the eastern and western portions, respectively, of a 75 km^2 inactive proglacial scoria-rich alluvial fan extending at its apex, 8 km from the snout of Sólheimajökull and the Þurragil

Fig. 11.18. Intraclasts within a larger intraclast exposed in a section downstream of Selfjall.

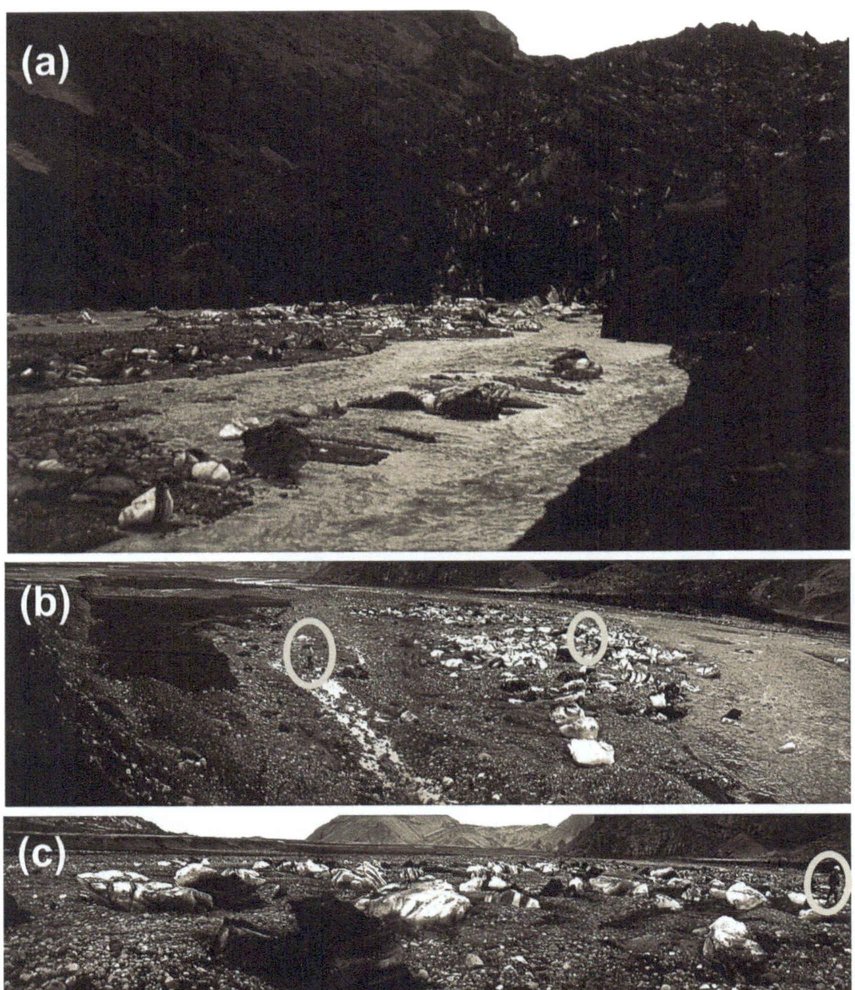

Fig. 11.19. (a) 2003 jökulhlaup exited Kötlujökull via a single conduit. Ice blocks were derived from conduit roof collapse. (b) and (c) Panoramas of 2003 jökulhlaup aftermath. Note large numbers of ice blocks grounded on proximal bar surfaces. Most ice blocks generated obstacle scour marks and no blocks were buried by flood sediment. People are circled for scale.

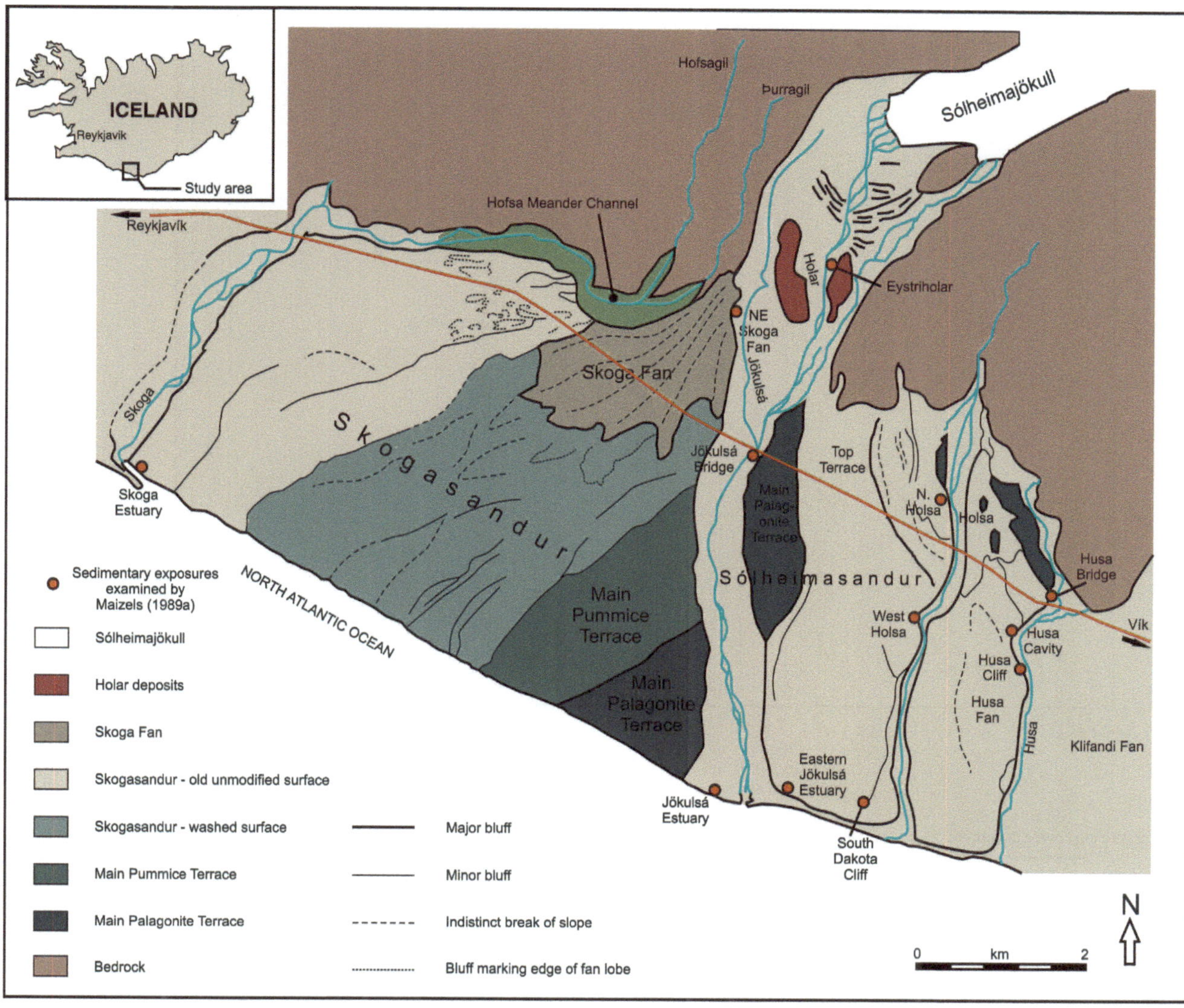

Fig. 11.20. Location of Skógasandur and adjacent areas showing the main morphological units of the Sólheimajökull sandur area. From west to east the main sandur zones include the Skógasandur, the active Jökulsá valley, Sólheimasandur, Hólsá Fan, Húsa Fan and Klifandi Fan. The northern part of Skógasandur is overlain by the Skóga pumice fan, which is truncated on its northern margin by the Hofsá meander channel. The southern margin of the sandur deposits is marked by continuous coastal cliff sections between 3 and 15 m high (after Maizels, 1989a).

and Hólsárgil bedrock gorges directly into the North Atlantic (Fig. 11.20). Active proglacial drainage is concentrated within the Jökulsá á Sólheimasandi which follows an incised course to the Jökulsá estuary (Fig. 11.20). Sólheimasandur and Skógasandur are type-sites for Icelandic volcanogenic floods and their sedimentary characteristics (Einarsson *et al.*, 1980; Maizels and Dugmore, 1985; Maizels, 1989a, 1989b, 1991, 1993, 1997; Le Heron and Etienne, 2005).

11.4.2. Jökulhlaup Magnitude and Frequency

Sólheimajökull is thought to have experienced at least eight major jökulhlaups from the KVS between 4500 BP and AD 1357 (Maizels and Dugmore, 1985; Maizels 1989a, 1989b). Maizels (1989a) used a range of palaeohydraulic reconstruction techniques to estimate a series of prehistoric jökulhlaup peak discharges between 2.1×10^3 and $3.3 \times 10^5\,m^3/s$ (Fig. 11.21). Fan-forming jökulhlaups are estimated to have had higher peak discharges than those reconstructed for later erosional channels on the basis of greater palaeoflow widths and stages (Maizels, 1989a). Jökulhlaups with the highest reconstructed magnitudes ($\sim 10^5\,m^3/s$) occurred roughly every 1,000 years (Fig. 11.21).

Although most historic Katlahlaups exit Kötlujökull, a series of floods inundated Sólheimasandur and Skógasandur in the early tenth century (Dugmore, 1987; Dugmore *et al.*, 2000) and a small component of floodwater associated with the 1860 Katla eruption was routed through Sólheimajökull (Hákonarson, 1860; Björnsson *et al.*, 2000; Larsen, 2000). The ice and water divide of Sólheimajökull continues into the southwestern rim of the Katla subglacial caldera (Lawler *et al.*, 1996; Björnsson *et al.*, 2000; Guðmundsson and Högnadóttir, 2006) (Fig. 11.2). Most recently, a small subglacial volcanic eruption in July 1999 generated a jökulhlaup at Sólheimajökull, with a characteristic rapid rate of discharge increase (Einarsson, 2000; Guðmundsson

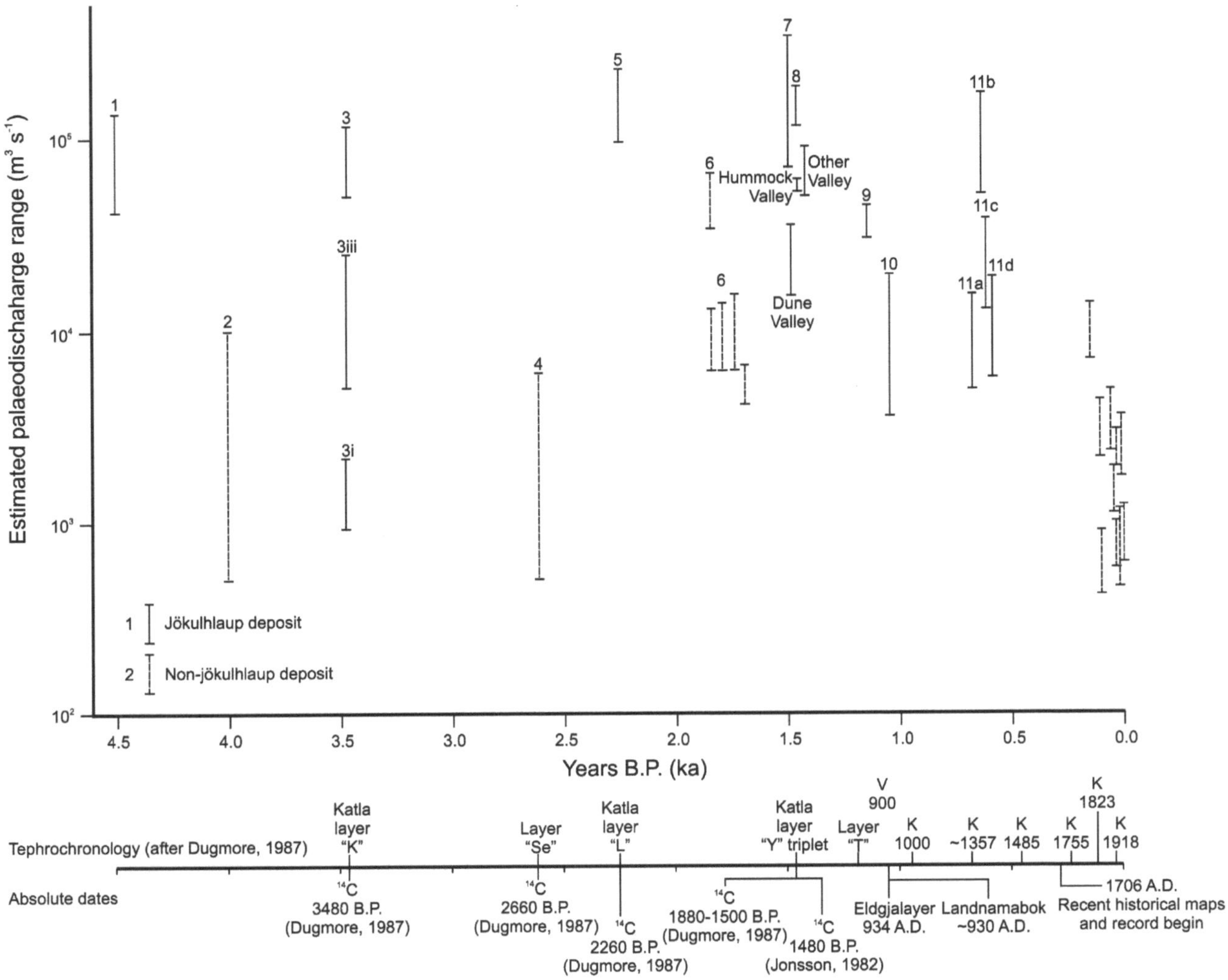

Fig. 11.21. Estimated long-term variations in jökulhlaup and non-jökulhlaup peak discharges, Sólheimasandur deposits. Estimates are based on minimum and maximum discharge predictions from models of Costa (1984). Sources of dating evidence are largely based on Maizels and Dugmore (1985) and Dugmore (1987). The discharge estimates indicate that the jökulhlaups recorded in the stratigraphic record were associated with flows averaging 10^3 m^3/s. Non-jökulhlaup sandur deposits were associated with peak flows ranging between only about 5×10^2 and 10^4 m^3/s (after Maizels, 1989a).

et al., 2000; Sigurðsson *et al.*, 2000; Russell *et al.*, 2000, 2002, submitted for publication; Roberts *et al.*, 2000a, 2000b, 2003). Reconstructed peak discharge for this event ranged from 4.4×10^3 m^3/s proximally (Russell *et al.*, 2000, 2002) to 1.7×10^3 m^3/s distally (Sigurðsson *et al.*, 2000). The jökulhlaup rose to peak discharge within 1 h of flood onset and had a total duration of approximately 6 h (Sigurðsson *et al.*, 2000; Roberts *et al.*, 2003). The July 1999 jökulhlaup was relatively small compared with prehistoric jökulhlaups that inundated Sólheimasandur and Skógasandur. Maizels (1989a) noted that jökulhlaups of magnitudes comparable to the 1999 event are at the lower limits of detection within the sedimentary and landform record.

Sólheimasandur also receives jökulhlaups from the drainage of an ice-dammed lake in Jökulsárgil, a tributary valley to the north of Sólheimajökull, the last large jökulhlaup from which occurred in 1936 (Thórarinsson, 1939). Retreat and thinning of the snout has since precluded a large ice-dammed lake from forming, although smaller lake volumes were impounded periodically between 1990 and 1991 (Tweed, 2000a, 2000b). Jökulsárgil was filled suddenly during the July 1999 Katlahlaup, which subsequently drained catastrophically beneath the western margin of Sólheimajökull (Sigurðsson *et al.*, 2000; Russell *et al.*, 2000, 2002, submitted for publication; Roberts *et al.*, 2000a, 2000b, 2003).

11.4.3. Geomorphological and Sedimentary Evidence of Jökulhlaups on Sólheimasandur and Skógasandur

Large bedrock gorges provide spectacular evidence of the erosional power of palaeo-jökulhlaups feeding water and sediment to the proglacial sandur (Guðmundsson and Högnadóttir, 2006) (Fig. 11.20). Despite detailed investigations of proglacial jökulhlaup geomorphology and sedimentology little attention has been given to the evidence of supraglacial, subglacial or ice-marginal jökulhlaup drainage routeways (Roberts *et al.*, 2003).

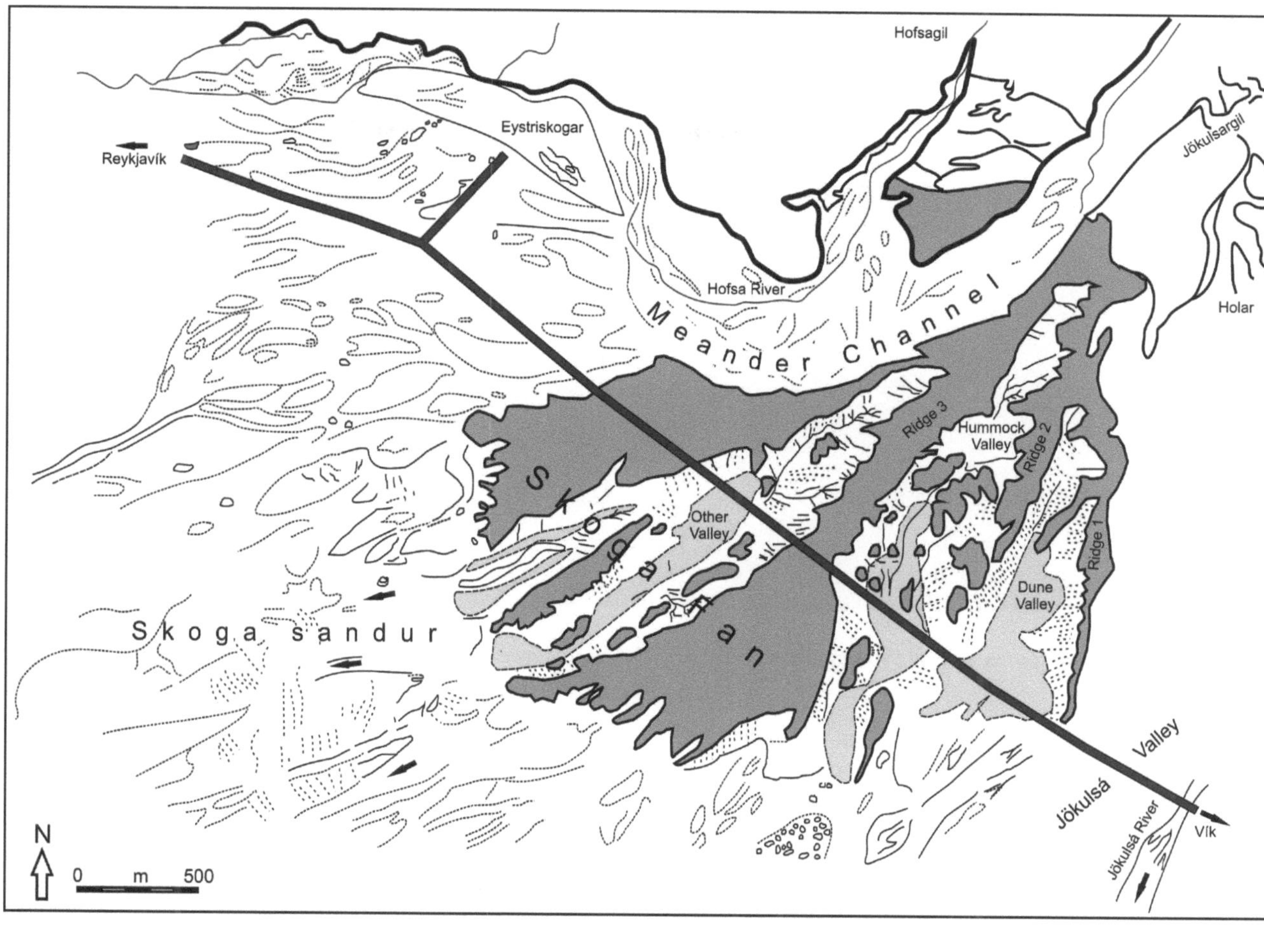

Fig. 11.22. Morphology of the Skógar pumice fan. The fan surface is deeply dissected by three major flood channels: (from west to east) Other Valley, Hummock Valley and Dune Valley. To the southwest lies Skógasandur which is characterised by a series of low, transverse boulder ridges, relic bars and channels, shallow depressions and boulder lags, representing the area of sandur washed over by floodwaters draining from the Skógar pumice fan flood channels. To the northwest lie unmodified areas of Skógasandur, exhibiting subdued surface topography, relict braid bars and channels and finer-grained surface material with few boulders. To the north lies the Hófsá meander channel, which emerges from two canyons cut into the edge of the massif and which truncates the northernmost part of the Skógar pumice fan (after Maizels, 1989a).

Sólheimasandur and Skógasandur display an overall fan-like morphology expanding from the Þuragil and Hólsárgil gorges with a relative relief of erosional and terraced landscape of 30 m (Maizels, 1989a, 1989b) (Fig. 11.22). The Skogar scoria fan has a well-defined lobate morphology dissected by a number of major channels containing streamlined erosional residuals, boulders and megaripples (Maizels, 1989a, 1989b). Megaripples with wavelengths of 6.6–10.5 m and heights of 0.16–0.25 m are composed of fine-grained scoria cores with basalt pebble lag and coarser grain size crests (Maizels, 1989a). Although Maizels (1989a) attributed the megaripples to fluvial deposition on the jökulhlaup waning stage, these bedforms are most probably of aeolian origin as they are comparable with aeolian bedform processes and forms elsewhere in Iceland (Mountney and Russell, 2004, 2006). Much of Sólheimasandur and Skógasandur are terraced, reflecting numerous erosional episodes during and between jökulhlaups (Maizels, 1987, 1989a, 1989b) (Figs. 11.20 and 11.22). Recent fluctuations and recession of Sólheimajökull are likely to have contributed to ongoing proglacial fluvial incision as found in other Icelandic proglacial fluvial systems (e.g. Krigström, 1962; Thompson and Jones, 1986; Marren, 2002).

Maizels (1989a) estimated that 85% of the total sandur sediment thickness is composed of jökulhlaup deposits. Using a combination of lichenometry and tephrochronology Maizels and Dugmore (1985) identified eleven jökulhlaup units that pre-date the Little Ice Age (Fig. 11.21). Sandur deposits are dominated by three lithofacies types: Type A, <4.6 m-thick massive, homogenous, well-sorted, black scoria gravels with palagonite pebble stringers; Type B1, massive, homogenous, well-sorted, black scoria gravels overlain unconformably by a thin unit of bedded black scoria sands and gravels (Figs. 11.10 and 11.23); Type B2, horizontally bedded black scoria sands overlying Type B1 or Type A; and Type C consisting of a range of graded sediments such as normal, cyclic normal, normal-inverse, and inverse (Maizels, 1989a, 1989b). Maizels (1989a, 1989b) interpreted Lithofacies Types A, B and C as the products of deposition from non-Newtonian flows, reflecting a transition from single or multiple flood surge deposits

Fig. 11.23. Photo of scoria gravels found proximally on Sólheimasandur.

of black scoria gravels to thinner units of planar and cross-bedded gravels. Sedimentary successions indicate the occurrence of jökulhlaups with a rapid rise to peak discharge, possibly accompanied by a series of pulses related to cyclic incision by large antidunes. As flows expanded across the sandur, flow deceleration and deposition took place allowing the 'freezing' of lobate flow fronts such as the Skógafan followed by a transition to channelised fluidal flows (Maizels, 1989a, 1989b, 1997) (Fig. 11.22). More recently, Duller (2007) presented clear evidence of large-scale structures such as channels and lateral accretion surfaces within distal jökulhlaup deposits on Skógasandur (Figs. 11.24 and 11.25). This suggests that some Katlahlaups draining Sólheimajökull were more fluidal in nature and more similar to the 1918 Katlahlaup which drained across Mýrdalssandur.

11.4.4. Impact of July 1999 Jökulhlaup

The 1999 jökulhlaup emanated from the western margin of Sólheimajökull via a series of ice fractures and conduits, 4 km from the glacier snout (Roberts *et al.*, 2000a, 2000, 2003; Russell *et al.*, 2000, 2002, submitted for publication) (Figs. 11.26 and 11.27a–c) and drained ice-marginally into an ice-dammed lake basin 3.7 km from the glacier snout (Sigurðsson *et al.*, 2000; Roberts *et al.*, 2000b, 2003). Drainage from the upper ice-dammed lake occurred via a spillway channel and a series of ice fractures and conduits subsequently fed a lower ice-dammed lake basin, Jökulsárgil (Fig. 11.26). Floodwater also exited supraglacially via a 250 m long fracture complex, 3 km up glacier of the snout (Roberts *et al.*, 2000a, 2000, 2003) (Figs. 11.26 and 11.27a,b). At the glacier snout, most water discharged via a 150 m wide ice-roofed tunnel on the western margin. A second tunnel exiting the centre of the glacier snout captured most waning stage and post-jökulhlaup flow (Figs. 11.26 and 11.27f). Smaller flows also exited the eastern margin of the snout, flowing across vegetated hillsides and parallel to the snout via a newly cut steep-sided channel.

Outflow from the uppermost western conduits cut through moraine ridges and transported boulders up to 5 m in diameter into a high-gradient ice-marginal channel (Fig. 11.27c). Boulders up to 10 m in diameter were scoured from older moraine and slope sediments and transported for short distances (Russell *et al.*, 2002; Roberts *et al.*, 2003). Temporary storage of jökulhlaup water within the upper ice-dammed lake basin allowed localised delta formation where floodwaters entered the lake from the high-gradient ice-marginal feeder channel (Russell *et al.*, 2002; Roberts *et al.*, 2003) (Fig. 11.26). Jökulhlaup outflow from the upper lake eroded a spillway channel through older sediments and, in places, exposed bedrock.

The jökulhlaup deposited a 1,200 m^2 boulder bar on top of pre-flood vegetated surfaces immediately in front of the western snout outlet with a thickness of 6 m proximally, reducing to less than 1 m within 500 m of the outlet (Figs. 11.26 and 11.27d–f. Boulders up to 10 m in diameter and ice blocks up to 15 m in diameter formed clusters, suggesting simultaneous rapid deposition of sediment and ice blocks (Fig. 11.27e). The almost structureless, inversely graded and matrix-rich nature of the surrounding deposits suggests extremely rapid deposition from a non-cohesive sediment-rich flow (Russell *et al.*, 2000, 2002) (Fig. 11.26b). Numerous kettle holes formed after the jökulhlaup following ablation of partially and completely buried ice blocks (Russell *et al.*, 2000, 2002). The number and size of boulders declined distally and laterally across the bar surface away from the jökulhlaup portal. The number and size of completely buried ice blocks decreased downstream with declining flood unit thickness. Waning stage bar-surface reworking was slight, with most erosion confined to the development of main channels which became progressively less incised with increasing distance downstream (Russell *et al.*, 2000, 2002).

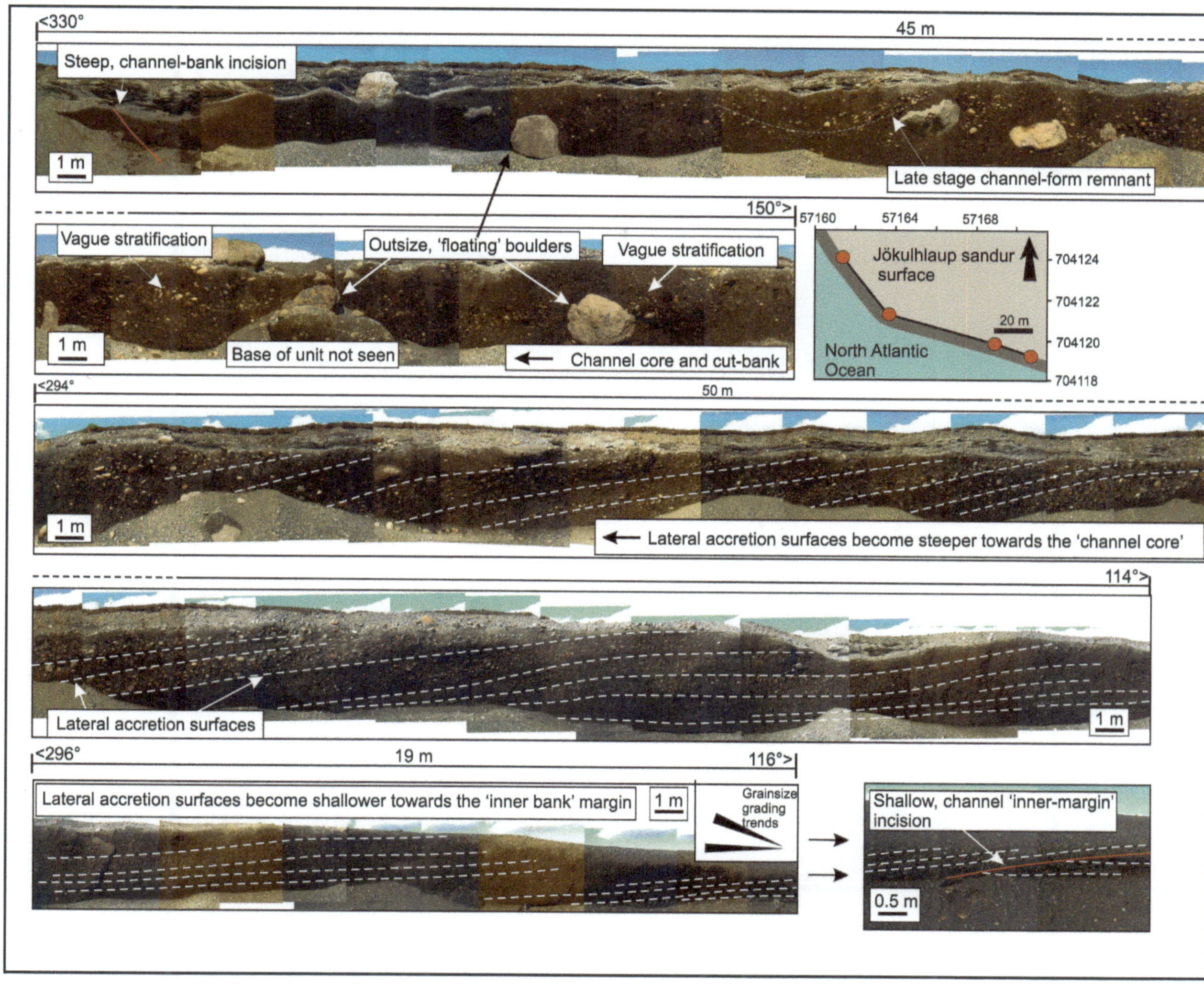

Fig. 11.24. Photopanoramic and overlay illustration highlighting architectural features of distal jökulhlaup deposits, Skógarsandur, southern Iceland (Duller, 2007). The inset map shows the section in plan view.

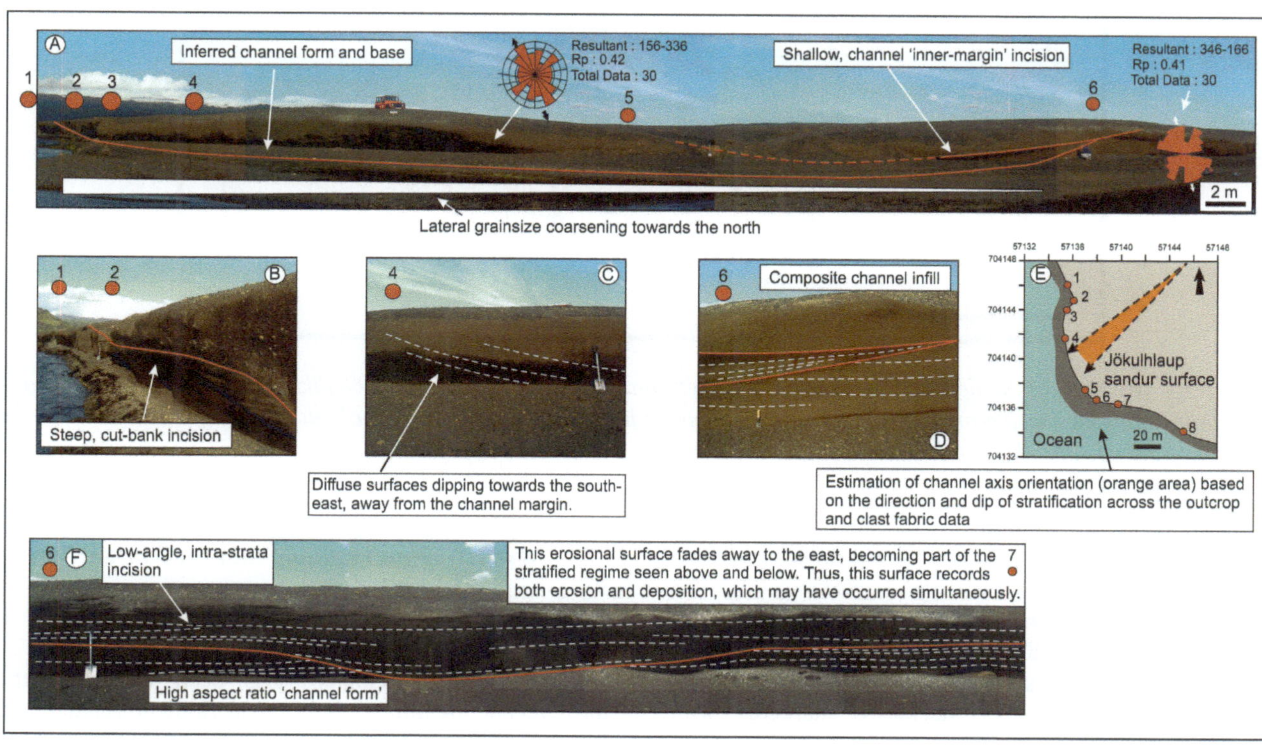

Fig. 11.25. Architectural features of distal jökulhlaup deposits, Skógarsandur, southern Iceland (Duller, 2007).

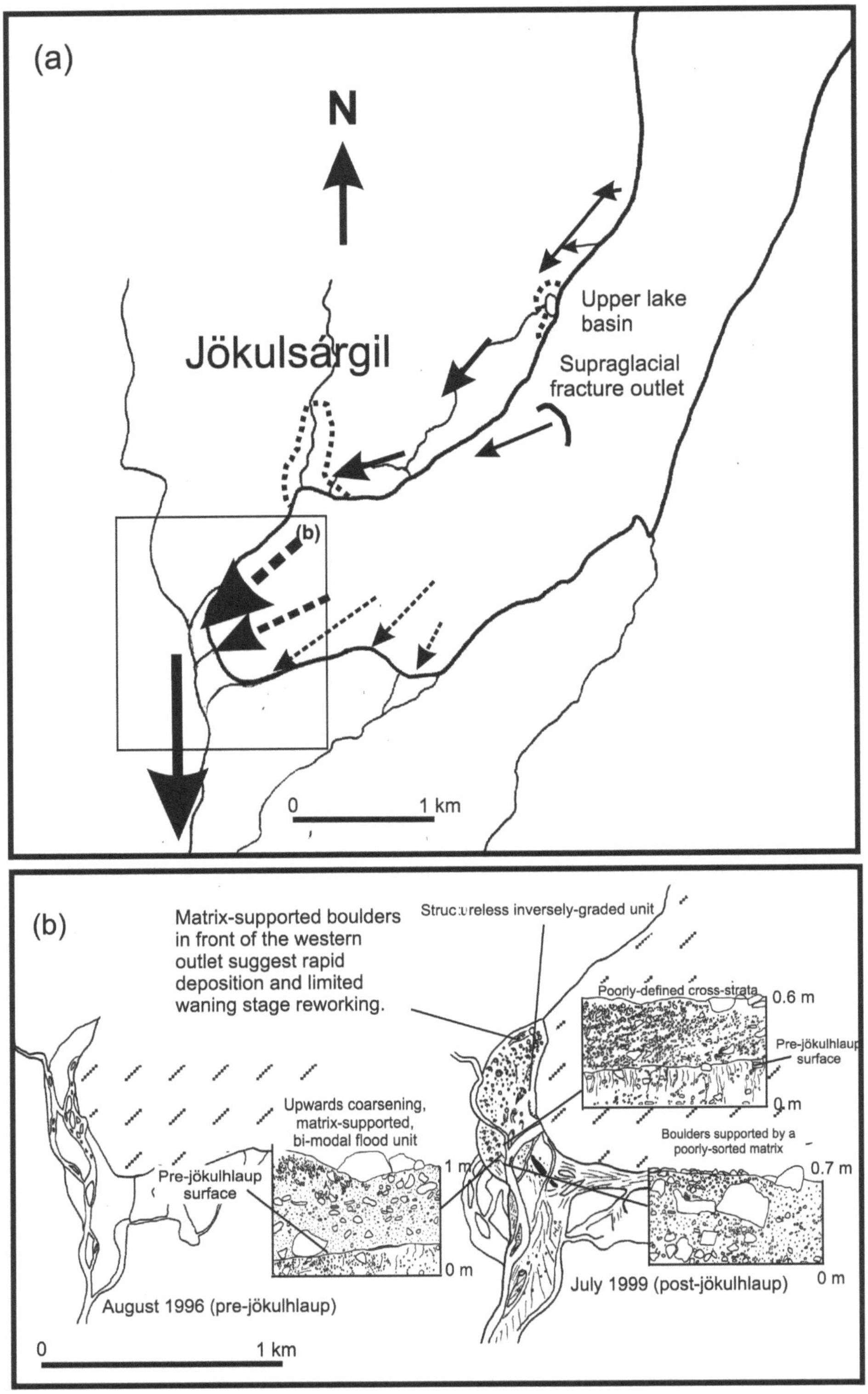

Fig. 11.26. (a) Map of Sólheimajökull, showing outlets and routeways of July 1999 jökulhlaup. (b) Pre- and post-jökulhlaup maps of the ice-proximal area of Sólheimasandur, indicating geomorphological change. The nature and location of July 1999 jökulhlaup deposits are indicated (after Russell et al., 2002).

Large volumes of sediment evacuated from the subglacial system acted as a major control on ice-proximal proglacial deposition (Russell *et al.*, 2000, 2002). The flood was largely confined to the well-established incised channel of Jökulsá á Sólheimasandi, where its impact was greatest at the glacier snout. The erosional and depositional impact of the flood decreased markedly with distance downstream (Russell *et al.*, 2000, 2002).

11.5. Jökulhlaups from the Northwestern Sector of Mýrdalsjökull Draining into the Markarfljót

11.5.1. Introduction

Knowledge of Katlahlaups has been based on the most recent jökulhlaup activity on Sólheimasandur and Mýrdalssandur; however, prehistoric jökulhlaups have

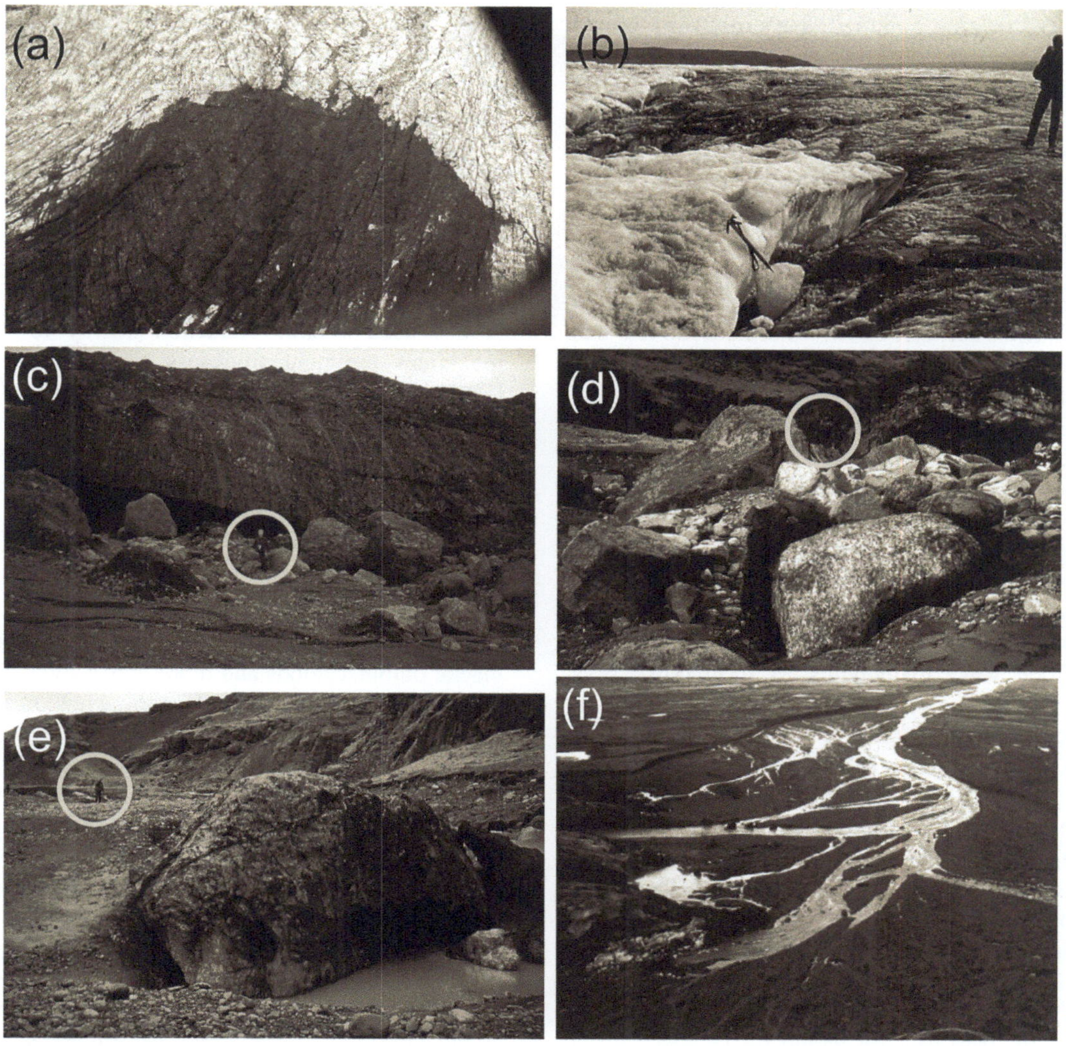

Fig. 11.27. (a) Supraglacial outlet of the July 1999 jökulhlaup at Sólheimajökull. Water ascended through nearly 400 m of glacier ice. Flow was from top to bottom. (b) Surface expression of jökulhlaup hydrofractures indicated in (a). Flow was from left to right. (c) Jökulhlaup outlet located 5 km up glacier from the glacier snout. Flows truncated a lateral moraine ridge and transported large boulders. Flow was from left to right. Person circled for scale. (d) Proximal portion of the July 1999 jökulhlaup bar deposited in front of the western outlet. Boulders were deposited on top of ice blocks, suggesting rapid deposition. Flow was directly towards the viewer. Note people circled for scale. (e) Large ice block obstacle mark directly in front of the western jökulhlaup outlet. Flows were from right to left. Note person circled for scale. (f) Post-1999 jökulhlaup view across the snout of Sólheimajökull showing western (bottom) and central outlets (top).

been known to drain also into the Markarfljót river system (Haraldsson, 1981; Sigurðsson, 1988; Björnsson *et al.*, 2000; Tómasson, 2002). More recently, the magnitude and frequency of prehistoric Katlahlaups draining from the northwest margin of Mýrdalsjökull, mainly via Entujökull, and also from Slettjökull, Merkurjökull, Krossárjökull and Tungnakvíslarjökull outlet glaciers has been reconstructed from geomorphological and stratigraphic evidence (Knudsen and Eggertsson, 2005; Smith and Haraldsson, 2005; Larsen *et al.*, 2005; Smith and Dugmore, 2006). The current period of unrest at Katla has raised concerns that a future Katla eruption might route very large jökulhlaups ($\sim 2.5–3 \times 10^5$ m^3/s) into the Markarfljót river system, impacting on farming communities and areas popular with tourists (Guðmundsson and Gylfason, 2005).

11.5.2. Markarfljót Jökulhlaup Magnitude and Frequency

Smith (2004) and Larsen *et al.* (2005) identified a minimum of 11 major Holocene jökulhlaups draining into the Markarfljót at 9000, 7900, 7500, 6600, 6100, 5100, 4400, 3500, 2000, 1600 and 1200 yrs BP. The largest of these jökulhlaups (4400 yrs BP) has a reconstructed peak discharge of $2.5–3 \times 10^5$ m^3/s (Gröndal *et al.*, 2005; Guðmundsson and Gylfason, 2005). Smith and Dugmore (2006) present detailed stratigraphic evidence of the last three Holocene jökulhlaups, dated using tephrochronology and ^{14}C AD 500, 700 and 900. Smith and Haraldsson (2005) and Knudsen and Eggertsson (2005) present evidence of the 1230 ^{14}C yrs BP Katlahlaup with an estimated peak discharge of 1×10^5 m^3/s from a site

(Drumbabót) containing buried tree trunks in growth position.

Katlahlaups with the largest estimated discharges believed to have been routed through Markarfljót from the snout of Entujökull have been modelled using a 2D hydraulic model, with specified input hydrographs (Hólm and Kjaran, 2005; Elíasson *et al.*, 2007). A symmetrical hydrograph with a 2 h rise to peak discharge of 3×10^5 m^3/s took 6 h to inundate the entire valley as far as the coastline. Modelled flow depths reached 20–25 m in the Markarfljót but were shown to decrease to less than 10 m by the time the flood wave reached the coast. A symmetrical hydrograph with a 2-h rise to peak discharge of 1×10^5 m^3/s took 10 h to inundate the entire valley as far as the coastline. Although the input hydrographs used are hypothetical, the model output is iteratively calibrated against relatively well-defined peak jökulhlaup water surface slope data, which provided slope–area calculations (Gröndal *et al.*, 2005; Hólm and Kjaran, 2005). Elíasson *et al.* (2007) observed that the modelled jökulhlaup hydrographs behaved in a wave-like fashion with the flood wave maintaining its form down flow. Subsequent modelling of the jökulhlaup as a translatory flood wave yields comparable results when compared to slope–area methods used to calculate the 1918 Katlahlaup discharge on Mýrdalssandur (Tómasson, 1996), and for which there is sufficient observational evidence of jökulhlaup flow behaviour to constrain model parameters (Elíasson *et al.*, 2007). On entering the sea, the largest Katlahlaups have been known to generate tsunami related to the sudden influx of a jökulhlaup frontal wave (Elíasson, 2008). Modelling of the passage of jökulhlaup hydrographs as translatory waves suggests that the largest predicted Katlahlaup tsunami would hit Heimaey in the Vestmannaeyjar Islands with a wave height of 2–3 m, therefore not representing a major hazard (Elíasson, 2008). The potential for the generation of large hazardous tsunami from submarine slope failure, triggered by large influxes of water and sediment, during Katlahlaups remains to be investigated.

11.5.3. Geomorphological and Sedimentary Evidence of Katlahlaups in the Markarfljót Valley

The Markarfljót valley contains a spectacular sequence of large-scale erosional landforms including bedrock gorges (Markarfljótsgljúfur), cataracts, washed lava surfaces, truncated moraines and eroded terraces (Krigström, 1962; Haraldsson, 1981; Tómasson, 2002; Guðmundsson *et al.*, 2005; Smith and Dugmore, 2006). Smith and Dugmore (2006) use the erosional evidence to identify Katlahlaups draining from more advanced glacier margin positions. Haraldsson (1981) linked the Markarfljót canyons to potential volcanogenic bursts from Entujökull and also raised the possibility of bursts generated by the drainage of ice-dammed lakes impounded by the advances of Entujökull. Boulder bars, imbricated boulders and large bedforms are consistent with jökulhlaup deposition within the bedrock reaches of the system (Smith and Dugmore, 2006). More distally, jökulhlaup deposits are commonly scoria rich and similar to those described for Sólheimasandur and Skógasandur (Knudsen and Eggertsson, 2005; Smith and Haraldsson, 2005; Smith and Dugmore, 2006).

Fluvial and sedimentary processes within the modern Markarfljót river system do not reflect the passage of large jökulhlaups or the large volumes of sediment deposited in the vast areas of Auster and Vestur Landeyjar at the mouth of the Markarfljót valley (e.g. Krigström, 1962; Bluck, 1974). Haraldsson and Palm (1980) used seismic refraction survey to establish a maximum sediment thickness of 250 m beneath the Landeyjar and sediment thicknesses of up to 75 m in the Markarfljót valley. It is clear that the present-day fluvial processes cannot account for large-scale erosional and depositional evidence within the Markarfljót valley.

11.6. Conclusions

The KVS has generated some of world's largest volcano-glacial outburst floods and represents a significant hazard to surrounding population. Historic eyewitness accounts of Katlahlaups and tephrochronological reconstructions provide one of the best records of jökulhlaup frequency and magnitude for any volcanic system. The study of Katlahlaup landforms and deposits has improved our understanding of sandur evolution, high-magnitude flood processes and ice block meltout structures. Interpretation of often fine-grained Katlahlaup deposits has improved our ability to identify the sedimentary signature of high-magnitude events in the absence of coarse-grained sediment. Recently, the use of large-scale sedimentary architecture and micro-fabrics has improved our understanding of Katlahlaup depositional processes and flow rheology. Sedimentary structures associated with the formation, migration and collapse of large-scale antidunes suggest that some Katlahlaups have a more turbulent and fluidal nature than previously thought. The sedimentary impact of supercritical flows (for 1918 deposits) has been presented here. However, if we are to assess the long-term role of jökulhlaups in terms of sediment supply and landscape change, the recognition of supercritical flow during these events is an important parameter for input into models of sediment transport in the proglacial environments. Katlahlaups are powerful conveyors of sediment to the Iceland shelf and offshore, beyond the shelf edge, supplying a large amount of sediment to the southern oceanic basin in the form of submarine flows.

Despite these recent advances in our knowledge of Katlahlaups, there are still significant gaps in our understanding of Katlahlaups. The causes and impacts of, often undetected, lower magnitude Katlahlaups have only just being realised. Such events may be important precursors to large-scale subglacial volcanic activity. As the probability of a major Katla subglacial eruption increases with time, further research on Katlahlaup impacts is vital to provide input to hydraulic modelling studies, which in turn can inform the Icelandic civil defence community.

The jökulhlaup landscapes and deposits surrounding Mýrdalsjökull provide useful, well-constrained analogues which can be applied elsewhere within the sedimentary record of proglacial and volcanic areas. Katlahlaups represent the only well-constrained modern analogue for large (mega) glacier outburst floods during the Quaternary. Katlahlaup generation mechanisms and proglacial hydrograph characteristics require non-steady-state sub-, en- and proglacial hydraulic processes. Wider recognition of hydraulic transience as a driver of landscape evolution will improve our interpretation of the landform and sediment record of Quaternary megafloods, resulting from subglacial eruptions and the drainage of large ice-dammed lakes. The study of Katlahlaup deposits allows better interpretation of high-magnitude events within systems dominated by fine-grained sediment and an understanding of the controls on sediment preservation.

Acknowledgements

We acknowledge fieldwork grants from the Earthwatch Institute and the assistance of Earthwatch volunteers' teams based in Vík from 2003 onwards. R.D. would like to thank Sheila Gomez and Emma Sykes for helpful assistance in the field. We thank Ívar Örn Benediktsson, Per Möller and Anders Schomacker for constructive comments on this manuscript.

References

Björnsson, H., 1976. Jökulhlaup og jarðskjálftar í Mýrdalsjökli í nóv. 1975. Jökull 26, 51.

Björnsson, H., 1993. Ýmis sjónarmið um eðli Kötlugos. Kötlustefna 1993, Raunvísindastofnun Háskóla Íslands 1993, RH-3-93.

Björnsson, H., 2002. Subglacial lakes and jökulhlaups in Iceland. Global and Planetary Change 35, 255–271.

Björnsson, H. & Einarsson, P., 1976. Skjálftar og jökulhlaup í Múlakvísl í ágúst 1975 og 1976. Jökull 26, 58.

Björnsson, H., Pálsson, F., & Guðmunsson, M.T., 1993. Mýrdalsjökull: yfirborð, botn og rennslisleiðir jökulhlaupa. In: Larsen, G. (Ed), Kötlustefna, rannsóknir á eldverkni undir Mýrdalsjökli. University of Iceland, Science Institute, RH-3-93, Reykjavik, 11–13.

Björnsson, H., Pálsson, F., & Guðmundsson, M.T., 2000. Surface and bedrock topography of the Mýrdalsjökull ice cap, Iceland: the Katla caldera, eruption sites and routes of jökulhlaups. Jökull 49, 29–46.

Bluck, B.J., 1974. Structure and directional properties of some valley sandur deposits in southern Iceland. Sedimentology 21, 533–554.

Cas, R.A.F. & Wright, J.V., 1987. Volcanic successions: modern and ancient. London, Allen and Unwin, 544 pp.

Cashman, K.V., Sturtevant, B., Papale, P., & Navon, O., 2000. Magmatic fragmentation. In: Sigurdsson, H., Houghton, B.F., Mcnutt, S.R., Rymer, H. (Eds), Encyclopedia of volcanoes. San Diego, Academic Press, 421–430.

Costa, J.E., 1984. Physical geomorphology of debris flows. In: Costa, J.E., Heister, P.J. (Eds), Developments and applications of geomorphology. Berlin, Springer Verlag, 268–317.

Dugmore, A.J., 1987. Holocene glacier fluctuations around Eyjafjallajökull, south Iceland: A tephrochronological study. PhD Thesis. Department of Geography, University of Aberdeen.

Dugmore, A.J., Newton, A.J., Larsen, G., & Cook, G.T., 2000. Tephrochronology, environmental change and the Norse settlement of Iceland. Environmental Archaeology 5, 21–34.

Duller, R.A., 2007. Depositional processes associated with volcaniclastic jökulhlaups, Mýrdalssandur, Iceland. Unpublished PhD Thesis. Keele University, 346 pp.

Duller, R.A., Mountney, N.P., Russell, A.J., & Cassidy, N.J., 2008. Architectural analysis of a volcaniclastic jökulhlaup deposit, southern Iceland: sedimentary evidence for supercritical flow. Sedimentology 55, 939–964.

Duller, R.A., Mountney, N.P., & Russell, A.J., in press. Particle fabric and sedimentation of structureless sands, southern Iceland. Journal of Sedimentary Research.

Eggertsson, S., 1919. Ýmislegt smávegis viðvíkandi Kötlugosinu 1918. Eimreiðin 25, 212–222.

Einarsson, G., 1979. Mýrdalssandur pumice deposit. Reykjavik, Iðntæknistofnun 79, 05.

Einarsson, P., 2000. Atburðarás í tengslum við hlaup í Jökulsá á Sölheimasandi í Júlí 1999, Geoscience Society of Iceland, February Meeting, 14.

Einarsson, E.H., Larsen, G., & Thórarinsson, S., 1980. The Sólheimar tephra layer and the Katla eruption of 1357. Acta Naturalia Íslandica 28, 1–24.

Elíasson, J., 2008. A glacial burst tsunami near Vestmannaeyjar, Iceland. Journal of Coastal Research 24, 13–20.

Elíasson, J., Larsen, G., Guðmundsson, M.T., & Sigmundsson, F., 2006. Probabilistic model for eruptions and associated flood events in the Katla caldera, Iceland. Computational Geosciences 10, 179–200.

Elíasson, J., Kjaran, S.P., Holm, S.L., Guðmundsson, M.T., & Larsen, G., 2007. Large hazardous floods as translatory waves. Environmental Modelling & Software 22, 1392–1399.

Everest, J. & Bradwell, T., 2003. Buried glacier ice in southern Iceland and its wider significance. Geomorphology 52, 347–358.

Gröndal, G.O., Larsen, G., & Elefsen, S., 2005. Stærðir forsögulegra hamfaraflóða í Markarfljóti – mæling á farvegum neðan Einhyrningsflata. In: Guðmundsson, M.T., Gylfason, A.G. (Eds), Hættumat vegna eldgosa og hlaupa frá vestanverðum Mýrdalsjökli og Eyjafjallajökli. Reykjavík, Ríkislögreglustjórinn, 99–104.

Guðmundsson, H.J., 1997. A review of the Holocene environmental history of Iceland. Quaternary Science Reviews 16, 81–92.

Guðmundsson, M.T. & Gylfason, Á.G. (Eds), 2005. Hættumat vegna eldgosa og hlaupa frá vestanverðum Mýrdalsjökli og Eyjafjallajökli. Reykjavík, Ríkislögreglustjórinn, 210 pp.

Guðmundsson, M.T. & Högnadóttir, Þ., 2006. Ísbráðnun og upptakarennsli jökulhlaupa vegna eldgosa í Kötluöskju og austanverðum Mýrdalsjökli. Skýrsla unnin fyrir Almannavarnadeild Ríkislögreglustjóra. Reykjavík, Jarðvísindastofnun Háskólans, 35 pp.

Guðmundsson, M.T., Högnadóttir, T., Björnsson, H., Pálsson, F., & Guðmundsson, G., 2000. Jarðhitinn í Mýrdalsjökuli og atburðirnir sumarið, Geoscience Society of Iceland, February Meeting, 13.

Guðmundsson, M.T., Elíasson, J., Larsen, G., Gylfason, A.G., Einarsson, P., Jóhannesson, T., Hákonardóttir, K.M., & Torfason, H., 2005. Yfirlit um hættu vegna eldgosa og hlaupa frávesturhluta Mýrdalsjökuls og Eyjafjallajökli (Overview of hazard due to eruptions and floods from Katla and Eyjafjallajökull, south Iceland). In: Guðmundsson, M.T., Gylfason, Á.G. (Eds), Hættumat vegna eldgosa og hlaupa frá vestanverðum Mýrdalsjökli og Eyjafjallajökli. Reykjavík, Ríkislögreglustjórinn, 11–45.

Hákonarson, M., 1860. Kötluhlaup í Jökulsá á Sólheimasandi. Íslendingur 1, 19th July.

Hannesson, P., 1934. Kötlugosið síðasta. Náttúrufræðingurinn 4, 1–4.

Haraldsson, H., 1981. The Markarfljót sandur area, southern Iceland: sedimentological, petrographical and stratigraphical studies. Striae 15, 65 pp.

Haraldsson, H. & Palm, H., 1980. A seismic investigation in the Markarfljót sandur area, southern Iceland. Striolae 2, 54 pp.

Heim, D., 1983. Glaziäre Entwässerung und Sanderbildung am Kötlujökull, Südisland. Polarforschung 53, 17–29.

Heim, D., 1992. Sandergenese und Gletcherentwässerung am Kötlujökull (Höfðabrekkajökull), Südisland. Polarforschung 62, 95–128.

Hólm,, S.L. & Kjaran,, S.P., 2005. Reiknilíkan fyrir útbreiðslu hlaupa frá Entujökli (Simulation of jökulhlaups from Entujökull, S-Iceland). In: Guðmundsson, M.T., Gylfason, Á.G. (Eds), Hættumat vegna eldgosa og hlaupa frá vestanverðum Mýrdalsjökli og Eyjafjallajökli. Reykjavík, Ríkislögreglustjórinn, 197–210.

Höskuldsson, A. & Sparks, R.S.J., 1997. Thermodynamics and fluid dynamics of effusive subglacial eruptions. Bulletin of Volcanology 59, 219–230.

Jakobsson, S.P., 1979. Petrology of recent basalts of the Eastern Volcanic Zone, Iceland. Acta Naturalia Islandica 26, 1–103.

Jóhannsson, G., 1919. Kötlugosið. Reykjavík, Bókaverzlun Ársæls Árnasonar.

Jónsson, J., 1982. Notes on the Katla volcanogenic debris flows. Jökull 32, 61–68.

Kjær, K.H., Sultan, L., Krüger, J., & Schomacker, A., 2004. Architecture and sedimentation of outwash fans in front of the Mýrdalsjökull ice cap, Iceland. Sedimentary Geology 172, 139–163.

Knudsen, Ó. & Eggertsson, Ó., 2005. Jökulhlaupaset við Þverá í Fljótshlíð. In: Guðmundsson, M.T., Gylfason, Á.G. (Eds), Hættumat vegna eldgosa og hlaupa frá vestanverðum Mýrdalsjökli og Eyjafjallajökli. Reykjavík, Ríkislögreglustjórinn, 113–122.

Krigström, A., 1962. Geomorphological studies of sandur plains and their braided rivers in Iceland. Geografiska Annaler 44, 328–346.

Krüger, J., 1994. Glacial processes, sediments, landforms, and stratigraphy in the terminus region of Mýrdalsjökull, Iceland. Two interdisciplinary case studies. Folia Geographica Danica 21, 1–233.

Krüger, J., 1997. Development of minor outwash fans at Kötlujökull, Iceland. Quaternary Science Reviews 16, 649–659.

Lacasse, C., Sigurðsson, H., Jóhannesson, H., Paterne, M., & Carey, S., 1995. Source of ash zone I in the North Atlantic. Bulletin of Volcanology 57, 18–32.

Lacasse, C., Sigurðsson, H., Carey, S., Paterne, M., & Guichard, F., 1996. North Atlantic deep-sea sedimentation of Late Quaternary tephra from the Iceland hotspot. Marine Geology 129, 207–235.

Larsen, G., 2000. Holocene eruptions within the Katla volcanic system, south Iceland: characteristics and environmental impact. Jökull 49, 1–28.

Larsen, G. & Ásbjörnsson, S., 1995. Volume of tephra and rock debris deposited by the 1918 jökulhlaups on western Mýrdalssandur, South Iceland. Abstracts, International Symposium on Glacial Erosion and Sedimentation, 20–25 August. International Glaciological Society, Reykjavík, 67.

Larsen, G., Guðmundsson, M.T., & Björnsson, H., 1998. Eight centuries of periodic volcanism at the centre of the Icelandic hotspot revealed by glacier tephrostratigraphy. Geology 26, 943–946.

Larsen, G., Newton, A.G., Dugmore, A.J., & Vilmundardóttir, E.G., 2001. Geochemistry, dispersal, volumes and chronology of Holocene silicic tephra layers from the Katla Volcanic System, Iceland. Journal of Quaternary Science 16, 119–132.

Larsen, G., Smith, K.T., Newton, A.J., & Knudsen, Ó., 2005. Jökulhlaup til vesturs frá Mýrdalsjökli: Ummerki um forsöguleg hlaup niður Markarfljót. In: Guðmundsson, M.T., Gylfason, Á.G. (Eds), Hættumat vegna eldgosa og hlaupa frá vestanverðum Mýrdalsjökli og Eyjafjallajökli. Reykjavík, Ríkislögreglustjórinn, 75–98.

Lawler, D.M., Björnsson, H., & Dolan, M., 1996. Impact of subglacial geothermal activity on meltwater quality in the Jökulsá á Sólheimasandi system, southern Iceland. Hydrological Processes 10, 557–578.

Le Heron, D.P. & Etienne, J.L., 2005. A complex subglacial clastic dyke swarm, Sólheimajökull, southern Iceland. Sedimentary Geology 181, 25–37.

Magnússon, Þ., 1626. Þorsteins Magnússonar um jökulbrunann fyrir austan 1625. Safn til Sögu Íslands IV, 1907–1915. Kaupmannahöfn og Reykjavík, 201–215.

Maizels, J.K., 1987. Modelling of paleohydrologic changes during deglaciation. Geographie Physique et Quaternaire 40, 263–277.

Maizels, J.K., 1989a. Sedimentology, palaeoflow dynamics and flood history of jökulhlaup deposits: palaeohydrology of Holocene sediment sequences in southern Iceland sandur deposits. Journal of Sedimentary Petrology 59, 204–223.

Maizels, J.K., 1989b. Sedimentology and palaeohydrology of Holocene flood deposits in front of a jökulhlaup glacier, South Iceland. In: Bevan, K., Carling, P. (Eds), Floods: hydrological, sedimentological and geomorphological implications: an overview. Chichester, John Wiley & Sons, 239–253.

Maizels, J.K., 1991. The origin and evolution of Holocene sandur deposits in areas of jökulhlaup drainage, south Iceland. In: Maizels, J.K., Caseldine, C. (Eds), Environmental change in Iceland: past and present. Dordrecht, Kluwer, 267–302.

Maizels, J.K., 1992. Boulder ring structures produced during jökulhlaup flows – origin and hydraulic significance. Geografiska Annaler 74A, 21–33.

Maizels, J.K., 1993. Lithofacies variations within sandur deposits: the role of runoff regime, flow dynamics and sediment supply characteristics. Sedimentary Geology 85, 299–325.

Maizels, J.K., 1997. Jökulhlaup deposits in proglacial areas. Quaternary Science Reviews 16, 793–819.

Maizels, J.K. & Dugmore, A.J., 1985. Lichenometric dating and tephrochronology of sandur deposits, Sólheimajökull area, southern Iceland. Jökull 35, 69–77.

Maizels, J.K. & Russell, A.J., 1992. Quaternary perspectives on jökulhlaup prediction. In: Gray, J.M. (Ed.), Applications of Quaternary research. Quaternary Research Association, London, 133–153.

Marren, P.M., 2002. Glacier margin fluctuations, Skaftafellsjökull, Iceland: implications for sandur evolution. Boreas 31, 75–81.

Mattsson, H. & Höskuldsson, A., 2003. Geology of the Heimaey volcanic centre, south Iceland: early evolution of a central volcano in a propagating rift? Journal of Volcanology and Geothermal Research 127, 55–71.

van der Meer, J.J.M., Kjær, K.H., & Krüger, J., 1999. Subglacial water-escape structures and till structures, Sléttjökull, Iceland. Journal of Quaternary Science 14, 191–205.

Mountney, N.P. & Russell, A.J., 2004. Sedimentology of aeolian sandsheet deposits in the Askja region of NE Iceland. Sedimentary Geology 166, 223–244.

Mountney, N.P. & Russell, A.J., 2006. Coastal aeolian dune development, Sólheimasandur, southern Iceland. Sedimentary Geology 192, 167–191.

Norðdahl, H., 1990. Late Weichselian and early Holocene de-glaciation history of Iceland. Jökull 40, 27–50.

Norðdahl, H. & Pétursson, H.G., 2005. Relative sea-level changes in Iceland: new aspects of the Weichselian deglaciation of Iceland. In: Caseldine, C., Russell, A., Harðardóttir, J., Knudsen, Ó. (Eds), Iceland: modern processes and past environments. Elsevier. Developments in Quaternary Science 5, 25–78.

Nummedal, D., Hine, A.C., & Boothroyd, J.C., 1987. Holocene evolution of the south-central coast of Iceland. In: Fitzgerald, D.M., Rosen, P.S. (Eds), Glaciated coasts. London, Academic Press, 115–150.

Olszewski, A. & Weckwerth, P., 1999. The morphogenesis of kettles in the Höfðabrekkujökull forefield, Mýrdalssandur, Iceland. Jökull 47, 71–88.

Rist, S., 1967. Jökulhlaups from the ice cover of Mýrdalsjökull on June 25, 1955 and January 20, 1956. Jökull 17, 243–248.

Rist, S., 1983. Floods and flood danger in Iceland. Jökull 33, 119–132.

Roberts, M.J., 2005. Jökulhlaups: a reassessment of floodwater flow through glaciers. Reviews of Geophysics 43(1). RG1002, 1–21.

Roberts, M.J., 2008. Constraints on the timing, source location, and routing of jökulhlaup at Icelandic test sites. Unpublished Poster.

Roberts, M.J., Russell, A.J., Tweed, F.S., & Knudsen, Ó., 2000a. Rapid sediment entrainment and englacial deposition during jökulhlaups. Journal of Glaciology 153, 349–351.

Roberts, M.J., Russell, A.J., Tweed, F.S., & Knudsen, Ó., 2000b. Ice fracturing during jökulhlaups: implications for englacial floodwater routing and outlet development. Earth Surface Processes and Landforms 25, 1429–1446.

Roberts, M.J., Tweed, F.S., Russell, A.J., Knudsen, Ó., & Harris, T.D., 2003. Hydrologic and geomorphic effects of temporary ice-dammed lake formation during jökulhlaups. Earth Surface Processes and Landforms 28, 723–737.

Russell, A.J., Tweed, F.S., & Knudsen, Ó., 2000. Flash flood at Sólheimajökull heralds the reawakening of an Icelandic subglacial volcano. Geology Today 16(3), 103–107.

Russell, A.J., Tweed, F.S., Knudsen, Ó., Roberts, M.J., Harris, T.D., & Marren, P.M., 2002. The geomorphic impact and sedimentary characteristics of the July 1999 jökulhlaup on the Jökulsá á Sólheimasandi, Mýrdalsjökull, southern Iceland. In: Snorasson, A., Finnsdóttir, H.P., Moss, M. (Eds), The extremes of the extremes: extraordinary floods. Proceedings of a symposium held at Reykjavík, Iceland, July 2000. IAHS Publication Number 271, 249–254.

Russell, A.J., Fay, H., Marren, P.M., Tweed, F.S., & Knudsen, Ó., 2005. Icelandic jökulhlaup impacts. In: Caseldine, C., Russell, A., Harðardóttir, J., Knudsen, Ó. (Eds), Iceland: modern processes and past environments. Elsevier. Developments in Quaternary Science 5, 153–203.

Russell, A.J., Gregory, A., Duller, R., Large, A., Blauvelt, D., Marren, P.M., Mountney, N.P., & Iturrizaga, L., 2008. The geomorphic and sedimentary signature of two recent jökulhlaups from Kötlujökull, Mýrdalsjökull, Iceland. IAVCEI 2008 General Assembly, Reykjavík, Iceland, Poster Session III Abstract Volume, 87.

Russell, A.J., Tweed, F.S., Roberts, M.J., Harris, T.D., Knudsen, Ó., Guðmundsson, M.T., & Marren, P.M. An unusual jökulhlaup resulting from subglacial volcanism, Sólheimajökull, Iceland. Quaternary Science Reviews, submitted for publication.

Sigurðsson, F., 1988. Fold og vötn að Fjallabaki. Árbók 1988, Ferðafélags Íslands 181–202.

Sigurðsson, O., Zóphóníasson, S., & Ísleifsson, E., 2000. Jökulhlaup úr Sólheimajökuli 18 Júlí 1999. Jökull 49, 75–80.

Smellie, J.L., 2000. Subglacial eruptions. In: Sigurdsson, H., Houghton, B.F., Mcnutt, S.R., Rymer, H. (Eds), Encyclopedia of volcanoes. San Diego, Academic Press, 403–418.

Smith, K.T., 2004. Holocene jökulhlaups, glacier fluctuations and palaeoenvironment, Mýrdalsjökull, South Iceland. Unpublished PhD Thesis. University of Edinburgh, 400 pp.

Smith, K.T. & Dugmore, A.J., 2006. Jökulhlaups circa Landnám: mid-to late first millennium AD floods in south Iceland and their implications for landscapes of settlement. Geografiska Annaler 88A, 165–176.

Smith, K.T. & Haraldsson, H., 2005. A late Holocene jökulhlaup, Markarfljót, Iceland: nature, impact and implications for future hazards. Jökull 55, 75–86.

Soosalu, H., Jónsdóttir, K., & Einarsson, P., 2006. Seismicity crisis at the Katla volcano, Iceland – signs of a cryptodome? Journal of Volcanology and Geothermal Research 153, 177–186.

Sturkell, E., Einarsson, P., Roberts, M.J., Geirsson, H., Guðmundsson, M.T., Sigmundsson, F., Pinel, V.G., Guðmundsson, G., & Stefánsson, R., 2008. Seismic and geodetic insights into magma accumulation at Katla subglacial volcano, Iceland: 1999 to 2005. Journal of Geophysical Research 113, B03212.

Sveinsson, G., 1919. Kötlugosið 1918 og afleiðingar þess. Reykjavík, Prentsmiðjan Gutenberg.

Thompson, A.P. & Jones, A., 1986. Rates and causes of proglacial river terrace formation in southeast Iceland: an application of lichonometric dating techniques. Boreas 15, 231–246.

Thórarinsson, S., 1939. The ice-dammed lakes of Iceland, with particular reference to their value as indicators of glacier oscillations. Geografiska Annaler 21A, 216–242.

Thórarinsson, S., 1957. The jökulhlaup from the Katla area in 1955 compared with other jökulhlaups in Iceland. Jökull 7, 21–25.

Thorarinsson, S., 1959. Um möguleika a thvi ad segja fyrir næsta Kötlugos. Jökull 9, 6–18.

Thórarinsson, S., 1975. Katla og annáll Kötlugosa. Árbók 1975, Ferðafélags Íslands 125–149.

Thordarson, T. & Höskuldsson, A., 2002. Iceland (Classic geology in Europe). Harpenden, Terra Publishing, 200 pp.

Tómasson, H., 1996. The jökulhlaup from Katla in 1918. Annals of Glaciology 22, 249–254.

Tómasson, H., 2002. Catastrophic floods in Iceland. In: Snorasson, A., Finnsdóttir, H.P., Moss, M. (Eds), the extremes of the extremes: extraordinary floods. Proceedings of a symposium held at Reykjavik, Iceland, July 2000. IAHS Publication Number 271, 121–126.

Tweed, F.S., 2000a. Jökulhlaup initiation by ice-dam flotation: the significance of Glacier Debris-Content. Earth Surface Processes and Landforms 25, 105–108.

Tweed, F.S., 2000b. An ice-dammed lake in Jökulsárgil: predictive modelling and geomorphological evidence. Jökull 48, 1–11.

Wisniewski, E., Andrzejewski, L., & Olszewski, A., 1999. Relief of the Höfðabrekkujökull forefield, South Iceland, in light of geomorphological mapping. Jökull 47, 59–70.

Subject Index

FSC
www.fsc.org
MIX
Papier | Fördert gute Waldnutzung
FSC® C083411

Kolibri 360 GmbH
Ferdinand-Jühlke-Straße 7
99095 Erfurt, Deutschland
produktsicherheit@kolibri360.de